Cengiz Kahraman (Ed.)

Fuzzy Engineering Economics with Applications

# Studies in Fuzziness and Soft Computing, Volume 233

**Editor-in-Chief**

Prof. Janusz Kacprzyk
Systems Research Institute
Polish Academy of Sciences
ul. Newelska 6
01-447 Warsaw
Poland
E-mail: kacprzyk@ibspan.waw.pl

---

Further volumes of this series can be found on our homepage: springer.com

Vol. 219. Roland R. Yager, Liping Liu (Eds.)
*Classic Works of the Dempster-Shafer Theory
of Belief Functions*, 2007
ISBN 978-3-540-25381-5

Vol. 220. Humberto Bustince,
Francisco Herrera, Javier Montero (Eds.)
*Fuzzy Sets and Their Extensions:
Representation, Aggregation and Models*, 2007
ISBN 978-3-540-73722-3

Vol. 221. Gleb Beliakov, Tomasa Calvo,
Ana Pradera
*Aggregation Functions: A Guide
for Practitioners*, 2007
ISBN 978-3-540-73720-9

Vol. 222. James J. Buckley,
Leonard J. Jowers
*Monte Carlo Methods in Fuzzy
Optimization*, 2008
ISBN 978-3-540-76289-8

Vol. 223. Oscar Castillo, Patricia Melin
*Type-2 Fuzzy Logic: Theory and
Applications*, 2008
ISBN 978-3-540-76283-6

Vol. 224. Rafael Bello, Rafael Falcón,
Witold Pedrycz, Janusz Kacprzyk (Eds.)
*Contributions to Fuzzy and Rough Sets
Theories and Their Applications*, 2008
ISBN 978-3-540-76972-9

Vol. 225. Terry D. Clark, Jennifer M. Larson,
John N. Mordeson, Joshua D. Potter,
Mark J. Wierman
*Applying Fuzzy Mathematics to Formal
Models in Comparative Politics*, 2008
ISBN 978-3-540-77460-0

Vol. 226. Bhanu Prasad (Ed.)
*Soft Computing Applications in Industry*, 2008
ISBN 978-3-540-77464-8

Vol. 227. Eugene Roventa, Tiberiu Spircu
*Management of Knowledge Imperfection in
Building Intelligent Systems*, 2008
ISBN 978-3-540-77462-4

Vol. 228. Adam Kasperski
*Discrete Optimization with Interval Data*, 2008
ISBN 978-3-540-78483-8

Vol. 229. Sadaaki Miyamoto,
Hidetomo Ichihashi, Katsuhiro Honda
*Algorithms for Fuzzy Clustering*, 2008
ISBN 978-3-540-78736-5

Vol. 230. Bhanu Prasad (Ed.)
*Soft Computing Applications in Business*, 2008
ISBN 978-3-540-79004-4

Vol. 231. Michal Baczynski,
Balasubramaniam Jayaram
*Soft Fuzzy Implications*, 2008
ISBN 978-3-540-69080-1

Vol. 232. Eduardo Massad,
Neli Regina Siqueira Ortega,
Laécio Carvalho de Barros,
Claudio José Struchiner
*Fuzzy Logic in Action: Applications
in Epidemiology and Beyond*, 2008
ISBN 978-3-540-69092-4

Vol. 233. Cengiz Kahraman (Ed.)
*Fuzzy Engineering Economics with
Applications*, 2008
ISBN 978-3-540-70809-4

Cengiz Kahraman (Ed.)

# Fuzzy Engineering Economics with Applications

 Springer

**Editor**

Professor Cengiz Kahraman
Istanbul Technical University
Management Faculty
Department of Industrial Engineering
34367 Macka, Istanbul
Turkey
E-mail: kahramanc@itu.edu.tr

ISBN 978-3-540-70809-4          e-ISBN 978-3-540-70810-0

DOI 10.1007/978-3-540-70810-0

Studies in Fuzziness and Soft Computing     ISSN 1434-9922

Library of Congress Control Number: 2008931159

© 2008 Springer-Verlag Berlin Heidelberg

This work is subject to copyright. All rights are reserved, whether the whole or part of the material is concerned, specifically the rights of translation, reprinting, reuse of illustrations, recitation, broadcasting, reproduction on microfilm or in any other way, and storage in data banks. Duplication of this publication or parts thereof is permitted only under the provisions of the German Copyright Law of September 9, 1965, in its current version, and permission for use must always be obtained from Springer. Violations are liable to prosecution under the German Copyright Law.

The use of general descriptive names, registered names, trademarks, etc. in this publication does not imply, even in the absence of a specific statement, that such names are exempt from the relevant protective laws and regulations and therefore free for general use.

*Typeset & Cover Design:* Scientific Publishing Services Pvt. Ltd., Chennai, India.

Printed in acid-free paper

9 8 7 6 5 4 3 2 1

springer.com

# Preface

Fuzzy set approaches are suitable to use when the modeling of human knowledge is necessary and when human evaluations are needed. Fuzzy set theory is recognized as an important problem modeling and solution technique. It has been studied extensively over the past 40 years. Most of the early interest in fuzzy set theory pertained to representing uncertainty in human cognitive processes. Fuzzy set theory is now applied to problems in engineering, business, medical and related health sciences, and the natural sciences. This book handles the fuzzy cases of classical engineering economics topics. It contains 15 original research and application chapters including different topics of fuzzy engineering economics.

When no probabilities are available for states of nature, decisions are given under uncertainty. Fuzzy sets are a good tool for the operation research analyst facing uncertainty and subjectivity. The main purpose of the first chapter is to present the role and importance of fuzzy sets in the economic decision making problem with the literature review of the most recent advances.

The second chapter includes two subchapters on fuzzy net present value analysis written by different authors and different points of view. The first subchapter presents arithmetic Operations over Independent Fuzzy Numbers and Economic Concepts Review. Then, it gives the techniques for Comparing and Ordering Fuzzy Numbers of Independent numbers and dependent numbers. It examines fuzzy case with Partial Correlation. The chapter also includes many numerical applications. The second subchapter introduces fuzzy net present value analysis. The present value approach is first analyzed by crisp values. After the basic definitions of fuzzy numbers are given, the Concept And Applications of fuzzy Present Value are handled. The other components of this chapter is the fuzzy and probabilistic approach to the present value, the fuzzy net present value maximization as an objective in project selection problems, applications to the valuation of projects with future options, interpretation of fuzzy present value and fuzzy net present value, fuzzy net present value as an objective in optimization problems from the industry, fuzzy net present value as an objective in spatial games, and fuzzy classification based on net present value.

Chapter 3 includes two subchapters on fuzzy equivalent annual worth analysis. Equivalent annual worth analysis is one of the most used analysis techniques for the evaluation of investments. The main process is to convert any cash flow to an equivalent uniform cash flow. Later, the decision is given taking the economic category of

VI    Preface

the problem into account. In the first subchapter, it is shown how you can apply this technique when the parameters are all fuzzy. Numerical examples are given for different analysis periods. In the second subchapter, some case studies are presented using fuzzy equivalent uniform annual worth analysis. It includes three case studies and each case is studied for both crisp and fuzzy cases. Trapezoidal fuzzy numbers and correlated, uncorrelated, and partially correlated cash flows are handled in the cases.

Chapter 4 is on fuzzy rate of return analysis and includes two subsections. The definition of the fuzzy internal rate of return is not unequivocal and requires some discussion. In the first subchapter, crisp rate of return analysis is first introduced. Then fuzzy rate of return analysis is handled. The problem of defining fuzzy internal rate of return together with certain applications is discussed. In the second subchapter, the problem of *IRR* estimation in fuzzy setting is considered in the framework of more general problem of fuzzy equations solving. The concept of restricted fuzzy *IRR* as the solution of the corresponding non-linear fuzzy equation is proposed and analyzed.

Benefit-cost analysis is a systematic evaluation of the economic advantages (benefits) and disadvantages (costs) of a set of investment alternatives. The objective of a benefit-cost analysis is to translate the effects of an investment into monetary terms and to account for the fact that benefits generally accrue over a long period of time while capital costs are incurred primarily in the initial years. In the next chapter, benefit cost ratio (BCR) analysis is analyzed under fuzzy environment. Fuzzy continuous payments and fuzzy discrete payments are summarized briefly.

In an uncertain economic decision environment, our knowledge about the defender's remaining life and its cash flow information usually consist of a lot of vagueness. To describe a planning horizon which may be implicitly forecasted from past incomplete information, a linguistic description like `approximately between 8 and 10 years' is often used. In Chapter 6, using fuzzy equivalent uniform annual cash flow analysis, a fuzzy replacement analysis for two operating systems is handled.

Depreciation is an income tax deduction that allows a taxpayer to recover the cost of property or assets placed in service. The seventh chapter includes the fuzzy after-tax cash flow analyses in case of fuzzy cash flows, fuzzy depreciation, fuzzy tax rate, and fuzzy minimum attractive rate of return with numerical examples.

The eighth chapter presents the ways of incorporating the parameter fuzzy inflation to the engineering economy analyses. Inflation is a financial parameter difficult to estimate. The fuzzy set theory gives us the possibility of converting linguistic expressions about inflation estimates to numerical values. In this chapter, discounted cash flow techniques including these fuzzy expressions are given. The obtained results show the interval of the worst and the best possible outcomes when fuzzy inflation rates are taken into account.

Sensitivity analysis is performed in case of uncertainty. In the ninth chapter, the authors are concerned with the concept, properties and algorithms of differentiation of the Choquet integral. The differentiation of the Choquet integral of a nonnegative measurable function is studied in the setting of sensitivity analysis. The differentiation of the Choquet integral is extended to the differentiation of the generalized t-conorm integral. The Choquet integral is applied to the credit risk analysis (long-term debt rating ) to make clear the significance of them.

Techniques for ranking simple fuzzy numbers are abundant in the literature. However, there is a lack of efficient methods for comparing complex fuzzy numbers that are generated by fuzzy engineering economic analyses. In the next chapter a probabilistic approach is taken instead of the usual fuzzy set manipulations. The Mellin transform is introduced to compute the mean and the variance of a complex fuzzy number. The fuzzy number with the higher mean is to be ranked higher. Two fuzzy cash flow analyses and a fuzzy multiple attribute decision analysis are illustrated in order to demonstrate the suitability of the probabilistic approach.

A decision tree is a method you can use to help make good choices, especially decisions that involve high costs and risks. Decision trees use a graphic approach to compare competing alternatives and assign values to those alternatives by combining uncertainties, costs, and payoffs into specific numerical values. A fuzzy decision tree is a generalization of the crisp case. Fuzzy decision trees are helpful for representing ill-defined structures in decision analysis. Chapter 11 presents investment analyses using fuzzy decision trees with examples.

Chapter 12 includes two subchapters on fuzzy multiple objective evaluation of investments. The first subchapter analyzes the methods for Multiobjective Decision Making in the fuzzy setting in context of Investment Evaluation Problem. The problems typical for Multiobjective Decision Making are indicated and new solutions of them are proposed. The problem of appropriate common scale for representation of objective and subjective criteria is solved using the simple subsethood measure based on $\alpha$-cut representation of fuzzy values. To elaborate an appropriate method for aggregation of aggregating modes, the authors use the synthesis of the tools of Type 2 and Level 2 Fuzzy Sets. The second subchapter proposes the use of fuzzy data mining process to support the analysis processes in order to discover useful properties that can help to improve investment decisions. The stock market analysis is a high demanded task to support investment decisions. The quality of those decisions is the key point in order to obtain profits and obtain new customers and keep old ones. The analysis of stock markets is high complex due to the amount of data analyzed and to the nature of those.

Chapter 13 presents two main results that are accomplished in the fields of financial management and strategic investment planning. Applying the obtained theoretical results, the authors discuss the development of two soft decision support models in detail, which use possibility distributions to describe and characterize the uncertainty about future payoffs.

Pricing of options, forwards or futures often requires using uncertain values of parameters in the model. For example future interest rates are usually uncertain. In Chapter 14, the authors use fuzzy numbers for these uncertain parameters to account for this uncertainty. When some of the parameters in the model are fuzzy, the price then also becomes fuzzy. The authors first discuss options: (1) the discrete binomial method; and then (2) the Black-Scholes model. Then they look at pricing futures and forwards.

Addressing uncertainty in Lorie-Savage and Weingartner capital rationing models is considered in the literature with different approaches. Stochastic and robust approach to Weingartner capital rationing problem are examples of non-fuzzy approaches. In Chapter 15, the authors provide examples of fuzzy approach to Lorie-Savage problem, and illustrate the models with numerical examples. The solution of the generic models requires evolutionary algorithms; however for the models with triangular or trapezoidal fuzzy numbers, branch-and-bound method is suggested to be sufficient.

The last chapter presents future directions in engineering economics. It has also some recommendations for the content of the engineering economics books in the future.

I hope that this book would provide a useful resource of ideas, techniques, and methods for present and further research in the applications of fuzzy sets in engineering economics. I am grateful to the referees whose valuable and highly appreciated works contributed to select the high quality chapters published in this book. I am also grateful to my research assistant, Dr. Ihsan Kaya, for his helps in editing this book.

Cengiz Kahraman
Editor
Istanbul Technical University
Department of Industrial Engineering
Maçka Istanbul Turkey

# Contents

**Fuzzy Sets in Engineering Economic Decision-Making**
*Cengiz Kahraman* ............................................... 1

**Fuzzy Present Worth Analysis with Correlated and Uncorrelated Cash Flows**
*Aleksandar Dimitrovski, Manuel Matos* ............................ 11

**Optimization with Fuzzy Present Worth Analysis and Applications**
*Dorota Kuchta* ................................................. 43

**Fuzzy Equivalent Annual-Worth Analysis and Applications**
*Cengiz Kahraman, İhsan Kaya* .................................... 71

**Case Studies Using Fuzzy Equivalent Annual Worth Analysis**
*Manuel Matos, Aleksandar Dimitrovski* ........................... 83

**Fuzzy Rate of Return Analysis and Applications**
*Dorota Kuchta* ................................................. 97

**On the Fuzzy Internal Rate of Return**
*P. Sewastjanow, L. Dymowa* ..................................... 105

**Fuzzy Benefit/Cost Analysis and Applications**
*Cengiz Kahraman, İhsan Kaya* ................................... 129

**Fuzzy Replacement Analysis**
*Cengiz Kahraman, Murat Levent Demircan* ........................ 145

**Depreciation and Income Tax Considerations under Fuzziness**
*Cengiz Kahraman, İhsan Kaya* ................................... 159

**Effects of Inflation under Fuzziness and Some Applications**
*Cengiz Kahraman, Tufan Demirel, Nihan Demirel* .................. 173

# X    Contents

**Fuzzy Sensitivity Analysis and Its Application**
*Toshihiro Kaino, Kaoru Hirota, Witold Pedrycz* ...................... 183

**A Probabilistic Approach to Fuzzy Engineering Economic Analysis**
*K. Paul Yoon* ...................................................... 217

**Investment Analyses Using Fuzzy Decision Trees**
*Cengiz Kahraman* .................................................. 231

**Fuzzy Multiobjective Evaluation of Investments with Applications**
*L. Dymova, P. Sevastjanov* ......................................... 243

**Using Fuzzy Multi-attribute Data Mining in Stock Market Analysis for Supporting Investment Decisions**
*Francisco Araque, Alberto Salguero, Ramon Carrasco, Luis Martinez* .... 289

**Soft Decision Support Systems for Evaluating Real and Financial Investments**
*Péter Majlender* ................................................... 307

**Pricing Options, Forwards and Futures Using Fuzzy Set Theory**
*James J. Buckley, Esfandiar Eslami* ................................. 339

**Fuzzy Capital Rationing Models**
*Esra Bas, Cengiz Kahraman* ........................................ 359

**Future Directions in Fuzzy Engineering Economics**
*Cengiz Kahraman* .................................................. 381

**Index** ............................................................ 383

**Author Index** .................................................... 389

# Fuzzy Sets in Engineering Economic Decision-Making

Cengiz Kahraman

İstanbul Technical University, Department of Industrial Engineering, 34367, İstanbul, Turkey

**Abstract.** When no probabilities are available for states of nature, decisions are given under uncertainty. Fuzzy sets are a good tool for the operation research analyst facing uncertainty and subjectivity. The main purpose of this chapter is to present the role and importance of fuzzy sets in the economic decision making problem with the literature review of the most recent advances.

## 1 Introduction

Engineering economic decision-making has strategic importance for the growth of a firm as such decisions commit its limited productive resources to its production system and as they strengthen and renew their resources. They consist of allocation of a firm's resources with plans for recouping the initial investment plus adequate profits (or other returns) from cash flows (or other benefits) generated during the economic life of an investment. Such decisions are hard to reverse without severely disturbing an organization economically and otherwise. Therefore, a Capital Budgeting Decision needs systematic and careful analysis. But, such analysis is a many-sided activity which includes among others the estimation and forecasting of current and future cash flows and the economic evaluation of alternative projects. Since in reality the cash flows estimation takes place in a non-deterministic environment, full of complex interplay of conflicting forces, an exact description about cash flows is virtually impossible. Therefore, a firm has to take such decisions in a fuzzy environment and this work makes an attempt to develop procedures and techniques so as to equip decision analyst to achieve a meaningful economic evaluation of projects (Park 2003, Sullivan et al. 2002, Blank and Tarquin 2007, Riggs et al. 1996).

The capital budgeting decisions are built on a three way classification of a decision making environment - certainty, risk and uncertainty. Hence, all techniques that already exist for capital budgeting decisions are meant for only these three decision making environments.

The fuzzy sets theory was specifically designed to mathematically represent uncertainty and vagueness and to provide formalized tools for dealing with the imprecision intrinsic to many problems. There are two main characteristics of fuzzy systems that give them better performance for specific applications:

- Fuzzy systems are suitable for uncertain or approximate reasoning, especially for the system with a mathematical model that is difficult to derive.
- Fuzzy logic allows decision making with estimated values under incomplete or uncertain information.

C. Kahraman (Ed.): Fuzzy Engineering Economics with Appl., STUDFUZZ 233, pp. 1–9, 2008.
springerlink.com                                                   © Springer-Verlag Berlin Heidelberg 2008

2    C. Kahraman

Traditionally in engineering economics, as well as in other fields, uncertainties have been represented by probabilities. The likelihood of possible futures is estimated and probability distributions are assigned. These estimations can be based on past data and experience in which case we are talking about objective probabilities or can be based on someone's subjective judgment in the case of so called subjective probabilities. Anyway, this approach contains an underlying assumption of repetition of events under some unchanged conditions end laws. This is questionable, at least for long range analysis. In addition, very often subjective probabilities actually represent imprecise knowledge, vague information, rather than relative frequency. This way they represent to what degree something can happen, not how often it will happen. Thus, it is a question of degree of membership and fuzzy sets that permit partial membership of their elements in the perpetuity from complete membership to nonmembership (Dimitrovski 2000).

The parameters in engineering economics are cash flows (first costs, annual costs, annual benefits, and salvage values), minimum attractive rate of return, and useful life. The values of these parameters must be estimated before using them in the discounted cash flow formulas. When the probability distribution of any parameter is not known, or the past data related to the considered parameter do not exist in our records, the fuzzy set theory is a good tool in capturing the uncertainty in estimating these parameters. These kinds of estimations can be expressed as "an annual maintenance cost around $4,200", "an annual benefit between $6,400 and $7,800", "an annual interest rate approximately 6%", and "a useful life of around 10 years". The fuzzy set theory lets us use these partial linguistic estimations and convert them to full numerical estimations to be able to put them into the classical discounted cash flow formulas.

The remaining sections of this chapter are organized as follows: The second chapter includes the main concepts of engineering economics in the fuzzy case. The third section presents a bibliography for fuzzy engineering economics.

## 2  Fuzzy Concepts in Engineering Economy

In this section, the main concepts of engineering economics will be introduced under fuzziness. The concepts include the time value of money, simple and compound interests, nominal and effective interests, continuous compounding, inflated interest rate, single payment present worth factor (SPPWF), single payment future worth factor (SPFWF), uniform payments present worth factor (UPPWF), uniform payments future worth factor (UPFWF), gradient payments present worth factor (GPPWF), gradient payments future worth factor (GrPFWF), geometric payments present worth factor (GePPWF), geometric payments future worth factor (GePFWF), capital recovery factor (CRF), and sinking fund factor (SFF).

### 2.1  The Time Value of Money

The time value of money is based on the premise that an investor prefers to receive a payment of a fixed amount of money today, rather than an equal amount in the future, all else being equal. The time value of money is represented by an interest rate. A

Fuzzy Sets in Engineering Economic Decision-Making    3

higher time value means a higher interest rate. For example, an interest rate of 10% per year means the amount of money will increase 10% each year exponentially.

In the fuzzy case, the time value of money is neither certain as above nor probabilistic. It might be expressed as "around 10% per year" or "between 9% and 11%". The first expression can be represented by a triangular fuzzy number while the second can be represented by a trapezoidal fuzzy number.

## 2.2 Simple and Compound Interest Rates

Interest is a fee paid on borrowed capital. When money is borrowed, interest is charged for the use of that money for a certain period of time. When the money is paid back, the principal (amount of money that was borrowed) and the interest is paid back. The amount to interest depends on the interest rate, the amount of money borrowed (principal) and the length of time that the money is borrowed.

In case of simple interest, it is calculated only on the principal, or on that portion of the principal which remains unpaid. The well-known crisp formula for simple interest is

$$S = I(1 + i_s \times n) \tag{1}$$

where $I$ is the amount of the borrowed money; $i_s$ is the simple interest rate per period; and $n$ is the number of interest periods.

In the fuzzy case, Eq. (1) transforms into Eq. (2):

$$f_j\left(y\big|\tilde{S}\right) = f_j\left(y\big|\tilde{I}\right)\left(1 + f_j\left(y\big|\tilde{i_s}\right) \times f_j\left(y\big|\tilde{n}\right)\right) \tag{2}$$

where $j=1, 2$ and $f_1(.)$ and $f_2(.)$ are the left and right representations of the triangular shaped fuzzy simple interest rate.

The conceptual difference between simple and compound interests is that the principal with compound interest changes with every time period, as any interest incurred over the period is added to the principal.

The crisp SPPW equation is given by

$$P = F(1 + i)^{-n} \tag{3}$$

With the fuzzy parameters including interest rate, cash flows, and project life, SPPW equation is given by

$$f_{n,j}\left(y\big|\tilde{P}\right) = f_j\left(y\big|\tilde{F}\right) \otimes \left(1 + f_k\left(y\big|\tilde{i_s}\right)\right)^{-f_j\left(y\big|\tilde{n}\right)} \tag{4}$$

for $j = 1,2$ where $k = j$ for negative $\tilde{F}$ and $k = 3 - j$ for positive $\tilde{F}$.

The fuzzy formula for SPFW equation is

$$f_{\tilde{n},j}\left(y\big|\tilde{F}\right) = f_k\left(y\big|\tilde{P}\right) \otimes \left(1 + f_k\left(y\big|\tilde{i_s}\right)\right)^{f_k\left(y\big|\tilde{n}\right)} \tag{5}$$

for $j = 1,2$   $k = 3 - j$ where for negative $\tilde{P}$ and $k = j$ for positive $\tilde{P}$.

4        C. Kahraman

## 2.3 Nominal and Effective Interest Rates

Nominal Interest Rate (NIR) is the periodic interest rate times the number of periods per year. NIR does not take the effect of the compoundings during a year into account whereas Effective Interest Rate (EIR) does. They are calculated by the following formulas:

$$NIR = r_m \times m \tag{6}$$

and

$$EIR = (1 + r_m)^m - 1 \tag{7}$$

Fuzzy NIR and EIR are defined as follows

$$f_{\tilde{m},j}\left(y\middle|\widetilde{NIR}\right) = f_j\left(y\middle|\tilde{i}_{\tilde{m}}\right) \otimes f_j\left(y\middle|\tilde{m}\right) \tag{8}$$

and

$$f_{\tilde{m},j}\left(y\middle|\widetilde{EIR}\right) = \left(1 + f_j\left(y\middle|\tilde{i}_{\tilde{m}}\right)\right)^{f_j(y|\tilde{m})} - 1 \tag{9}$$

or

$$f_{\tilde{m},j}\left(y\middle|\widetilde{EIR}\right) = \left(1 + f_j\left(y\middle|\widetilde{NIR}\right) \oslash f_j\left(y\middle|\tilde{m}\right)\right)^{f_j(y|\tilde{m})} - 1 \tag{10}$$

where $j = 1,2$.

Lets give a numerical example: If $\tilde{m} = (10,11,12)$, $\tilde{i}_m = (2\%, 2.2\%, 2.4\%)$, then

$$\widetilde{NIR} = (10, 11, 12) \otimes (2\%, 2.2\%, 2.4\%) = (20\%, 24.2\%, 28.8\%)$$

and

$$\widetilde{EIR} = (1.02, 1.022, 1.024)^{(10,11,12)} - 1 = (21.9\%, 27.05\%, 32.92\%)$$

## 2.4 Continuous Compounding

Continuous compounding is the process of accumulating the time value of money forward in time on a continuous, or instantaneous, basis. Interest is earned constantly, and at each instant, the interest that accrues immediately begins earning interest on itself. It is the limiting case of discrete compounding while m approaches infinity. For the discrete case, let's calculate the future value of a present amount, $P$, after $n$ years including $m$ periods per year.

$$F = P\left(1 + \frac{r}{m}\right)^{n \times m} \tag{11}$$

where $r$ is the nominal interest rate.

If $m$ approaches infinity, we obtain

$$F = P \, Lim_{n \to \infty} \left(1 + \frac{NIR}{m}\right)^{n \times m} = Pe^{rn} \tag{12}$$

In the fuzzy case, SPFW equation is represented by

$$f_{\tilde{n},j}\left(y\middle|\tilde{F}\right) = f_k\left(y\middle|\tilde{P}\right) \otimes e^{f_k\left(y\middle|\tilde{n}\right) \otimes f_k\left(y\middle|\tilde{r}\right)} \tag{13}$$

for $j = 1,2$ where $k = 3 - j$ for negative $\tilde{P}$ and $k = j$ for positive $\tilde{P}$.

SPPW equation is given by

$$f_{\tilde{n},j}\left(y\middle|\tilde{P}\right) = f_j\left(y\middle|\tilde{F}\right) \otimes e^{-f_k\left(y\middle|\tilde{n}\right) \otimes f_k\left(y\middle|\tilde{r}\right)} \tag{14}$$

for $j = 1,2$ where $k = j$ for negative $\tilde{F}$ and $k = 3 - j$ for positive $\tilde{F}$.

## 2.5 Inflated Interest Rate

Inflation is the increase in the amount of money necessary to obtain the same amount of product or service before the inflated price was present. It is vital to take into account the effects of inflation in an economic analysis. The basic inflation relationship between future and today's dollars is

$$\text{Future Dollars} = \text{Today's dollars} \times (1+f)^n \tag{15}$$

Real or inflation-free interest rate, $i$, represents the actual or real gain received/charged on investments or borrowing. The relation between inflation rate ($e$), inflated interest rate ($i_f$), and real interest rate ($i$) is given by

$$i_e = i + e + i \times e \tag{16}$$

In the fuzzy case, the inflated interest rate is given by

$$f_j\left(y\middle|\tilde{i}_e\right) = f_j\left(y\middle|\tilde{e}\right) \oplus f_j\left(y\middle|\tilde{i}\right) \oplus f_j\left(y\middle|\tilde{i}\right) \otimes f_j\left(y\middle|\tilde{e}\right) \tag{17}$$

or the real interest rate,

$$f_j\left(y\middle|\tilde{i}\right) = [\,f_j\left(y\middle|\tilde{i}_e\right) \ominus f_k\left(y\middle|\tilde{e}\right)] \oslash \left(1 + f\left(y\middle|\tilde{e}\right)\right) \tag{18}$$

where $j=1,2$ and $k=3-j$.

With a fuzzy tax rate, the inflated interest rate becomes

$$f_j\left(y\middle|\tilde{i}_e\right) = \left[f_j\left(y\middle|\tilde{e}\right) \oplus f_j\left(y\middle|\tilde{i}\right) \oplus f_j\left(y\middle|\tilde{i}\right) \otimes f_j\left(y\middle|\tilde{e}\right)\right] \oslash \left(1 - f_k\left(y\middle|\tilde{i}\right)\right) \tag{19}$$

where $j=1,2$ and $k=3-j$.

## 2.6 Interest Factors

The other fuzzy interest factors are summarized below:

Uniform payments present worth factor (UPPWF):

$$f_{P/A,j}\left(y\big|\tilde{i},\tilde{n}\right)=\left\{\left[1+f_k\left(y\big|\tilde{i}\right)\right]^{f_j\left(y\big|\tilde{n}\right)}-1\right\}\oslash\left\{f_k\left(y\big|\tilde{i}\right)\otimes\left[1+f_k\left(y\big|\tilde{i}\right)^{f_j\left(y\big|\tilde{n}\right)}\right]\right\} \quad (20)$$

where $j=1,2$ and $k=3\text{-}j$.

Uniform payments future worth factor (UPFWF):

$$f_{F/A,j}\left(y\big|\tilde{i},\tilde{n}\right)=\left\{\left[1+f_j\left(y\big|\tilde{i}\right)\right]^{f_j\left(y\big|\tilde{n}\right)}-1\right\}\oslash f_j\left(y\big|\tilde{i}\right) \quad (21)$$

Gradient payments present worth factor (GPPWF):

$$f_{P/G,j}\left(y\big|\tilde{i},\tilde{n}\right)=\left(1/f_k\left(y\big|\tilde{i}\right)\right)\otimes$$
$$\left[\left[\left(1+f_k\left(y\big|\tilde{i}\right)\right)^{f_j\left(y\big|\tilde{n}\right)}\oslash f_k\left(y\big|\tilde{i}\right)\right]-f_j\left(y\big|\tilde{n}\right)\right]\left(1/\left(1+f_k\left(y\big|\tilde{i}\right)\right)\right) \quad (22)$$

Gradient payments future worth factor (GrPFWF):

$$f_{F/G,j}\left(y\big|\tilde{i},\tilde{n}\right)=\left(1/f_j\left(y\big|\tilde{i}\right)\right)\otimes\left[\left[\left(1+f_j\left(y\big|\tilde{i}\right)\right)^{f_j\left(y\big|\tilde{n}\right)}\oslash f_j\left(y\big|\tilde{i}\right)\right]-f_j\left(y\big|\tilde{n}\right)\right] \quad (23)$$

Geometric payments present worth factor (GePPWF):

$$f_{P/C_1,j}\left(y\big|\tilde{i},\tilde{g},\tilde{n}\right)=$$
$$\left[1-\left(1+f_k\left(y\big|\tilde{i}\right)\right)^{-f_j\left(y\big|\tilde{n}\right)}\otimes\left(1+f_k\left(y\big|\tilde{g}\right)^{f_j\left(y\big|\tilde{n}\right)}\right)\right]\oslash\left[f_k\left(y\big|\tilde{i}\right)\ominus f_k\left(y\big|\tilde{g}\right)\right] \quad (24)$$

where $C_1$ is the first cash of the geometric cash flows.

Geometric payments future worth factor (GePFWF):

$$f_{F/C_1,j}\left(y\big|\tilde{i},\tilde{g},\tilde{n}\right)=$$
$$\left[\left(1+f_k\left(y\big|\tilde{i}\right)\right)^{f_j\left(y\big|\tilde{n}\right)}\ominus\left(1+f_k\left(y\big|\tilde{g}\right)^{f_j\left(y\big|\tilde{n}\right)}\right)\right]\oslash\left[f_k\left(y\big|\tilde{i}\right)\ominus f_k\left(y\big|\tilde{g}\right)\right] \quad (25)$$

Capital recovery factor:

$$f_{A/P,j}\left(y\big|\tilde{i},\tilde{n}\right)=\left[f_j\left(y\big|\tilde{i}\right)\otimes\left(1+f_j\left(y\big|\tilde{i}\right)\right)^{f_k\left(y\big|\tilde{n}\right)}\right]\oslash\left[\left(1+f_j\left(y\big|\tilde{i}\right)\right)^{f_k\left(y\big|\tilde{n}\right)}-1\right] \quad (26)$$

Sinking fund factor:

$$f_{A/F,j}\left(y\big|\tilde{i},\tilde{n}\right)=f_k\left(y\big|\tilde{i}\right)\oslash\left[\left(1+f_k\left(y\big|\tilde{i}\right)\right)^{f_k\left(y\big|\tilde{n}\right)}-1\right] \quad (27)$$

## 3  Bibliography for Fuzzy Engineering Economic Analysis

In the following, most of the works on fuzzy engineering economics are listed. Many databases were examined and the followings were found. There may still be some other works those I could not find out. Here they are:

- Antonio Terceno , Jorge De Andrés , Glòria Barberà , Tomás Lorenzana, Using fuzzy set theory to analyse investments and select portfolios of tangible investments in uncertain environments, International Journal of Uncertainty, Fuzziness and Knowledge-Based Systems, v.11 n.3, p.263-281, 2003.
- Babusiaux, D., Pierru, A., Capital budgeting, project valuation and financing mix: Methodological proposals, Europian Journal of Operational Research, 135, pp. 326-337, 2001.
- Buckley, J.J., The fuzzy mathematics of finance, Fuzzy Sets and Systems, v.21 n.3, p.257-273, 1987.
- Buckley, J.J., Solving fuzzy equations in economics and finance, Fuzzy Sets and Systems, v.48 n.3, p.289-296, 1992.
- Calzi, M.L., Towards a general setting for the fuzzy mathematics of finance, Fuzzy Sets and Systems, v.35 n.3, p.265-280, 1990.
- Carlsson, C., Fuller, R., On fuzzy capital budgeting problem, in: Proceedings of the International ICSC Symposium on Soft Computing in Financial Markets, Rochester, New York, USA, ICSC Academic Press, June 22-25, 1999.
- Chansangavej, Ch., Mount-Campbell, C.A., Decision criteria in capital budgeting under uncertainties: implications for future research, Int. J. Prod. Economics 23, pp. 25-35, 1991.
- Carmichael, D.G., Balatbat, M.C.A., Probabilistic DCF Analysis and Capital Budgeting and Investment—a Survey, The Engineering Economist: A Journal Devoted to the Problems of Capital Investment, Volume 53, Issue 1, 2008, Pages 84 – 102.
- Chiu, C.-Y., Park, C.S., "Fuzzy Cash Flow Analysis Using Present Worth Criterion," The Engineering Economist, 39, (2), pp.113-138, 1994.
- Chiu, C.-Y., Park, C. S., Capital budgeting decisions with fuzzy projects, The Engineering Economist: A Journal Devoted to the Problems of Capital Investment, Vol.43, No.2, pp.125-150, 1998.
- Dimitrovski, A.D., Matos, M.A., Fuzzy engineering economic analysis, IEEE Transactions on Power Systems, Vol. 15, No. 1, pp. 283 – 289, 2000.
- Gupta, C.P., A note on the transformation of possibilistic information into probabilistic information for investment decisions, Fuzzy Sets and Systems, Vol.56, n.2, p.175-182, 1993.
- Iwamura, K., Liu, B., Chance Constrained Integer Programming Models for Capital Budgeting in Fuzzy Environments, The Journal of the Operational Research Society, Vol. 49, No. 8, pp. 854-860, 1998.
- Kahraman, C., Fuzzy versus probabilistic benefit/cost ratio analisis for public work projects, Int. J. Appl. Math. Comp. Sci., v. 11, N 3, pp.705-718, 2001.
- Kahraman, C., Beskese, B., Ruan, D., Measuring flexibility of computer integrated manufacturing systems using fuzzy cash flow analysis, Information Sciences— Informatics and Computer Science: An International Journal, v.168 n.1-4, p.77-94, 2004.

- Kahraman, C., Bozdag, C.E., Fuzzy Investment Analyses Using Capital Budgeting And Dynamic Programming Techniques, in Computational Intelligence in Economics and Finance, (edited by Shu-Heng Chen and P. Paul Wang), Springer-Verlag, Chapter 3, pp. 94-128, 2003 .
- Kahraman, C., Gülbay, M., Ulukan, Z., Applications of Fuzzy Capital Budgeting Techniques, in Fuzzy Applications in Industrial Engineering (Editor: C. Kahraman), pp. 177 – 203, Studies in Fuzziness and Soft Computing, Springer Berlin / Heidelberg, Volume 201, 2006.
- Kahraman C., Ruan D., Bozdag, C.E., Optimisation of Multilevel Investments using Dynamic Programming based on Fuzzy Cash Flows, Fuzzy Optimization and Decision Making Journal, Kluwer, Vol. 2, No. 2, pp. 101-122, 2003.
- Kahraman, C., Ruan, D., Tolga, E., Capital Budgeting Techniques Using Discounted Fuzzy Versus Probabilistic Cash Flows, Information Sciences, Vol. 142, No. 1-4, pp. 57-76, 2002.
- Kahraman, C., Tolga, E., Ulukan, Z., Justification of manufacturing technologies using fuzzy benefit/cost ratio analysis, International Journal of Production Economics, 66 (1), pp.45 –52, 2000.
- Kahraman, C.,Ulukan, Z., Continuous compounding in capital budgeting using fuzzy concept, in: Proceedings of the 6th IEEE International Conference on Fuzzy Systems (FUZZ-IEEE '97), Bellaterra, Spain, pp.1451 –1455, 1997.
- Kahraman, C., Ulukan,Z., Investment Analysis Using Grey and Fuzzy Logic, FLINS 2006, Proceedings, pp. 283-290, 29-31 August 2006, Genova-Italy.
- Kahraman, C., Ulukan, Z., Fuzzy cash flows under inflation, in: M. Mares et al eds., Proceedings of Seventh IFSA World Congress, June 25-29, 1997, Prague, Academia Prague, Vol. IV, pp.104-108, 1997.
- Kahraman, C., Ulukan, Z., Gülbay, M., Investment analyses under fuzziness using possibilities of probabilities, Proceedings, Vol. III, pp. 1721-1724, 11th IFSA World Congress, 2005, Beijing, China.
- Karsak, E.E., Measures Of Liquidity Risk Supplementing Fuzzy Discounted Cash Flow Analysis, The Engineering Economist: A Journal Devoted to the Problems of Capital Investment, Volume 43, Issue 4, 1998, Pages 331 – 344.
- Kuchta, D., Fuzzy capital budgeting, Fuzzy Sets and Systems, v.111 n.3, pp.367-385, 2000.
- Kuchta, D., A Fuzzy Model For R&D Project Selection With Benefit, Outcome And Resource Interactions, The Engineering Economist: A Journal Devoted to the Problems of Capital Investment, Volume 46, Issue 3, 2001, Pages 164 – 180.
- Liang, P., Song, F.,Computer-aided risk evaluation system for capital investment, Omega 22 (4), pp. 391- 400, 1994.
- Liou, T-S., Chen, C-W., Fuzzy Decision Analysis for Alternative Selection Using a Fuzzy Annual Worth Criterion, The Engineering Economist: A Journal Devoted to the Problems of Capital Investment, Volume 51, Issue 1, 2006, Pages 19 – 34.
- Nachtmann,H., Needy, K.L., Fuzzy Activity Based Costing: A Methodology For Handling Uncertainty In Activity Based Costing Systems, The Engineering Economist: A Journal Devoted to the Problems of Capital Investment, Volume 46, Issue 4, 2001, Pages 245 – 273.

- Omitaomu, O.A., Badiru, A., Fuzzy Present Value Analysis Model For Evaluating Information System Projects, The Engineering Economist: A Journal Devoted to the Problems of Capital Investment, Volume 52, Issue 2, 2007, Pages 157 – 178.
- Sevastjanov, P., Dimova, L., and Sevastianov, D., Fuzzy Capital Budgeting: Investment Project Evaluation and Optimization, in Fuzzy Applications in Industrial Engineering, Springer, C. Kahraman (editor), Volume 201, pp.205-228, 2006.
- Sevastianov, P., Sevastianov, D., Risk and capital budgeting parameters evaluation from the fuzzy sets theory position (in Russian), Reliable software, 1, pp. 10-19, 1997.
- Sorenson, G.E., Lavelle, J.P., A Comparison of Fuzzy Set and Probabilistic Paradigms for Ranking Vague Economic Investment Information Using a Present Worth Criterion, The Engineering Economist: A Journal Devoted to the Problems of Capital Investment, Volume 53, Issue 1, 2008, Pages 42 – 67.
- Tolga, E., Demircan, M.L., Kahraman, C., Operating System Selection using Fuzzy Replacement Analysis and Analytic Hierarchy Process, International Journal of Production Economics, Vol. 97, pp. 89-117, 2005.
- Xiaoxia Huang, Chance-constrained programming models for capital budgeting with NPV as fuzzy parameters, Journal of Computational and Applied Mathematics, v.198 n.1, p.149-159, 2007.
- Wang, M.-J., Liang, G.-S., Benefit/cost analysis using fuzzy concept, The Engineering Economist 40 (4) (1995) 359}376.
- Ward, T.L., Discounted fuzzy cash flow analysis, in 1985 Fall Industrial Engineering Conference Proceedings, pp.476 –481, 1985.

# 4 Conclusions

The basic terms of engineering economics have been examined under fuzziness in this chapter. This chapter is an introduction to fuzzy engineering economics. Fuzzy interest factors, fuzzy simple and compound interests, fuzzy nominal and effective interest rates, and fuzzy discrete and continuous compounding are the basic terms of engineering economic analysis. The equations given in this chapter can be used in developing the fuzzy present worth of gradient cash flows or geometric cash flows, or any other cash flows. The obtained results should be evaluated carefully since they involve more information than a single number gives.

# References

Blank, L.T., Tarquin, A.J.: Basics of engineering economy. McGraw-Hill, New York (2007)

Dimitrovski, A.D.: Fuzzy engineering economic analysis. IEEE Transactions on Power Systems 15(1), 283–289 (2000)

Park, C.S.: Fundamentals of engineering economics. Prentice Hall, Englewood Cliffs (2003)

Riggs, J.L., Bedworth, D.D., Randhawa, S.U.: Engineering economics. McGraw-Hill, New York (1996)

Sullivan, W.G., Wicks, E.M., Luxhoj, J.: Engineering economy. Prentice Hall, Englewood Cliffs (2002)

# Fuzzy Present Worth Analysis with Correlated and Uncorrelated Cash Flows

Aleksandar Dimitrovski[1] and Manuel Matos[2]

[1] Washington State University, School of EE&CS, P.O. Box 642752, Pullman, WA 99164, USA, 4200-465 Porto, Portugal

[2] INESC Porto & Universidade do Porto, Faculdade de Engenharia, Rua Dr Roberto Frias 378

**Abstract.** This chapter first presents arithmetic operations over independent fuzzy numbers and economic concepts review. Then, it gives the techniques for comparing and ordering fuzzy numbers of independent numbers and dependent numbers. It examines fuzzy case with partial correlation. The chapter also includes many numerical applications.

## 1 Introduction

The motivation to develop fuzzy models for financial calculations came from the recognition that discount rates and future returns and costs are affected by uncertainty that, in most cases, could hardly be modeled by probabilistic models, due to lack of statistical information. In fact, what is available, in general, are vague or qualitative expert declarations about rates and estimated returns, sometimes expressed in the form of intervals or "three point estimates" that are directly transformable in rectangular or triangular fuzzy membership functions. Fuzzy financial calculations, then, have the task of propagating this uncertainty in the data through the financial models and produce fuzzy versions of the usual financial indicators (like PW, AW, etc) able to show the Decision Maker *what may possibly happen* if the specific project under consideration is carried out.

A different point of view is expressed by Botterud et al. (2005), in the area of Power Systems planning, who use probabilistic models and argue that descriptions based on fuzzy sets "do not take into account the autocorrelation, and the gradual unfolding of new information, that is required to capture the option value of more information". This criticism shows an important path for fuzzy financial models development, but we are still aware that most deterministic calculations simply forget uncertainty and work with just a "best estimate", while some probabilistic models rely on assumed statistical distributions and "typical" parameters, without a real data support. Note that, if this support from past experience exists, there is no objection on the use of probabilistic models, at least if the user is convinced that the future will be a "smooth" continuity from the past.

One way or another, the number of new published material in the area of fuzzy financing calculations has been increasing through the years. Without any claim of completeness, we selected some references that may help the reader get insight into

C. Kahraman (Ed.): Fuzzy Engineering Economics with Appl., STUDFUZZ 233, pp. 11–41, 2008.
springerlink.com © Springer-Verlag Berlin Heidelberg 2008

application areas. Some of them are briefly commented below, just to manifest the diversity of the contributions.

Beskese et al. (2004) review alternative ways of quantifying flexibility in manufacturing systems through the use of fuzzy models that include present worth calculations. Capital budgeting attracted the attention of Kahraman et al (2002) and, more recently, of Huang (2007), with an interesting proposal of a fuzzy chance-constrained model. An extensive study on operating system selection, using a fuzzy version of AHP and fuzzy financial calculations was presented by Tolga et al. (2005). Rebiasz (2007) addressed the issue of project risk appraisal, mixing fuzzy and probabilistic models. Sheen contributed to the area with a number of different approaches, mainly addressing electricity demand-side management issues (Sheen 2005a, 2005c, 2005d) but also regarding cogeneration alternatives in the petrochemical industry (Sheen 2005b).

Regarding the specific aspects of fuzzy financial calculations, it should be pointed out that not all the models described in the literature take into account the possible dependence between fuzzy numbers, a subject that will be developed in the next sections of this chapter. In many cases, this is not an issue, either because the assumption of independence is correct or the approximation is acceptable. However, there are situations where the uncertainty of the results may be artificially exaggerated by this simplification, as will be demonstrated in the sequel.

## 2 Economic Concepts Review

Every technically feasible solution has to be analyzed for its economic efficiency. Only then a decision can be made which of the solutions, if any, are justified and worth doing. Here, we give a brief review of some the basic concepts needed for the economic analysis.

### 2.1 Time Value of Money

The value of money depends on time and this is a fundamental economic concept. A certain amount of money today is worth more in the future and a certain amount of money in the future is worth less today. Thus, money has an inherent property to grow. This is usually described as the cost of money and is given as a percentage rate over a specified period, typically one year in engineering economics. The cost of money may include not only the real cost, but also an adjustment for inflation and premiums to cover different uncertainties. The combined rate is usually referred to as the *discount rate* (or simply *interest rate*). For example, if the discount rate is $i$ %/year, an amount of money $P$ now will be worth $P(1+i)$ after one year. The same amount of money will be worth $P(1+i)^n$ after $n$ years, if the discount rate is kept constant over the period.

### 2.2 Cash Flow

Different solutions (plans) incur different monetary transactions scattered over time. A graphical representation of these transactions in the form of cash flow diagram is

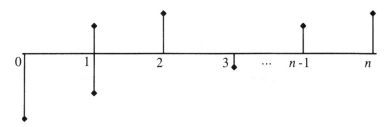

**Fig. 1.** Cash flow diagram

used to help describe the problem. Although the transactions may occur at any point in time, they are usually lumped at the end of the periods used to divide the time axis. In engineering economics, the period is typically one year, but this depends on the dynamics of the problem at hand. Figure 1 shows an example of cash flow diagram with positive arrows describing revenues (benefits) and negative arrows describing costs.

In order to compare solutions with different cash flows, because of the time value of money, they first need to be made financially equivalent. Several, well known, expressions are useful in the process of equivalence.

### 2.3 Equivalence Formulae

We already gave the expression for finding the future value $F$ of a present amount $P$, at the end of year $n$, if the discount rate is $d$: $F = P(1+i)^n$. The ratio $F/P$ is given by:

$$(F/P, i, n) = (1+i)^n \tag{1}$$

and is called the compound interest factor.

The inverse factor, used to find the present worth of a future amount, is given by:

$$(P/F, i, n) = 1/(1+i)^n \tag{2}$$

and is called the discount factor or the present worth factor.

Very often, the cash flow contains a series of equal payments called *annuities*. This is shown on Figure 2.

It is easy to find the equivalent present worth $P$ of such a series of annuities, each with value $A$. The ratio $P/A$ is given by:

$$(P/A, i, n) = \frac{(1+i)^n - 1}{i \cdot (1+i)^n} \tag{3}$$

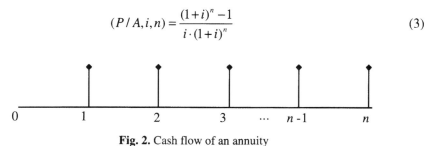

**Fig. 2.** Cash flow of an annuity

and is called the *present worth annuity factor*. Note that the case discussed here is when the first payment occurs at the end of the first year. A similar expression can be derived if the first payment occurs at the beginning of the first year.

The inverse factor, used to find the annuity of a present value, is given by:

$$(A/P, i, n) = \frac{i \cdot (1+i)^n}{(1+i)^n - 1} \quad (4)$$

and is called the *capital recovery factor*, in reference to the series of equal future payments needed to repay the borrowed capital (for example, a house mortgage loan).

Another pattern that usually occurs in cash flows is geometric series. It consists of a series of payments which increase by a factor $(1+f)$ each period. Thus, $A_{k+1} = A_k(1+f)$, $k \geq 1$. This is shown on Figure 3. Such series is used to describe costs that escalate each year, for example, due to inflation.

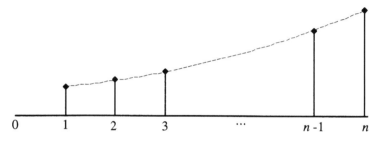

**Fig. 3.** Cash flow of a geometric series

With some algebra, it is not difficult to find the equivalent present worth $P$ of such a series of annuities that starts with value $A_1$. The ratio $P/A_1$ is given by:

$$(P/A_1, i, f, n) = \frac{1 - \left(\frac{1+f}{1+i}\right)^n}{(i-f)} \quad (5)$$

and is called the geometric series present worth factor.

Again, note that the case discussed here is when the first payment occurs at the end of the first year. It is easy to find the equivalent expression when $A_1$ occurs at the beginning of the first year.

All these factors can be combined appropriately to derive new equivalence expressions. For example, multiplying (4) by (5), we can obtain the ratio $A/A_1$:

$$(A/A_1, i, f, n) = \frac{i \cdot [(1+i)^n - (1+f)^n]}{[(1+i)^n - 1] \cdot (i-f)} \quad (6)$$

which is the geometric series annuity factor.

Another frequently used factor can be derived by multiplying (2) and (4) to obtain the ratio $A/F$:

$$(A/F, i, n) = \frac{i}{(1+i)^n - 1} \tag{7}$$

which is called the *sinking fund factor*, in reference to the annuity required to set aside in an account ("sinking fund") so that an amount $F$ is available at the end of year $n$.

## 2.4 Techniques of Comparison

Once the cash flows of different alternatives are recognized and determined, several techniques can be employed to measure their efficiency and rank them accordingly. The most frequently used ones are:

- Payback period: determines the number of years required to recover the initial investments;
- Present worth: converts all cash flows to a single equivalent sum at present time;
- Equivalent annual worth: converts all cash flows to an equivalent annuity;
- Future worth: converts all cash flows to a single equivalent future sum at the end of the period of consideration;
- Rate of return: finds the interest rate that yields zero present worth;
- Benefit/Cost ratio: finds the ratio of the equivalent present worth of benefits and costs, or the equivalent annual worth of benefits and costs.

All of the above techniques, except the first one, are known as exact analysis techniques and will give the same optimal alternative. Each of them can be applied to any given problem. However, depending on the particular problem and associated data, there can be certain advantages to applying one technique or another.

This chapter and the following one deal with present worth and equivalent annual worth techniques in their fuzzy form. Therefore, we define only these two formally here, but the others follow analogously.

The net present worth (present value) for alternative $k$, $\text{NPW}_k$, that has a general cash flow diagram like the one shown on Figure 1, is the sum of all its annual cash flows, discounted to the present time:

$$\text{NPW}_k = \sum_{j=0}^{n} A_{kj} / (1+i)^j \tag{8}$$

where

- $A_{kj}$ is the net cash flow for alternative $k$ at the end of period $j$;
- $n$ is the number of periods (years) in the planning horizon;
- $i$ is the discount rate.

Alternatives are ranked according to their NPW with the best one having the highest NPW. This, of course, has to be a positive number for the corresponding alternative to have a return rate bigger than the discount rate and be attractive. Otherwise, the money is better spent elsewhere in the economy.

The equivalent net annual worth for alternative $k$, $EAW_k$, can be obtained simply by multiplying the $NPW_k$ with the capital recovery factor given in (4):

$$\text{EAW}_k = \frac{i \cdot (1+i)^n}{(1+i)^n - 1} \sum_{j=0}^{n} A_{kj} / (1+i)^j \tag{9}$$

Again, alternatives are ranked according to their EAW with the best one having the highest EAW. Like with NPW, this has to be a positive number for the alternative to be attractive, or the opportunity cost of the money will not be covered.

There are special problems where the revenues are fixed and the alternatives differ in their costs. In such cases it is easier to consider only the costs in (8) or (9), where they are typically taken as positive numbers. Then, the best alternative is the one that has the lowest NPW or EAW.

Similarly, there are special problems where the costs are fixed and the alternatives differ in their revenues only. Then, it is easier to simply omit the costs in (8) or (9) and perform the analysis as usual.

In general, all the variables in (8) or (9) can be uncertain and defined as fuzzy numbers. But, even if only one of them is fuzzy, the NPW or EAW will turn fuzzy, too. Then the analysis becomes much more complicated as shown in the sequel.

### 2.5 Arithmetic Operations over Fuzzy Numbers That Are Not Independent

Arithmetic operations over fuzzy numbers follow the rules established by the *Extension Principle* (Zadeh 1965), synthesized in equation (10) for a generic operation $\theta$, where $\tilde{Z} = \theta(\tilde{X}, \tilde{Y})$.

$$u(z) = \sup_{z = \theta(x,y)} \left\{ \min\left( u_{\tilde{X}}(x), u_{\tilde{Y}}(y) \right) \right\} \tag{10}$$

When $\tilde{X}$ and $\tilde{Y}$ are independent and take standard shapes like trapezoidal numbers, this leads most of the times to straightforward rules for the ordinary arithmetic operations, for instance:

$$\tilde{Z} = \tilde{X} + \tilde{Y} \quad [z_1; z_2; z_3; z_4] = [x_1+y_1; x_2+y_2; x_3+y_3; x_4+y_4] \tag{11}$$

$$\tilde{Z} = \tilde{X} - \tilde{Y} \quad [z_1; z_2; z_3; z_4] = [x_1-y_4; x_2-y_3; x_3-y_2; x_4-y_1] \tag{12}$$

As an example, if a specific activity leads to an estimated cost of $\tilde{C} = [100; 120; 130; 150]$ and an estimated return of $\tilde{R} = [100; 120; 130; 150]$, the net return would be $\tilde{N} = \tilde{R} - \tilde{C} = [-50; -10; 10; 50]$ if $\tilde{C}$ and $\tilde{R}$ are independent.

Now, consider the case where $\tilde{R}$ is a reimbursement of the cost $\tilde{C}$, so for each specific value of the (uncertain) cost, the reimbursement has exactly the same value.

This corresponds to a complete dependence between the two variables, so equation (12) is no longer applicable and the general equation (10) must be used instead, in the form:

$$u(z) = \sup_{z=\theta(x,y)} \{\min(u_{\tilde{X}}(x), u_{\tilde{Y}}(y|x))\} \qquad (13)$$

leading to $\tilde{R} - \tilde{C} = 0$, a crisp result that means "in any circumstance, the final net return is null". The joint possibility distributions for the two situations, shown in Figure 4 and Figure 5, turn evident the difference between them.

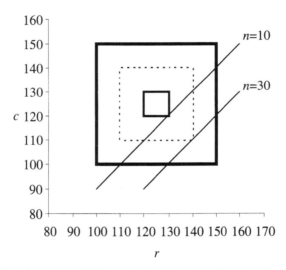

**Fig. 4.** Joint possibility distribution (independent variables)

In Figure 4, the situation of independent variables is addressed, so all the combinations of the possible values of $\tilde{R}$ and $\tilde{C}$ are considered. The outer square corresponds to the limit of the domain where $\min(u_{\tilde{C}}, u_{\tilde{R}}) = 0$, while the inner square corresponds to the zone where $\min(u_{\tilde{C}}, u_{\tilde{R}}) = 1$. The intermediate square (dotted) corresponds to $\min(u_{\tilde{C}}, u_{\tilde{R}}) = 0.5$. The locus of all the combinations of $\tilde{C}$ and $\tilde{R}$ that lead to $n = 10$ and $n = 30$ are also depicted. In the first case, the maximum degree of membership is 1 ($r = 130$, $c = 120$), while in the second the maximum degree of membership is 0.5 ($r = 140$, $c = 110$).

Figure 5 shows the situation when $\tilde{R}$ and $\tilde{C}$ are completely correlated ($r = c$, where $r$ and $c$ are possible instances of $\tilde{R}$ and $\tilde{C}$, respectively), so for all the situations where $r \neq c$ we get $\min(u_{\tilde{C}}, u_{\tilde{R}}) = 0$. Therefore, in this case only the result $n = 0$ (shown in the picture) has a degree of membership greater than zero ($u_{\tilde{N}}(0) = 1$), leading to a crisp result already commented.

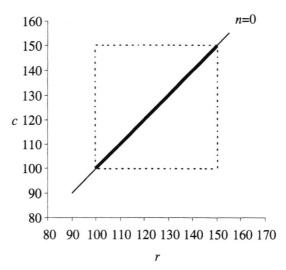

**Fig. 5.** Joint possibility distribution (dependent variables)

This issue is especially important when dealing with expressions where the same fuzzy number appears more than once, like

$$\tilde{Z} = \frac{\tilde{X}}{\tilde{Y}} = \frac{\tilde{A} - \tilde{B}}{\tilde{A} + \tilde{B}} \qquad (14)$$

One could think that performing first the subtraction and the sum and then the division (in all the cases using the general rules of operation) could be a correct procedure, but that is not the case. In fact, even if $\tilde{A}$ is uncertain, it never takes different values in the numerator and denominator. To turn the example simple, let us consider $\tilde{A}$ and $\tilde{B}$ to be intervals, say [21 25] and [10 13], respectively. Now, we could obtain, using ordinary operations for independent numbers:

$$\tilde{X} = \tilde{A} - \tilde{B} = [8 \quad 15]$$

$$\tilde{Y} = \tilde{A} + \tilde{B} = [31 \quad 38]$$

and

$$\tilde{Z}_{inaccurate} = \frac{\tilde{X}}{\tilde{Y}} = \frac{[8 \quad 15]}{[31 \quad 38]} = \left[\frac{8}{38} \quad \frac{15}{31}\right] = [0.21 \quad 0.48]$$

This is inaccurate, since we are considering $\tilde{A}$ to be simultaneously equal to 21 and 25 (8=**21**-13, 38 = **25** + 13) on the left side, and the same on the other side. The correct value of $\tilde{Z}$ would then be given by:

$$\tilde{Z} = \left[\min_{\substack{a \in [21\ 25] \\ b \in [10\ 13]}} \left(\frac{a-b}{a+b}\right) \quad \max_{\substack{a \in [21\ 25] \\ b \in [10\ 13]}} \left(\frac{a-b}{a+b}\right)\right] = \left[\frac{21-13}{21+13} \quad \frac{25-10}{25+10}\right] = [0.24 \quad 0.43] \qquad (15)$$

Comparing the two results, it is easy to conclude that the inaccurate value shows an additional uncertainty that is not real, corresponding to impossible situations of the combined instances of $\tilde{A}$ and $\tilde{B}$.

This aspect is of major importance in the calculation of the Present Value of projects, since some of the uncertain variables (like interest rates) influence both costs and returns. On the other hand, comparison between different projects must also take into account the dependences motivated by the use of common uncertain parameters.

In formal terms, the recommended strategy - used in (15) - consists of rewriting the arithmetic operation using only independent variables and applying directly the *Extension Principle*. If the dependent $\tilde{X}$ and $\tilde{Y}$ can be expressed in terms of independent (primary) variables $\tilde{A}, \tilde{B},\dots$, we may substitute the operation $\tilde{Z} = \theta(\tilde{X},\tilde{Y})$ by $\tilde{Z} = \theta(\tilde{A}, \tilde{B},\dots)$ and determine, for each $\alpha$-level, the interval $\tilde{Z}(\alpha) = \left[ z_L(\alpha) \quad z_R(\alpha) \right]$ through:

$$z_L(\alpha) = \min_{\substack{a \in \tilde{A}(\alpha) \\ b \in \tilde{B}(\alpha) \\ \dots}} \theta(a,b,\dots) \tag{16}$$

and

$$z_R(\alpha) = \max_{\substack{a \in \tilde{A}(\alpha) \\ b \in \tilde{B}(\alpha) \\ \dots}} \theta(a,b,\dots) \tag{17}$$

This strategy is also adequate for situations of partial dependence, where rewriting the problem generally simplifies the calculations and eliminates the risk of obtaining inaccurate results. For instance, consider $\tilde{X} = \tilde{A} = [20 \quad 30]$ and $\tilde{Y} = 2\tilde{A} + [13 \quad 17] = [53 \quad 77]$, so $\tilde{Y}$ and $\tilde{X}$ are partially dependent. Taking into account the dependence, the value of $\tilde{Z} = \tilde{Y} - \tilde{X}$ is obtained by

$$\tilde{Z} = \tilde{Y} - \tilde{X} = 2.\tilde{A} + [13 \quad 17] - \tilde{A}$$

or, after some simplification,

$$\tilde{Z} = \tilde{A} + [13 \quad 17] = [33 \quad 47]$$

Note that this is very different from the inaccurate result [53 77] − [20 30] = [23 57] that would be obtained ignoring the partial dependence between $\tilde{Y}$ and $\tilde{X}$.

When addressing models with many variables and complex dependences, these are generally described by a vector of equations $\mathbf{g}(x, y, \dots) = \mathbf{0}$. In that case, the previous approach must be extended, leading to the following mathematical programs, to be solved for each $\alpha$-level:

$$
z_L(\alpha) = \quad \min \quad \theta(x, y, \ldots)
$$
$$
\text{subj to:} \quad x \in \tilde{X}(\alpha)
$$
$$
y \in \tilde{Y}(\alpha) \tag{18}
$$
$$
\ldots
$$
$$
\mathbf{g}(x, y) = 0
$$

and

$$
z_R(\alpha) = \quad \max \quad \theta(x, y, \ldots)
$$
$$
\text{subj to:} \quad x \in \tilde{X}(\alpha)
$$
$$
y \in \tilde{Y}(\alpha) \tag{19}
$$
$$
\ldots
$$
$$
\mathbf{g}(x, y) = \mathbf{0}
$$

Note that, in the case of complex models, the solutions of these mathematical programming problems are not necessarily constituted by combinations of the maximum and minimum values of $x$, $y$, etc. Conversely, in more simple situations like the previous examples, it is generally easy to find by inspection the correct extremes of $\tilde{X}(\alpha)$, $\tilde{Y}(\alpha)$ that should be used. However, the dependence is commonly ignored in practice, possibly because it is not recognized.

## 2.6 Example of Financial Calculations

In order to apply the preceding considerations to the theme of this chapter, consider the $(P/A, i, n)$ factor, that is, the expression that relates the present value $P$ with a uniform series $A$, at a known interest rate $i$ in a horizon of $n$ years:

$$
(P/A, i, n) = \frac{(1+i)^n - 1}{i \cdot (1+i)^n}
$$

Now, consider that the rate $i$ is estimated to lie between 7 % and 10 % and $n = 10$ years. It is easy to see that the minimum value for $P/A$ corresponds to $i = 10\%$, while the maximum is obtained for $i = 7\%$, giving $P/A = [6.145 \ 7.024]$ (note the "compression" of uncertainty, due to the nonlinear nature of the equation). This is the correct answer, in accordance with the *Extension Principle*. But, if the two terms of the fraction are calculated separately:

$$
(1+i)^n - 1 = [0.9672 \ 1.59371]
$$
$$
i.(1+i)^n = [0.1377 \ 0.2594]
$$

it may lead to the result that ignores the dependence

$$
P/A_{inaccurate} = \begin{bmatrix} 0.9672 & 1.5937 \\ 0.2594 & 0.1377 \end{bmatrix} = [3.729 \ 11.574]
$$

when different instances of $i$ were used at the same time to calculate both the left and right values of the uncertain $P/A$. Therefore, the correct interval $P/A$ when $i = [i_{min}\ i_{max}]$ is:

$$(P/A,i,n) = \left[ \frac{(1+i_{max})^n - 1}{i_{max} \cdot (1+i_{max})^n} \quad \frac{(1+i_{min})^n - 1}{i_{min} \cdot (1+i_{min})^n} \right] \tag{20}$$

If the project horizon is also uncertain, $n = [n_{min}\ n_{max}]$, a similar reasoning leads to:

$$(P/A,i,n) = \left[ \frac{(1+i_{max})^{n_{min}} - 1}{i_{max} \cdot (1+i_{max})^{n_{min}}} \quad \frac{(1+i_{min})^{n_{max}} - 1}{i_{min} \cdot (1+i_{min})^{n_{max}}} \right] \tag{21}$$

For instance, assuming the same uncertain rate as before and $n = [10\ 15]$ years, we would find:

$$P/A = \left[ \frac{(1+0.1)^{10} - 1}{0.1\,(1+0.1)^{10}} \quad \frac{(1+0.07)^{15} - 1}{0.07\,(1+0.07)^{15}} \right] = [6.1446\quad 9.1079]$$

Now, if the fuzzy information is represented by triangular or trapezoidal numbers, we just need to repeat the previous process for different $\alpha$ levels, with the granularity required by the problem. It should be pointed out that the calculations do not preserve the shape of the data, due to the nonlinearity of the mathematical expression of $P/A$. However, triangular or trapezoidal approximations can be easily obtained that do not introduce too much error. In such cases, it is sufficient to calculate the extreme $\alpha$-cuts ($\alpha = 0$ and $\alpha = 1$) and assume linearity to complete the fuzzy $P/A$ membership function.

To illustrate, set $i = [7\%;\ 8.5\%;\ 10\%]$ and $n = [10;\ 12;\ 13;\ 15]$. Since $\alpha = 0$ leads to the case described above,

$$P/A_{\alpha=0} = [6.1446\quad 9.1079]$$

we only need to calculate for $\alpha = 1$, where $i = 8.5\%$ and $n = [12\ 13]$. Using again (21), we obtain:

$$P/A_{\alpha=1} = \left[ \frac{(1+0.085)^{12} - 1}{0.085\,(1+0.085)^{12}} \quad \frac{(1+0.085)^{13} - 1}{0.085\,(1+0.085)^{13}} \right] = [7.3447\quad 7.6910]$$

The complete representation of the fuzzy $P/A$ is shown on Figure 6, where we depicted the exact membership function (bold) and the trapezoidal approximation.

The following comparison of the exact and approximated 0.5-cuts shows that the error is negligible (less than 0.5% in both extremes):

$$P/A_{\alpha=0.5}^{exact} = [6.7255\quad 8.3653] \qquad P/A_{\alpha=0.5}^{approx} = [6.7446\quad 8.3994]$$

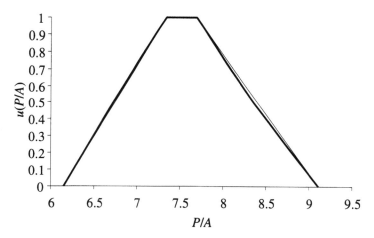

**Fig. 6.** Example of *P/A* exact and approximate values

## 3 Comparing and Ordering Fuzzy Numbers

### 3.1 Independent Numbers

In many decision situations, the final task consists of choosing among alternatives represented by a fuzzy attribute, or ordering a set of such alternatives. We will address this issue more thoroughly later in this text, but will review now the calculation of crisp equivalents, a straightforward approach for comparing and ordering fuzzy numbers.

Perhaps the most popular way of collapsing a fuzzy number in a single number is the Removal, defined as the average of the external and internal areas defined by an origin and the fuzzy number. Figure 7 illustrates the definition for a trapezoidal fuzzy

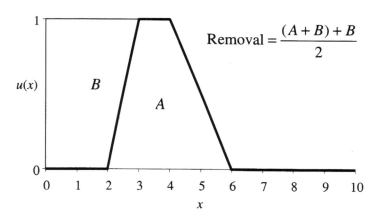

**Fig. 7.** Removal definition (origin=0)

number, when the origin is 0 (which is normally the case, so we will take it as the default situation).

It is easy to see that, for a trapezoidal number:

$$Removal_{trap} = \frac{a+b+c+d}{4} \tag{22}$$

The expressions for triangular numbers and intervals are immediate:

$$Removal_{tri} = \frac{a+2b+c}{4} \tag{23}$$

$$Removal_{int} = \frac{a+b}{2} \tag{24}$$

A more complex alternative is the Center of Gravity (COG) also known as *centroid*, defined for a general fuzzy number whose support is the interval $[a\ b]$ as:

$$COG = \frac{\int_a^b x \cdot u(x)\,dx}{\int_a^b u(x)\,dx} \tag{25}$$

The expressions for the three main types of fuzzy numbers follow:

$$COG_{trap} = \frac{1}{3} \cdot \frac{(a^2 + b^2 + ab) - (c^2 + d^2 + cd)}{(a+b) - (c+d)} \tag{26}$$

$$COG_{tri} = \frac{a+b+c}{3} \tag{27}$$

$$COG_{int} = \frac{a+b}{2} \tag{28}$$

So, the removal appears to be a good (and simpler) approximation of the COG that coincides with it if the fuzzy number is symmetrical (which is always the case for intervals). Note that, if a trapezoidal number is symmetrical, both the expressions for the COG and Removal reduce to $(a+d)/2 = (b+c)/2$.

Now, using either the Removal or the COG, it is possible to induce an order over a set of fuzzy numbers. If needed, the median of trapezoidal numbers $(b + c)/2$ can be used to untie (if the numbers are not symmetrical). As a final resource to untie, the amplitude of the fuzzy number support interval can be used. The sequence of criteria (removal, median, amplitude) follows a suggestion of (Kaufmann and Gupta 1988), but the alternative of using the COG should not be neglected. For instance, when comparing the two triangular numbers for cost $\tilde{A} = [6;\ 10;\ 18]$ and $\tilde{B} = [5;\ 11;\ 17]$, the removal is the same (11) while the COG favors slightly $\tilde{B}$ ($COG(\tilde{A}) = 11.33$, $COG(\tilde{B}) = 11$).

## 3.2 Dependent Fuzzy Numbers

In the case of dependent fuzzy numbers, using the preceding approach to compare two fuzzy numbers $\tilde{A}$ and $\tilde{B}$ may lead to inaccurate results, so it is recommended to first perform the operation $\tilde{A} - \tilde{B}$ and then, if necessary, calculate COG ($\tilde{A} - \tilde{B}$).

Ordering more than two dependent fuzzy numbers is a more complex task, since the COG of the differences are not necessarily additive: COG($\tilde{A} - \tilde{B}$) + COG($\tilde{B} - \tilde{C}$) may be different from COG($\tilde{A} - \tilde{C}$). This circumstance may produce intransitive situations.

## 3.3 Decision Issues

The calculation of the fuzzy present worth (value) is a means to help deciding about projects in the presence of uncertainty. The basic reasoning of the methodology, in the crisp case, can be stated as the following rule: "the project is worthwhile if NPW > 0", meaning that the balance of costs and returns, taking into account the specified discount rate through the years, is positive. This is a simple test, but not so simple when the uncertainty leads to a fuzzy NPW that may take negative values.

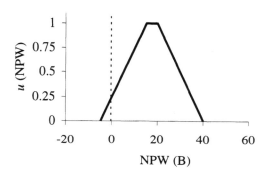

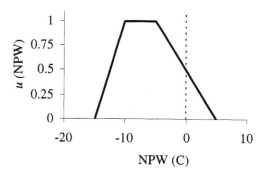

**Fig. 8.** Two cases of partly negative NPW ($\alpha_B$=0.25, $\alpha_C$=1)

Consider, for instance, the case of projects A, B and C, with $\widehat{NPW}_A$ = [5; 10; 20; 25], $\widehat{NPW}_B$ = [-5; 15; 20; 40] and $\widehat{NPW}_C$ = [15; -10; -5; 5]. Project A clearly passes the test, since its NPW is positive in all situations. Let us concentrate on projects B and C, whose NPWs are depicted in Figure 8.

Although both projects B and C have degrees of membership to negative values of their respective NPW, we see easily that this only happens for a small part of the support of $NPW_B$, while most of the $NPW_C$ support is negative. Of course, we may decide to reject both projects, if we want to be strict about not accepting NPW < 0, but to make a distinction between the two situations is always useful.

In fact, the maximum degree of membership of $NPW_B$ that leads to NPW < 0 is 0.25, as it is apparent from Figure 8 (and can be deducted analytically). We may say that the *exposure* of $NPW_B$ regarding negative values is 0.25, or that its *robustness* is 0.75 (=1-0.25). So, even if we cannot say that $NPW_B$ is positive we have a measure of the possibility of not having a negative NPW. On the other hand, the maximum degree of membership of $NPW_C$ that leads to NPW < 0 is 1 (maximum exposure), so the robustness of $NPW_C$ regarding negative values is 0.

The case of Project B is interesting also because it shows that aggregate measures sometimes do not tell the whole story. In this case, $COG_B$ = 17.5, which is even better that $COG_A$ = 15 (of course, there is no surprise with $COG_C$ = -6). Therefore, in an uncertain environment, it is always advisable to characterize a project, not only by a central measure (like COG, or Removal) that shows the general tendency of the data, but also by a risk indicator (like the *robustness index* just described). Afterwards, it is a matter of the risk attitude of the decision maker, which may, or may not, prefer losing some opportunities in order to exclude the possibility of a negative NPW.

### 3.4 NPW Example 1

Let us consider a simple example without many complicated time value calculations of money. This way, the arithmetic complexities will not obscure the effects of introducing fuzzy variables in the analysis.

A factory is faced with two alternatives denoted A and B. Alternative A is to make an investment and purchase new equipment that will result in decreased operating costs. Alternative B is to continue operation with the current equipment i.e. donothing alternative. Both alternatives are assumed to have the same revenues and a book life of twenty years.

### 3.4.1 Crisp Case

Table 1 shows the costs associated with each alternative for the crisp case. The '$' symbol is used to denote a generic monetary unit, not necessarily the US currency.

**Table 1.** Costs for two alternatives ($) - crisp case

| Table . | Initial cost | Operating costs |
|---|---|---|
| Alternative A | 600,000 | 50,000 |
| Alternative B | 0 | 120,000 |

## 26 A. Dimitrovski and M. Matos

**Table 2.** Present value of cumulative costs ($) – crisp case

| Year | Alternative A | Alternative B |
|------|--------------|--------------|
| 1 | 646,296 | 111,111 |
| 2 | 689,163 | 213,992 |
| 3 | 728,855 | 309,252 |
| 4 | 765,606 | 397,455 |
| 5 | 799,636 | 479,125 |
| 6 | 831,144 | 554,746 |
| 7 | 860,319 | 624,764 |
| 8 | 887,332 | 689,597 |
| 9 | 912,344 | 749,627 |
| 10 | 935,504 | 805,210 |
| 11 | 956,948 | 856,676 |
| 12 | 976,804 | 904,329 |
| 13 | 995,189 | 948,453 |
| 14 | 1,012,212 | 989,308 |
| 15 | 1,027,974 | 1,027,137 |
| 16 | 1,042,568 | 1,062,164 |
| 17 | 1,056,082 | 1,094,597 |
| 18 | 1,068,594 | 1,124,626 |
| 19 | 1,080,180 | 1,152,432 |
| 20 | 1,090,907 | 1,178,178 |

The initial cost consists of purchase costs and installation costs. The operating costs include raw material, direct and indirect labor, and maintenance. The discount rate is assumed to be 8%/year. Inflation, costs escalation, and tax implications are neglected.

Because of the assumption of equal revenues in both alternatives, we only need to deal with costs. In such case, as usual, costs are represented as positive numbers and the objective is their minimization.

Table 2 shows present values (present worth) of cumulative costs for both alternatives during their lifetime. This is also shown on Figure 9. As can be seen from these data, starting from year 16 which is the break-even year, the alternative A yields smaller cumulative costs than alternative B. The total costs at the end of the project are 1,090,907 and 1,178,178, for alternatives A and B respectively. Thus, alternative A is preferable and the investment in the new equipment pays off over time.

### 3.4.2 Fuzzy Case

Let us now take into account a few likely uncertainties in the data given above, by defining them with trapezoidal FN. We will use ±10% variation in crisp costs to define the 0-cut intervals of the fuzzy numbers, and ±5% for the 1-cuts. Thus, the crisp costs

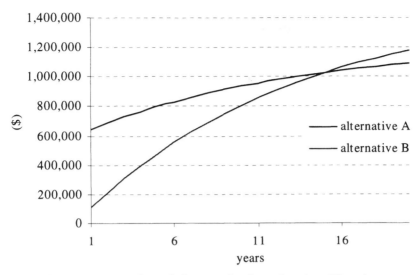

**Fig. 9.** Present values of cumulative costs for alternatives A and B – crisp case

are the median values of the corresponding fuzzy numbers. Table 3 shows the costs associated with each alternative in this case.

We will also assume that the discount rate is uncertain, modeled with the trapezoidal FN: [7%; 7.5%; 8.5%; 9%]/year.

**Table 3.** Costs for two alternatives ($) - fuzzy case

|  | Initial cost |
|---|---|
| Alternative A | [540,000; 570,000; 630,000; 660,000] |
| Alternative B | 0 |
|  | Operating costs |
| Alternative A | [45,000; 47,500; 52,500; 55,000] |
| Alternative B | [108,000; 114,000; 126,000; 132,000] |

Table 4 shows the present values for alternatives A and B in this case. The results are now in the form of fuzzy numbers and the table shows the lower and upper bounds of both 0-cuts and 1-cuts. These are also the characteristic points of the trapezoidal approximations of the fuzzy numbers.

The present values of the total costs for both alternatives are also shown on Figure 10 as trapezoidal fuzzy numbers. These FNs correspond to the data from the last year in Table 4.

At first sight, when comparing the outcomes from the two alternatives shown on Figure 10, it may seem again that alternative A is preferred over alternative B, even with all uncertainties included in the analysis. This conclusion could be backed up by one definition of FN comparison which states that one FN is smaller (larger) than the

**Table 4.** Present value of cumulative costs ($) – fuzzy case

| | Alternative A | | | | Alternative B | | | |
|---|---|---|---|---|---|---|---|---|
| Year | $\alpha = 0$ | $\alpha = 1$ | $\alpha = 1$ | $\alpha = 0$ | $\alpha = 0$ | $\alpha = 1$ | $\alpha = 1$ | $\alpha = 0$ |
| 1 | 581,284 | 613,779 | 678,837 | 711,402 | 99,083 | 105,069 | 117,209 | 123,364 |
| 2 | 619,160 | 654,128 | 724,267 | 759,441 | 189,984 | 201,907 | 226,241 | 238,658 |
| 3 | 653,908 | 691,316 | 766,528 | 804,337 | 273,380 | 291,159 | 327,666 | 346,410 |
| 4 | 685,787 | 725,591 | 805,840 | 846,297 | 349,890 | 373,418 | 422,015 | 447,112 |
| 5 | 715,034 | 757,180 | 842,409 | 885,511 | 420,082 | 449,233 | 509,781 | 541,226 |
| 6 | 741,866 | 786,295 | 876,427 | 922,160 | 484,479 | 519,109 | 591,425 | 629,183 |
| 7 | 766,483 | 813,129 | 908,072 | 956,411 | 543,559 | 583,511 | 667,372 | 711,386 |
| 8 | 789,067 | 837,861 | 937,508 | 988,421 | 597,760 | 642,867 | 738,020 | 788,211 |
| 9 | 809,786 | 860,655 | 964,892 | 1,018,338 | 647,487 | 697,573 | 803,740 | 860,011 |
| 10 | 828,795 | 881,664 | 990,364 | 1,046,297 | 693,107 | 747,994 | 864,874 | 927,113 |
| 11 | 846,234 | 901,027 | 1,014,060 | 1,072,427 | 734,961 | 794,464 | 921,743 | 989,825 |
| 12 | 862,233 | 918,873 | 1,036,102 | 1,096,848 | 773,358 | 837,294 | 974,645 | 1,048,435 |
| 13 | 876,911 | 935,320 | 1,056,607 | 1,119,671 | 808,586 | 876,769 | 1,023,856 | 1,103,210 |
| 14 | 890,377 | 950,480 | 1,075,681 | 1,141,001 | 840,904 | 913,151 | 1,069,633 | 1,154,402 |
| 15 | 902,731 | 964,451 | 1,093,424 | 1,160,935 | 870,554 | 946,683 | 1,112,217 | 1,202,245 |
| 16 | 914,065 | 977,328 | 1,109,929 | 1,179,566 | 897,756 | 977,588 | 1,151,830 | 1,246,958 |
| 17 | 924,463 | 989,197 | 1,125,283 | 1,196,977 | 922,712 | 1,006,072 | 1,188,679 | 1,288,745 |
| 18 | 934,003 | 1,000,135 | 1,139,565 | 1,213,250 | 945,608 | 1,032,324 | 1,222,957 | 1,327,799 |
| 19 | 942,755 | 1,010,217 | 1,152,852 | 1,228,458 | 966,612 | 1,056,520 | 1,254,844 | 1,364,299 |
| 20 | 950,785 | 1,019,508 | 1,165,211 | 1,242,671 | 985,883 | 1,078,820 | 1,284,506 | 1,398,410 |

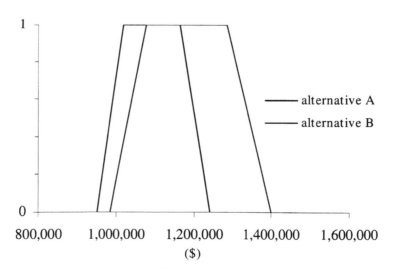

**Fig. 10.** Final present values of costs for alternatives A and B – fuzzy case

other FN if, for each α-cut, the lower and upper bounds of the first FN are smaller (larger) than the respective bounds of the second FN.

Although this definition seems logical, it is not always applicable as it implies correlation between the two variables represented with FNs. It implies that the two variables behave in a same manner. When one of the variables takes a certain value from its support set, the other variable takes a value in proportion to that one from its support set. In other words, the variables "travel" along their membership functions in unison.

If we assume that the variables representing present values of the cumulative costs are independent from each other, then we can not say right away which alternative is better. Alternative B may take a value on the left side of its membership function, while alternative A, at the same time, may take a value on the right side of its membership function. In such scenario, alternative B would have smaller overall costs and, hence, would be preferable. Similarly, there are scenarios when the opposite happens and alternative A is preferable.

The difference between the two FNs from Figure 10, for the two cases of correlated and uncorrelated costs, are shown on Figure 11. When the costs are correlated, the difference between the costs in alternative A and alternative B is always a negative number. In this case, no matter the scenario, alternative A always has smaller total costs and is preferable. If the difference had been always a positive number, then alternative B would have been always preferable.

When the costs are uncorrelated, their difference becomes a fuzzy number around the origin which includes both negative and positive numbers. Choosing an alternative in this case is not straightforward anymore. It becomes a more complicated decision making process that involves additional analysis and criteria.

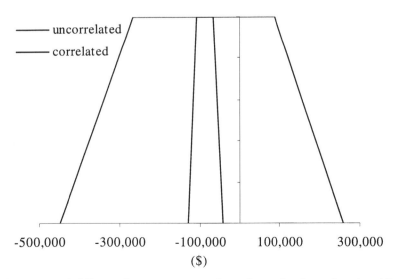

**Fig. 11.** Final difference between present values of costs for alternatives A and B

For the sake of this example, we have assumed that the costs in both alternatives can be completely independent. However, this is not accurate! Even if all the other parameters are independent, the discount rate is common to both alternatives, affecting their future values the same way. That is, if a particular instance of the discount rate is going to happen for alternative A, the same value will happen if alternative B was selected. This means that it is inaccurate to calculate the difference between the costs of the alternatives assuming they are independent FNs. That corresponds to an impossible situation when both the extreme values of the discount rate would happen simultaneously in the future, one for each alternative.

In order to take into account this dependence it is more convenient to directly calculate the cumulative PV of difference between alternatives, instead of the cumulative PVs of the alternatives first. When put in a single expression, the dependence between the fuzzy variables can be resolved using the approach stated in section 2.5.

The differences between the total costs when such dependence is recognized are shown on Figure 12. The result for the correlated case is the same as before because, together with the correlation of the costs, a common discount rate was also implicitly assumed. The result for the uncorrelated case, however, is quite different. The resulting FN is much narrower as a result of the removal of the artificial uncertainty introduced before by inaccurate arithmetic. Nevertheless, it is still a fuzzy number around the origin and choosing the right alternative is not clear-cut.

One approach to choosing between alternatives that have "overlapping" membership functions and whose difference is a FN around the origin, is to use one of the comparison criteria introduced before. The positive and the negative part of the fuzzy difference, taken with absolute values, are compared as two FNs. For example, if the COG criterion is used, it is calculated separately for both parts of the membership function, left and right of the origin. The comparison between these two crisp numbers determines which way the fuzzy difference goes and this, in turn, determines which alternative is preferable. Using the comparison approach in this particular

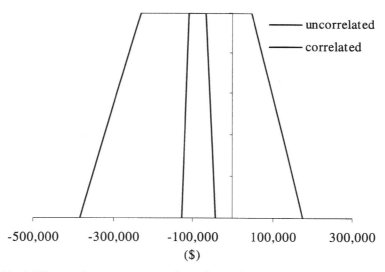

**Fig. 12.** Final difference between present values of costs for alternatives A and B – dependence recognized

example, no matter which criterion is used, it is obvious that alternative A will still be preferable in the uncorrelated case. In other words, the overall possibility that the costs in alternative A will be less than the costs in alternative B is bigger than the possibility of the opposite.

The COG criterion is similar to the *expected value* in probabilistic analysis, with possibility function replacing the probability density function. Because the sum of all possibilities is generally different than one, unlike the sum of all probabilities, the calculated "*expected possibility*" is normalized by that value. Hence, the nominator in the definition of COG.

## 3.5 Fuzzy Case with Partial Correlation

One may argue that it is not only the discount rate that is common to both alternatives and that causes dependence between the alternatives. Other hidden dependencies may occur that will exaggerate the uncertainty for the uncorrelated case shown in Figure 12. For instance, the operating costs for alternative A are likely to be correlated to its initial cost. Furthermore, the operating costs for both alternatives are likely to share some common characteristics and to be affected in a similar manner by the same causes in the future (fuel price, economic environment, social issues, etc.).

The last dependence can be modeled as partial, i.e. consisting of two parts. The first part is totally correlated and the second part represents some sort of *fuzzy noise*, an independent fuzzy number around zero. For this example, if we assume the fuzzy noise to be [-24,000; -12,000; 12,000; 24,000] as shown in Figure 13, then by scaling it with a coefficient $k \leq 1$, we can obtain different levels of dependence. The marginal cases shown on Figure 12 can be obtained by setting $k = 0$, for the case of total correlation, and $k = 1$, for the case of independent operating costs – uncorrelated case.

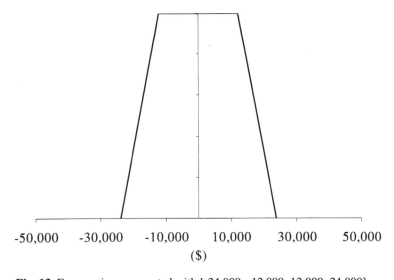

**Fig. 13.** Fuzzy noise represented with [-24,000; -12,000; 12,000; 24,000]

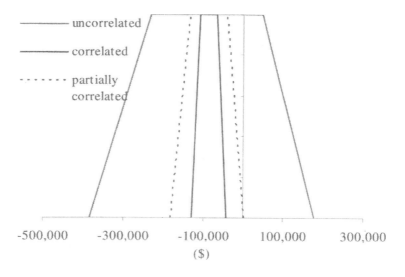

**Fig. 14.** Final difference between present values of costs for alternatives A and B – different levels of correlation

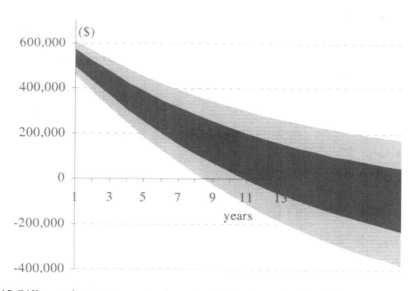

**Fig. 15.** Difference between present values of costs for alternatives A and B – uncorrelated case

Figure 14 shows once again total cost differences, this time including also the case of partial correlation between operating costs with scaling factor $k = 0.2$.

It is interesting to see how the cost differences develop over time. Figure 15 – Figure 17 show this for the three cases of uncorrelated, correlated, and partially correlated operating costs, respectively. The 0-cuts of the resulting FN are shown as gray and the 1-cuts are shown as black intervals.

Fuzzy Present Worth Analysis with Correlated and Uncorrelated Cash Flows    33

The uncorrelated case shown on Figure 15 exhibits an ever-growing uncertainty in the total cost difference. It starts with the initial uncertain cost for alternative A that is added to and compounded over time with the difference in operating costs between the alternatives. On the other hand, in the correlated case shown on Figure 16 the uncertainty stays constant over time. Although individual uncertainty for each alternative grows as it compounds over time, because of the total correlation, the uncertainty

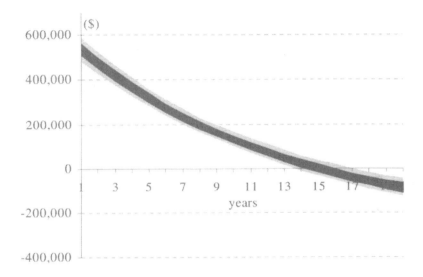

**Fig. 16.** Difference between present values of costs for alternatives A and B – correlated case

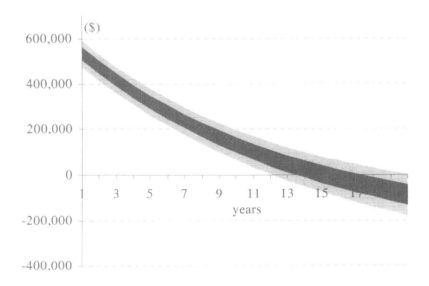

**Fig. 17.** Difference between present values of costs for alternatives A and B – partially correlated case with $k = 0.2$

of the cost differences between the alternatives are pretty much the same. The final result is a much narrower FN.

Partially correlated cases lie between the two marginal cases. The one shown on Figure 17 with $k = 0.2$ still produces a clear outcome, where alternative A is always preferable. Increasing $k$ beyond that value will include 0 within the final result and, hence, will induce some trade-off in the decision making.

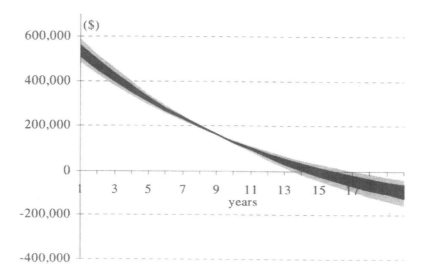

**Fig. 18.** Difference between present values of costs for alternatives A and B – correlated case with incorrect application of the extension principle

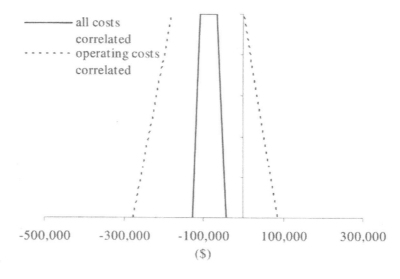

**Fig. 19.** Final difference between present values of costs for alternatives A and B – correlated and uncorrelated initial cost

Fuzzy Present Worth Analysis with Correlated and Uncorrelated Cash Flows     35

It is worth noting that for proper calculation in the correlated case it is not sufficient to just use the points that correspond to the same position on the membership functions. In this particular example, to find the difference between costs over time it is not sufficient to find the difference between the points from the same row and column in Table 4. Those points are obtained for a particular value of the discount rate that gives the extreme points when looking at the costs individually. However, when looking at the cost difference, there may be a different value of the discount rate that yields smaller (bigger) value. This is illustrated on Figure 18 where the cost differences are obtained incorrectly, without proper application of the Extension Principle.

So far, in the cases with correlation, it has been assumed for alternative A that its operating costs are correlated to its initial cost. Figure 19 compares the total cost difference between the alternatives with and without this assumption. The cases shown are when the operating costs are still kept correlated. Thus, one case is when all the costs are correlated and it is the same one shown on Figure 11, Figure 12 and Figure 14. As expected, removing the assumption of correlation between the initial and operating costs increases the uncertainty. The change is significant, although it would probably not have changed the preferred choice in this particular example. The change will get smaller in cases of partial correlation. It should be clear though, in general, the existence or absence of any such dependence can make the difference and it is important to recognize it in the analysis.

## 3.6  NPW Example 2

The previous example dealt with cost minimization of alternatives with equal revenues. Let us now consider another case that typically occurs in engineering economic analysis, selection among alternatives with different revenues and different costs. In this case, the objective is maximization of the net present value (worth). Net present value, as usual, is the difference between the present values of total revenues and total costs.

A company is faced with three alternatives denoted A, B, and C. Each of the alternatives includes initial capital expenditure and a stream of uniform revenues during its lifetime. All alternatives are assumed to have equal book life of twenty years and no salvage value at its end. Again, we keep the example simple enough, without many complicated time-value calculations of money so that the effects of introducing fuzzy variables in the analysis are clear.

### 3.6.1  Crisp Case

Table 5 shows the initial costs and net annual revenues associated with each alternative for the crisp case. The net annual revenues are obtained after subtracting the annual costs from the total annual revenues. The discount rate is assumed to be 9%/year. Inflation, costs escalation, and tax implications are again neglected.

Table 6 shows net present values (present worth) for the three alternatives during their lifetime. This is also shown on Figure 20. As can be seen from there, all alternatives have a positive net present value at the end of their lifetime. Thus, they all generate a better return rate than the assumed discount rate of 9%/year. The alternative C yields the largest net present value and, since the goal is return maximization, it is the preferable alternative in this case.

**Table 5.** Cash flows for the alternatives ($) – crisp case

|  | Initial cost | Net annual revenues |
|---|---|---|
| Alternative A | 50,000 | 6,000 |
| Alternative B | 100,000 | 11,000 |
| Alternative C | 200,000 | 23,000 |

**Table 6.** Net present values ($) – crisp case

| Year | Alternative A | Alternative B | Alternative C |
|---|---|---|---|
| 1 | -44495 | -89908 | -178899 |
| 2 | -39445 | -80650 | -159540 |
| 3 | -34812 | -72156 | -141780 |
| 4 | -30562 | -64363 | -125486 |
| 5 | -26662 | -57214 | -110538 |
| 6 | -23084 | -50655 | -96824 |
| 7 | -19802 | -44638 | -84242 |
| 8 | -16791 | -39117 | -72699 |
| 9 | -14029 | -34052 | -62109 |
| 10 | -11494 | -29406 | -52394 |
| 11 | -9169 | -25143 | -43481 |
| 12 | -7036 | -21232 | -35303 |
| 13 | -5079 | -17644 | -27801 |
| 14 | -3283 | -14352 | -20919 |
| 15 | -1636 | -11332 | -14604 |
| 16 | -125 | -8562 | -8811 |
| 17 | 1262 | -6020 | -3496 |
| 18 | 2534 | -3688 | 1379 |
| 19 | 3701 | -1549 | 5853 |
| 20 | 4771 | 414 | 9957 |

### 3.6.2  Fuzzy Case

Let us now introduce uncertainty in some of the data from above. The future net annual revenues are something which can never be known for certain and we will model them as FNs. Again, we will use ±10% variation in the crisp values to define the 0-cut intervals of the fuzzy numbers, and ±5% for the 1-cuts. So, the crisp revenues are the median values of the corresponding fuzzy numbers. As for the initial costs, we will assume that they are precisely known and the same as before. In other words, the company is about to start with the endeavor right away and knows exactly the initial costs of each alternative. Table 7 shows the costs for each alternative in this case.

We will assume for now that the discount rate is also uncertain, modeled with the trapezoidal FN around the crisp 9%/year, [8%; 8.5%; 9.5%; 10%]/year.

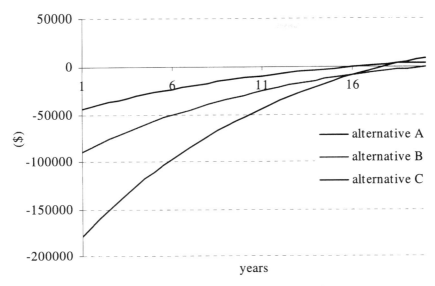

**Fig. 20.** Net present values of the alternatives – crisp case

**Table 7.** Cash flows for the alternatives ($) - fuzzy case

|  | Initial cost | Net annual revenues |
| --- | --- | --- |
| Alternative A | 50,000 | [5,300; 5,700; 6,300; 6,600] |
| Alternative B | 100,000 | [9,900; 10,450; 11,550; 12,100] |
| Alternative C | 200,000 | [20,700; 21,850; 24,150; 25,300] |

Table 8 shows the net present values for the alternatives in this case. Again, the lower and upper bounds of both 0-cuts and 1-cuts are shown. The data from the end of the lifetime are also shown on Figure 21 as approximate trapezoidal FNs.

These FNs show that there is no clear choice among the alternatives. All of them, depending on the realization of the uncertain variables in the future, can yield both positive and negative outcomes. Thus, all the net PVs are FNs around the origin, but the amount of uncertainty they contain differs significantly. As Figure 21 shows, the net PV of alternative A is almost completely contained in the net PV of alternative B, and this one is almost completely contained in the net PV of alternative C. In other words, as the possibility of making a larger return increases, so does the possibility of making a larger loss. Which one is to be chosen depends entirely on the decision maker willingness to accept more or less risk.

When calculating the risk it is important to find the amount by which the alternatives' outcomes differ. As we observed from Figure 10 and Figure 11, this is not simply the difference between the corresponding points from the two membership functions. That will imply a complete correlation between the alternatives. Neither is it simply the fuzzy difference between the two FNs. That will imply completely unrelated cases, while we know that they share at least a common discount factor.

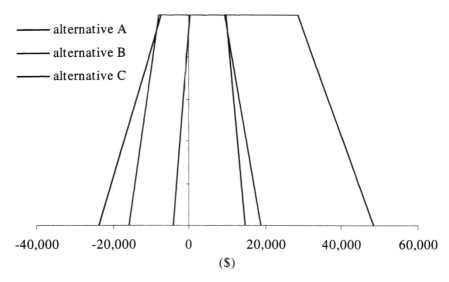

**Fig. 21.** Final net present values of the alternatives – fuzzy case

**Table 8.** Net present values ($) – fuzzy case

|  | Alternative A |  |  |  | Alternative B |  |  |  | Alternative C |  |  |  |
|---|---|---|---|---|---|---|---|---|---|---|---|---|
| Year | $\alpha=0$ | $\alpha=1$ | $\alpha=1$ | $\alpha=0$ | $\alpha=0$ | $\alpha=1$ | $\alpha=1$ | $\alpha=0$ | $\alpha=0$ | $\alpha=1$ | $\alpha=1$ | $\alpha=0$ |
| 1 | -45,091 | -44,795 | -44,194 | -43,889 | -91,000 | -90,457 | -89,355 | -88,796 | -181,182 | -180,046 | -177,742 | -176,574 |
| 2 | -40,628 | -40,041 | -38,842 | -38,230 | -82,818 | -81,741 | -79,544 | -78,422 | -164,074 | -161,823 | -157,228 | -154,883 |
| 3 | -36,571 | -35,699 | -33,910 | -32,991 | -75,380 | -73,782 | -70,501 | -68,817 | -148,522 | -145,180 | -138,320 | -134,799 |
| 4 | -32,883 | -31,734 | -29,364 | -28,140 | -68,618 | -66,513 | -62,167 | -59,923 | -134,384 | -129,982 | -120,894 | -116,203 |
| 5 | -29,530 | -28,114 | -25,174 | -23,648 | -62,471 | -59,875 | -54,486 | -51,688 | -121,531 | -116,102 | -104,833 | -98,984 |
| 6 | -26,482 | -24,807 | -21,312 | -19,489 | -56,883 | -53,813 | -47,406 | -44,063 | -109,846 | -103,427 | -90,031 | -83,041 |
| 7 | -23,711 | -21,787 | -17,753 | -15,638 | -51,803 | -48,277 | -40,881 | -37,003 | -99,224 | -91,851 | -76,388 | -68,279 |
| 8 | -21,191 | -19,029 | -14,473 | -12,072 | -47,184 | -43,221 | -34,867 | -30,466 | -89,567 | -81,279 | -63,814 | -54,610 |
| 9 | -18,901 | -16,511 | -11,450 | -8,771 | -42,986 | -38,603 | -29,325 | -24,413 | -80,788 | -71,625 | -52,225 | -41,954 |
| 10 | -16,819 | -14,211 | -8,664 | -5,713 | -39,169 | -34,387 | -24,216 | -18,808 | -72,807 | -62,808 | -41,543 | -30,235 |
| 11 | -14,927 | -12,110 | -6,095 | -2,883 | -35,699 | -30,536 | -19,508 | -13,619 | -65,552 | -54,756 | -31,699 | -19,384 |
| 12 | -13,206 | -10,192 | -3,728 | -262 | -32,544 | -27,019 | -15,169 | -8,813 | -58,957 | -47,403 | -22,626 | -9,337 |
| 13 | -11,642 | -8,440 | -1,547 | 2,165 | -29,677 | -23,807 | -11,169 | -4,364 | -52,961 | -40,688 | -14,263 | -34 |
| 14 | -10,220 | -6,840 | 464 | 4,412 | -27,070 | -20,874 | -7,483 | -245 | -47,510 | -34,555 | -6,556 | 8,579 |
| 15 | -8,927 | -5,379 | 2,317 | 6,493 | -24,700 | -18,196 | -4,086 | 3,570 | -42,554 | -28,954 | 547 | 16,555 |
| 16 | -7,752 | -4,045 | 4,025 | 8,419 | -22,545 | -15,749 | -955 | 7,102 | -38,049 | -23,840 | 7,094 | 23,940 |
| 17 | -6,684 | -2,827 | 5,599 | 10,203 | -20,587 | -13,515 | 1,931 | 10,372 | -33,954 | -19,169 | 13,128 | 30,777 |
| 18 | -5,712 | -1,714 | 7,050 | 11,854 | -18,806 | -11,475 | 4,591 | 13,400 | -30,231 | -14,903 | 18,690 | 37,109 |
| 19 | -4,829 | -698 | 8,387 | 13,384 | -17,187 | -9,612 | 7,042 | 16,204 | -26,846 | -11,007 | 23,815 | 42,971 |
| 20 | -4,027 | 231 | 9,619 | 14,800 | -15,716 | -7,911 | 9,302 | 18,800 | -23,769 | -7,449 | 28,540 | 48,399 |

Figure 22 shows the differences between the net PV of alternatives A and C for these two cases and for the accurately calculated one.

If we use one of the previously mentioned comparison criteria, for example COG, alternative C will be preferable. This can be seen on Figure 22. However, this criterion is

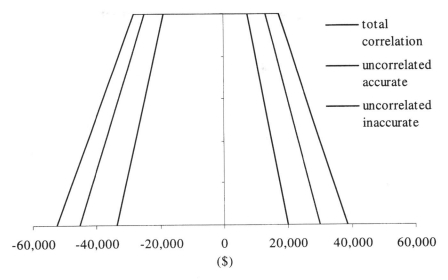

**Fig. 22.** Final difference between net present values of alternatives A and C – correlated and uncorrelated cases

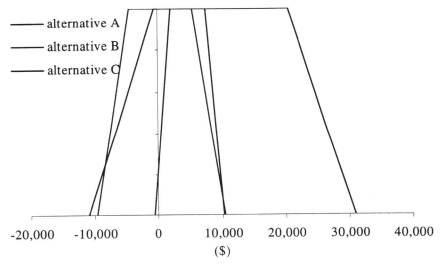

**Fig. 23.** Final net present values of the alternatives – fuzzy case with fixed MARR

an averaging criterion by its definition and may hide some unacceptable risks to the decision maker, as shown on Figure 21.

Instead of forecasting an appropriate discount rate, very often analysts set in advance its desired value which is then called minimum acceptable rate of return (MARR). The alternatives are then checked against this target. The calculated NPW should be $\geq 0$ at the MARR. This is translated to EAW $\geq 0$ at the MARR, if the equivalent annual worth method is used. Alternatives that do not meet the target are eliminated. This approach may somewhat help the fuzzy analysis as it eliminates one of the sources of uncertainty and introduces an additional criterion.

Figure 23 shows the results with a fixed MARR of 9%. Thus, the only source of uncertainty now left is the net annual revenues. In this case, alternative A practically doesn't bear any risk as it almost always yields a positive NPW for the given MARR. The other two alternatives are more risky, but it should be clear that alternative B is inferior to alternative C. The risk of loss is practically the same in both, but alternative C may yield far better return if the future conditions are favorable. Again, the final decision between A and C rests with the decision maker and his willingness to accept a certain level of risk.

# References

Beskese, A., Kahraman, C., Irani, Z.: Quantification of flexibility in advanced manufacturing systems using fuzzy concept. International Journal of Production Economics 89(1), 45–56 (2004)

Botterud, A., Ilic, M.D., Wangensteen, I.: Optimal investments in power generation under centralized and decentralized decision making. IEEE Transactions on Power Systems 20(1), 254–263 (2005)

Buckley, J.J.: The Fuzzy Mathematics of Finance. Fuzzy Sets and Systems 21, 257–273 (1987)

Carlsson, C., Fuller, R.: A fuzzy approach to real option valuation. Fuzzy Sets And Systems 139(2), 297–312 (2003)

Carlsson, C., Fuller, R., Heikkila, M., et al.: A fuzzy approach to R&D project portfolio selection. International Journal Of Approximate Reasoning 44(2), 93–105 (2007)

Dimitrovski, A.D., Matos, M.A.: Fuzzy Engineering Economic Analysis. IEEE Transactions on Power Systems 15(1), 283–289 (2000)

Dubois, D., Prade, H.: Fuzzy Sets and Systems–Theory and Applications. Academic Press Inc., Orlando (1980)

Huang, X.X.: Chance-constrained programming models for capital budgeting with NPV as fuzzy parameters. Journal of Computational and Applied Mathematics 198(1), 149–159 (2007)

Huang, X.X.: Optimal project selection with random fuzzy parameters. International Journal of Production Economics 106(2), 513–522 (2007)

Kahraman, C., Ruan, D., Tolga, E.: Capital budgeting techniques using discounted fuzzy versus probabilistic cash flows. Information Sciences 142(1-4), 57–76 (2002)

Kaufmann, A., Gupta, M.M.: Fuzzy Mathematical Models in Engineering and Management Science. Elsevier, Amsterdam (1988)

Matthews, J., Broadwater, R., Long, L.: The Application of Interval Mathematics to Utility Economic Analysis. IEEE Transactions on Power Systems 5(1), 177–181 (1990)

Merrill, H.M., Wood, A.J.: Risk and Uncertainty in Power System Planning. In: Proceedings of the 10th Power Systems Computation Conference, Graz (1990)

Miranda, V.: Fuzzy Reliability Analysis of Power Systems. In: Proceedings of the 12th Power Systems Computation Conference, Dresden (1996)

Mohanty, R.P., Agarwal, R., Choudhury, A.K., et al.: A fuzzy ANP-based approach to R&D project selection: a case study. International Journal of Production Research 43(24), 5199–5216 (2005)

Rebiasz, B.: Fuzziness and randomness in investment project risk appraisal. Computers & Operations Research 34(1), 199–210 (2007)

Riggs, J.L., West, T.M.: Essentials of Engineering Economics. McGraw-Hill Book Co., New York (1986)

Rubinson, T.: Multi-period risk management using fuzzy logic. International Journal of Uncertainty Fuzziness and Knowledge-Based Systems 4(5), 449–466 (1996)

Saade, J.J.: A Unifying Approach to Defuzzification and Comparison of the Outputs of Fuzzy Controllers. IEEE Transactions on Fuzzy Systems 4(3), 227–237 (1996)

Sheen, J.N.: Fuzzy financial profitability analyses of demand side management alternatives from participant perspective. Information Sciences 169(3-4), 329–364 (2005a)

Sheen, J.N.: Fuzzy evaluation of cogeneration alternatives in a petrochemical industry. Computers & Mathematics with Applications 49(5-6), 741–755 (2005b)

Sheen, J.N.: Fuzzy financial analyses of demand-side management alternatives. IEE Proceedings-Generation Transmission and Distribution 152(2), 215–226 (2005c)

Sheen, J.N.: Fuzzy Financial Decision-Making: Load Management Programs Case Study. IEEE Transactions on Power Systems 20(4), 1808–1817 (2005d)

Terceno, A., De Andres, J., Barbera, G.: Using fuzzy set theory to analyse investments and select portfolios of tangible investments in uncertain environments. International Journal of Uncertainty Fuzziness and Knowledge-Based Systems 11(3), 263–281 (2003)

Tolga, E., Demircan, M.L., Kahraman, C.: Operating system selection using fuzzy replacement analysis and analytic hierarchy process. International Journal Of Production Economics 97(1), 89–117 (2005)

Zadeh, L.A.: Fuzzy Sets. Information and Control 8, 338–353 (1965)

Zimmermann, H.-J.: Fuzzy Set Theory - and Its Applications, 2nd edn. Kluwer Academic Publishers, Boston (1991)

# Optimization with Fuzzy Present Worth Analysis and Applications

Dorota Kuchta

Wroclaw University of Technology, Institute of Industrial Engineering and Management, Smoluchowskiego, 25, 50-371 Wroclaw, Poland

**Abstract.** This chapter introduces fuzzy net present value analysis. The present value approach is first analyzed by crisp values. After the basic definitions of fuzzy numbers are given, the concept and applications of fuzzy present value are handled. The other components of this chapter is the fuzzy and probabilistic approach to the present value, the fuzzy net present value maximization as an objective in project selection problems, applications to the valuation of projects with future options, interpretation of fuzzy present value and fuzzy net present value, fuzzy net present value as objective in optimization problems from the industry, fuzzy net present value as an objective in spatial games, and fuzzy classification based on net present value.

## 1 Introduction

In this section the fuzzy present value and net present value will be presented, together with their selected applications. The section is organized as follows. We start with the notion of crisp Present value and crisp net present value, introducing all the necessary notation. We also present in the crisp case several selected applications of the present value concept, which will be discussed in the fuzzy case later on. Then we introduce the notation and basic information about fuzzy numbers. It is only then that we pass over to the case of fuzzy present value and fuzzy net present value. We present the definitions and interpretations of the concepts and then discuss the selected applications.

## 2 Introductory Information - Crisp Present Value and Fuzzy Numbers

### 2.1 Crisp Present Value

We consider a project $P$ or another type of investment that a company might be interested to undertake. For this project the following data has to be given:

a) $n$ - the duration of the project in assumed time units (for simplicity reasons, here they will be years), we will assume them to be integer (in some cases it might be a fraction also, for details see (Kuchta 2002));

b) $COF_i$ $(\geq 0)$ - cash outflow at the end of the $i$th year $(i=0,1,...,n)$ and $CIF_i$ $(\geq 0)$ - cash inflow at the end of the $i$th year $(i=1,...,n)$. Both yearly cash flows (in- and outflows) will be often summarized as a single cash flow $CF_i$, occurring at the end of the $i$th year $(i=0,1,...,n)$, where $CF_0 = -COF_0$, $CF_i = CIF_i - COF_i$ $(i=1,...,n)$.

---

C. Kahraman (Ed.): Fuzzy Engineering Economics with Appl., STUDFUZZ 233, pp. 43–69, 2008.
springerlink.com                                      © Springer-Verlag Berlin Heidelberg 2008

44    D. Kuchta

c) $r$ (>0) - the required rate of return or cost of capital, also called the discount rate, it is also used to model the risk linked to the estimated cash flow (usually the higher the risk that the cash flow will not occur in the estimated quantity, the higher $r$).

The cash outflows and inflows are calculated taking into account various components and factors (e.g. Horngen et al. 1994). The initial investment $COF_0$ is the sum of e.g. the initial investment in machines, cars, any other equipment and the initial working capital investment, minus the current disposal price of old equipment, which the new one will replace. The final inflow $CIF_n$ takes into account e.g. the terminal disposal price of the equipment and the recovery of working capital. The other out- and inflows are calculated on the basis of the overhead cost, depreciation, material cost, labour cost, sales, quantities of items to be sold and bought and their prices, taxes etc. There may be also some special ways of calculating the in- and outflows for particular cases. E.g. in (Sheen 2005a, 2005b, 2005c), where a special case of seeking for measures to optimize the electricity delivery is considered, the cash flows in the subsequent years are a function of the cash flow in the first year and inflation:

$$CF_i = CF_1 (1+d)^{n-1} \quad (i=1,...,n) \tag{1}$$

where $d$ stands for the yearly inflation rate. Quite often the in- and outflows generated by the project are understood in a more general sense, standing for all the advantages and disadvantages that come into being through the project in question. For example, in (Greendut 1995) the personal energy of the entrepreneur is considered and the present value of the energy spent during the whole project is calculated. More and more often social cost and benefits are considered (see e.g. (Dompere 1995)), like reduction in diseases, better living conditions etc.

Having discussed the entry date that is necessary to calculate the Present value and the Net Present Value, we can over to those notions themselves.

**Definition 1**

The present worth (or present value) $PV(CF_i)$ of a cash flow $CF_i$ (or cash outflow or inflow) ($i=0, 1,...,n$) is defined as

$$PV(CF_i) = \frac{CF_i}{(1+r)^i}$$

**Definition 2**

Net Present value of a project is defined as:

$$NPV(P) = CF_0 + \sum_{i=1}^{n} PV(CF_i)$$

Let us remark that this way of calculating the present values is based on the assumption that the decision makers prefers later outflows and earlier inflows, that is why the higher the $i$, the flows are divided by a higher value $(1+r)^i$. In some cases we have to differentiate the discount rates in different years, because the same $r$ cannot be applied. The reason of that lies sometimes in changes of financial market and sometimes in the need to take into account the higher risk linked with later inflows (the later the flow, the less accurate the forecast is). In some very special cases the values of $r$ may chosen in a completely different way. E.g. the flows of the entrepreneur's energy in (Greenhut 1995) may be discounted in an exactly reverse way: it may happen that the decision maker prefers to spend his energy now instead of doing it later. In (Sáez and Requena 2007) a thorough discussion can be found with respect to the discounting rate in case of long-term projects, where social cost and benefits have to be considered. E.g. some authors claim that a lower discount rate should be used for years which lie more ahead in the future, because the future generations and their benefits and costs should be of the highest importance to us while enterprising this type of projects.

In general, the interpretation of the NVP consists in selecting such projects whose NPV is the largest, and the minimal acceptance threshold is zero: normally no projects with NVP<0 are accepted. Sometimes projects are considered which generate only cost (outflows), in such cases the NPV is calculated according to the following formula:

$$NPV(P) = COF_0 + \sum_{i=1}^{n} PV(COF_i) \tag{2}$$

and the project with the smallest NPV is selected.

In the following only those applications of the NPV will be discussed where the fuzzy approach is used.

Very often the NPV is used as objective function in various optimization problems, by itself or as one of several criteria in multicriteria problems. Here is a simple example of such a problem, which is a project selection problem:

**Example 1**

Let us consider a company which has $k$ independent projects available for investment, each one of them will last not more than $n$ years. Let $NPV_i$ $(i = 1,...k)$ denote the Net Present value of the $i$th project, $CF_{ij}$ $(i = 1,...,n; j = 1,...k)$ - cash flows caused by the $i$th project in the $j$th year, and $B_j$ $(j = 1,...,n)$ - cash budget of the company available for the $j$-th year. The problem consists in choosing such a selection of projects which can be undertaken (the sum of their cash outflows for each year will be not greater than the corresponding budget) and which will maximize the global NPV.

If we define 0-1 variables $x_i$ $(i = 1,...,n)$, where $x_i = 1$ would mean the acceptance of project $i$ and $x_i = 0$ - its rejection, the following 0-1 linear programming problem will give us the solution of our problem:

$$\sum_{i=1}^{k} NPV_i x_i \rightarrow \max$$
$$\sum_{i=1}^{k} CF_{ij} x_i \leq B_j \ (j = 1,...,n) \tag{3}$$

Let us consider another example of an optimization problem where NPV constitutes the objective:

46    D. Kuchta

**Example 2 (Liu and Sahinidis 1997)**

The authors consider a long-range planning problem for a chemical process involving a network of chemical processes which are interconnected by chemicals. The latter include raw materials, intermediates and products that may be purchased from and/or sold to different markets. The decision to be taken consists in the selection of processes from among various available technologies and of the moments when the process should be expanded, as well as of the production levels for the processes. The decision maker wants to maximize the NPV of the project over a long range horizon, consisting of a finite number of time periods during which the process, the demands of chemicals and the costs can vary. The NPV, constituting the objective function in (Liu and Sahinidis 1997), is the difference between the sales revenue and the sum of investment, operating and raw material costs, expressed in their present values according to Definition 1:

$$NPV = -\sum_{i=1}^{NP}\sum_{t=1}^{NT}\left(\alpha_{it}E_{it} + \beta_{it}y_{it}\right)-$$
$$-\sum_{i=1}^{NP}\sum_{t=1}^{NT}\partial_{it}W_{it} + \sum_{j=1}^{NC}\sum_{l=1}^{NM}\sum_{t=1}^{NT}\left(\gamma_{jlt}S_{jlt} - \Gamma_{jlt}P_{jlt}\right) \tag{4}$$

where

- NP denotes the number of processes, NC - the number of chemicals, NM - of markets, NT - of periods;
- parameters $\alpha_{it}$, $\beta_{it}$ denote the variable/fixed cost of investment for the $i$-th process in the $t$-th period, parameters $\delta_{it}$ stand for the unit operating cost for the $i$-th process in the $t$-th period, parameters $\gamma_{ilt}$, $\Gamma_{ilt}$ denote the sales/purchase prices of the $j$-th chemical in the $l$-th market during the $t$-th period $\left(i = 1,.., NP; j = 1,..., NC; l = 1,..., NM; t = 1,..., NT\right)$, all the parameters are given in their present values (Definition 1)
- decision variables $E_{it}$ denote the capacity expansion of the plant of the $i$-th process in the $t$-th period, decision variables $y_{it}$ are binary variables equal to 1 if there is an expansion of the $i$-th process in the $t$-th period and 0 otherwise, decision variables $W_{it}$ stand for the operating level of the $i$-th process in the $t$-th period, decision variables $P_{ilt}$, $S_{ilt}$ denote the amount of the $j$-th product purchased from/sold to the $l$-th market during the $t$-th period $\left(i = 1,.., NP; j = 1,..., NC; l = 1,..., NM; t = 1,..., NT\right)$.

In the constraints the limits for capacity expansions and other feasibility conditions are taken into account. The entire model can be found in (Liu and Sahinidis 1997).

The following example (Zhou and Gong 2004) concerns a case where the NPV constitutes the objective, but also is included in the constraints. The example comes form the forest economics area.

## Example 3

Various forest management programs are considered (they differ from each other e.g. in respect to the time of harvesting and the time of regenerating) for a number of forest stands types (the stands are grouped according to the dominant tree species, site quality and stand age). The problem consists in deciding which management program to use for which stand, so that the total NPV of the whole project is maximal and some environmental concerns are taken into account.

The objective function takes the following form:

$$\sum_{i=1}^{NSP} \sum_{k=1}^{NMP} NPV_{ik} x_{ik} \to \max \tag{5}$$

where $NPV_{ik}$ is the Net Present value of the $k$-th management project if applied to the $i$-th stand type, $x_{ik}$ is a binary variable equal to 1 if the $k$-th management project is applied to the $i$-th stand type and to 0 otherwise, $NSP$ stands for the number of stand types and $NMP$ for the number of management programs. The constraints represent various environmental requirements. The constraints and the objective function (5) form together an integer linear programming problem.

Once this problem is solved and an optimal solution of the problem with objective (5) is obtained (let us denote the maximal value of the objective function (5) as $NPV_{\max}$ and the set of the indices corresponding to the basic variables in the optimal solution as $B$), a further step is required. It is so because in the considered problem not all environmental goals are taken into account. Some of them are difficult to measure or to formalize. In order to find solutions which would to some extent satisfy all the goals, also those immeasurable ones, the optimal solution of the problem with objective (5) is taken as a basis and a collection of new solutions is generated, for which the satisfaction of all the goals is verified by experts. The solutions which are generated are possibly no more optimal but fairly good with respect to objective (5) and, if possible, significantly different with respect to the decision variables values from the optimal solution of the problem (5). They are determined by means of solving the following problem:

$$\sum_{(i,k)\in B} x_{ik} \to \min$$

$$\sum_{i=1}^{NSP} \sum_{k=1}^{NMP} NPV_{ik} x_{ik} \geq NPV_{TV} \tag{6}$$

with the other constraints remaining unchanged, where $NPV_{TV}$ is a target value for the global Net Present Value, set by the decision maker as a percentage of $NPV_{\max}$. The collection of solutions is then analyzed by the decision maker with respect to the criteria not taken into account in the mathematical model.

Like in the previous example, very often the NPV is only one among many decision criteria. In the above example the other criteria could not be specified in the mathematical model and as a result they were taken into account in a completely informal way. In numerous other applications the NPV is combined into one model with other criteria, some of them may be formal, some them less formal:

48     D. Kuchta

**Example 4 (Chan et al. 2000)**

The authors use the NPV together with numerous other criteria to select technologies (it is assumed that for each technology its NPV can be calculated). This is the only criterion which can be expressed in a mathematical form: all the other criteria are subjective (e.g. flexibility). Then the AHP method is used to find the respective weightings of individual criteria, the weightings of each alternative under the individual criteria and finally the overall rating of the alternatives.

**Example 5 (Bogataj and Bogataj 2001)**

The authors consider a system composed of two centers (towns) $A$ and $B$, where $B$ is on a higher level than $A$, which means that only $B$ has at its disposal the components needed for a certain final product and $B$ manufactures the product anyway (but the human resources are more expensive in $B$ than in $A$). Both centers and the whole area to which they belong need the final product in question, also $A$ can produce it and sell, but if $A$ manufactures the product, it has to buy the components from $B$. The whole area divides itself (dynamically, according to the travel cost, shortage probability etc., in each of the centers) into two disjoint market areas: that of $A$ and that of $B$ - for example, the market area of $A$ is the area whose inhabitants travel in the first place to $A$ if they want to buy the product, and they go to B only in case they do not get the product in $A$. But of course, if $A$ does not manufacture the product, its market area is empty and the whole area is the market area of $B$. The centers should take the decision about their role in the system. $A$ can choose among two decisions: to buy the final product from $B$ or to buy only components and manufacture the product, selling it to its market area. $B$, which will manufacture the product and sell it to its market area anyway, should decide whether it will offer $A$ the final product (at a reduced price, so that $A$ can resell it in its market area) or the components. Both centers make their decisions taking into account various parameters, also those imposed by the other center. It is assumed that each center strives at minimizing the $NPV$ of all the payments associated with the decision in question. The $NPV$ is the function of the set-up cost in the production of the final product in both centers, of the set-up cost of the component, the ordering intervals for the final product and the components, the transportation cost, the demand from the area for the final product, the probability that there is a shortage of the final product the moment when the customer arrives, the selling price of the final product etc. Each center aims at satisfying its own interests, which are to a certain extent contradictory, but on the other hand they function in the framework of a common system and the division of the area into two market areas is not rigid. It is the customers who decide about it and $A$ and $B$ lead a certain game, where the rewards are the corresponding $NPV$. The game theory is used to determine the final solution (the equilibrium point), and this type of games is called spatial games.

There are of course much more practical applications of the Present value and of the Net Present Value, but here only those were presented which will be discussed later on as an illustration of the use of the fuzzy concept.

## 2.2  Basic Definitions - Fuzzy Numbers

Fuzzy numbers are a widely accepted tool to represent values which are unknown yet (in this context, fuzzy numbers constitute an alternative to probability distribution) or

Optimization with Fuzzy Present Worth Analysis and Applications

to model the way the decision maker understands certain concepts (e.g. "a high profit", "a low cost"). Here, we will use them in both roles, although mainly as expressions of yet unknown values, which will be known only in the future. It will be assumed that, for the moment, it is only known that certain realizations of the unknown values are possible to a higher degree and some to a smaller one.

We will use here the following definition of fuzzy number (more general definitions also exist, but the one we use is enough for our purposes):

**Definition 3**

A fuzzy number $\tilde{A}$ is a sextuplet of the following form:

$$\tilde{A} = \left(a_1, a_2, a_3, a_4, f_1^A(\lambda), f_2^A(\lambda)\right)$$

or shorter

$$\tilde{A} = \left(a_1, a_2, a_3, a_4, f_1^A, f_2^A\right)$$

where

- $a_1, a_2, a_3, a_4$ are real numbers such that $a_1 \leq a_2 \leq a_3 \leq a_4$ ;
- $f_1^A$ is a continuous non-decreasing real function defined on the interval [0,1], such that $f_1^A(0) = a_1, f_1^A(1) = a_2$ ;
- $f_2^A$ is a continuous non-decreasing real function defined on the interval [0,1], such that $f_2^A(0) = a_3, f_2^A(1) = a_4$

Each fuzzy number $\tilde{A}$ is closely linked to its membership function $\mu_A$, defined as follows:

**Definition 4**

Let $\tilde{A} = \left(a_1, a_2, a_3, a_4, f_1^A, f_2^A\right)$ be a fuzzy number. Its membership function is defined as:

$$\mu_A(x) = \begin{cases} 0 \text{ for } x \leq a_1 \text{ i } x \geq a_4 \\ f_1^A(x) \text{ for } x \in (a_1, a_2) \\ 1 \text{ for } x \in [a_2, a_3] \\ f_2^A(x) \text{ for } x \in (a_3, a_4) \end{cases}$$

The value $\mu_A(x)$ expresses the possibility that $x$ will be the realization of the unknown value $\tilde{A}$ .

**Definition 5**

Let $\tilde{A} = \left(a_1, a_2, a_3, a_4, f_1^A, f_2^A\right)$ be a fuzzy number. If $f_1^A(x), f_2^A(x)$ are linear, $\tilde{A} = \left(a_1, a_2, a_3, a_4, f_1^A, f_2^A\right)$ is a so called trapezoidal fuzzy number, if additionally $a_2 = a_3$, it is called a triangular fuzzy number.

50    D. Kuchta

In many applications only trapezoidal (triangular) fuzzy numbers are used and are sufficient to formulate fairly good models. As in the case of trapezoidal and triangular fuzzy numbers the functions $f_1^A(x), f_2^A(x)$ are determined unequivocally by the rest of the parameters, the following notations are used for such fuzzy numbers: $\tilde{A} = (a_1, a_2, a_3, a_4)$ for trapezoidal fuzzy numbers, $\tilde{A} = (a_1, a_2, a_4)$ for triangular fuzzy numbers.

**Definition 6**

A fuzzy number $\tilde{A} = (a_1, a_2, a_3, a_4, f_1^A, f_2^A)$ is greater or equal 0 if and only if $a_1 \geq 0$.

**Definition 7**

Let $\tilde{A}$ be a fuzzy number. The $\lambda$-cut $A^\lambda$ of $\tilde{A}$ $(\lambda \in [0,1])$ is a closed real interval defined as:

$$A^\lambda = \left[ \left(f_1^A\right)^{-1}(\lambda), \left(f_2^A\right)^{-1}(\lambda) \right] = \left[ \underline{A}^\lambda, \overline{A}^\lambda \right]$$

A fuzzy number may be considered to be a set of its $\lambda$-cuts. In numerous practical applications it is not necessary to know all the $\lambda$-cuts, the knowledge of $\lambda$-cuts for selected values of $\lambda$ is usually sufficient.

It is easy to prove that the following relationship holds:

$$A^\lambda = \{x : \mu_A(x) \geq \lambda\}$$

On fuzzy numbers arithmetical operations can be performed and the outcome will be also a fuzzy number. The definitions of arithmetical operations on fuzzy numbers we consider here are defined on the $\lambda$-cuts. The outcomes are real intervals - the $\lambda$-cuts of the resulting fuzzy number. For each $\lambda$ and each arithmetical operation we have the following definition:

**Definition 8**

Let, for a given $\lambda \in [0,1]$, $A^\lambda$ and $B^\lambda$ be the $\lambda$-cuts of, respectively, $\tilde{A}$ and $\tilde{B}$, and let * be an arithmetical operation defined on real numbers and giving a real number as an outcome. Then we define the $\lambda$-cut of the fuzzy outcome $\tilde{C}$ of the fuzzy extension of operation * as follows:

$$C^\lambda = \{z : \exists s, t \ z = s * t, s \in A^\lambda, t \in B^\lambda\}$$

In e.g. (Kuchta 2000) formulae for basic arithmetical operations on fuzzy numbers are given – they are easy to perform in practice, as they use only the outcomes of respective operations on real numbers and the min and max operators applied to real numbers.

## Definition 9 (Kuchta 2000)

Let $s>0$ be a positive real number and $\tilde{A} = \left(a_1, a_2, a_3, a_4, f_1^A(\lambda), f_2^A(\lambda)\right)$ and $\tilde{B} = \left(b_1, b_2, b_3, b_4, f_1^B(\lambda), f_2^B(\lambda)\right)$ two fuzzy numbers. Then:

$$s\tilde{A} = \left(sa_1, sa_2, sa_3, sa_4, f_1^{sA}(\lambda), f_2^{sA}(\lambda)\right)$$

$$\tilde{A} + \tilde{B} = \left(a_1 + b_1, a_2 + b_2, a_3 + b_3, a_4 + b_4, f_1^A(\lambda) + f_1^B(\lambda), f_2^A(\lambda) + f_2^B(\lambda)\right)$$

$$\tilde{A}\tilde{B} = \begin{pmatrix} \min\{a_1b_1, a_1b_4, a_4b_1, a_4b_4\}, \min\{a_2b_2, a_2b_3, a_3b_2, a_3b_3\}, \\ \max\{a_2b_2, a_2b_3, a_3b_2, a_3b_3\}, \max\{a_1b_1, a_1b_4, a_4b_1, a_4b_4\}, \\ \min\{f_1^A f_1^B, f_1^A f_2^B, f_2^A f_1^B, f_2^A f_2^B\}, \max\{f_1^A f_1^B, f_1^A f_2^B, f_2^A f_1^B, f_2^A f_2^B\} \end{pmatrix}$$

If $0 \notin [b_1, b_4]$ then $\tilde{A} / \tilde{B} = \tilde{A}\left(1/b_1, 1/b_2, 1/b_3, 1/b_4, 1/f_1^B(\lambda), 1/f_2^B(\lambda)\right)$.

In the case of trapezoidal and triangular fuzzy numbers, for sake of simplicity and also because it is often sufficient for practical purposes, Definition 9 without functions $f_1^A(\lambda), f_1^B(\lambda), f_2^A(\lambda), f_2^B(\lambda)$ is used.

## Definition 10 (Kuchta 2000)

Let $s>0$ be a positive real number and $\tilde{A} = \left(a_1, a_2, a_3, a_4\right)$ and $\tilde{B} = \left(b_1, b_2, b_3, b_4\right)$ two fuzzy trapezoidal numbers. Then:

$$s\tilde{A} = \left(sa_1, sa_2, sa_3, sa_4\right)$$

$$\tilde{A} + \tilde{B} = \left(a_1 + b_1, a_2 + b_2, a_3 + b_3, a_4 + b_4\right)$$

$$\tilde{A}\tilde{B} = \begin{pmatrix} \min\{a_1b_1, a_1b_4, a_4b_1, a_4b_4\}, \min\{a_2b_2, a_2b_3, a_3b_2, a_3b_3\}, \\ \max\{a_2b_2, a_2b_3, a_3b_2, a_3b_3\}, \max\{a_1b_1, a_1b_4, a_4b_1, a_4b_4\}, \end{pmatrix}$$

If $0 \notin [b_1, b_4]$ then $\tilde{A} / \tilde{B} = \tilde{A}\left(1/b_1, 1/b_2, 1/b_3, 1/b_4\right)$

Such a definition is simplified and the outcomes of the simplified operations do differ from the ones of exact operations according to original Definition 9 (also for trapezoidal and triangular fuzzy numbers), but the difference is usually negligible, which is justified - in the context of the fuzzy NPV - in (Terceño et al. 2003) and (Sheen 2005). Definition 8 and, consequently, Definitions 9 and 10 are sometimes not useful. Definition 8 is based on the assumption that all combinations of $s$ and $t$ are possible. Often it is not the case, some couples will never occur and this can be said in advance. In such a case the fuzzy number resulting from applying Definition 8 is unnecessarily large (i.e. the interval $(a_1, a_4)$ and possibly also $(a_2, a_3)$ contain values which in fact should not be there, as their occurrence in practice is impossible). This may lead to false interpretations of the results. That is why in (Dubois and Prade 1985) a modification of Definition 8 has been proposed, where a relation $R$ is defined on the set of $A^\lambda \otimes B^\lambda$, or more generally on $\{s \in (a_1, a_4)\} \otimes \{t \in (b_1, b_4)\}$, indicating which pairs *(s,t)* are possible and to which degree. This information is taken into account in the calculation of the resulting fuzzy numbers, so that no unnecessarily large intervals are computed. Relation $R$ is defined in an interactive way.

In (Lesage 2001) we can find a proposal of how this can be done. Lesage considers three types of $R$:

- non-interactive: no specific relation is assumed, all pairs $(s,t)$ such that $s \in (a_1, a_4)$ and $t \in (b_1, b_4)$ possible
- increasing: if $s \in (a_1, a_4)$ increases, $t \in (b_1, b_4)$ increases
- decreasing: if $s \in (a_1, a_4)$ increases, $t \in (b_1, b_4)$ decreases

Example of an increasing relation and of a decreasing one are given in Fig. 1: the matrices represent two axes: the horizontal one corresponds to $(a_1, a_4)$ and the vertical one to $(b_1, b_4)$, where it is assumed that only some discrete subsets of $(a_1, a_4)$ and $(b_1, b_4)$ are considered:

$$\begin{bmatrix} 0 & 0 & 1 & 1 \\ 0 & 1 & 1 & 0 \\ 0 & 1 & 0 & 0 \\ 1 & 0 & 0 & 0 \end{bmatrix} \quad \begin{bmatrix} 1 & 0 & 0 & 0 \\ 0 & 1 & 1 & 0 \\ 0 & 1 & 1 & 0 \\ 0 & 0 & 0 & 1 \end{bmatrix}$$

**Fig. 1.** Examples of relations between two fuzzy numbers - an increasing one and a decreasing one

The "1" regions in both matrices from Fig. 1 correspond to those couples which the decision maker considers to be possible. The other couples are omitted when Definition 8 is applied. The "1" regions can be made wider (till all the matrix elements are equal to 1 and the non-interactive case is obtained) or thinner (till a 1-1 function is defined between $(a_1, a_4)$ and $(b_1, b_4)$).

Fuzzy numbers cannot be compared with each other in a such direct manner as the crisp numbers: for example, it is not clear which fuzzy number from the following couple should be considered to be greater and to which degree: $(2,3,4,5)$ and $(1,2,3,6)$. However in all the applications of the Present value we usually have to decide which one from a collection of fuzzy numbers is the largest. There are several methods of deciding about this. An overview, made in the context of capital budgeting, is given e.g. in (Kahraman 2001). Here we will present only those methods of comparing fuzzy numbers which are used in the applications of Present value and Fuzzy Present Value.

However, before going over to comparing fuzzy numbers among themselves, let us mention that you should also compare fuzzy numbers with crisp ones and this can be done in various ways. Quite often it is required to determine what are the chances that the yet unknown realization of a fuzzy number will be smaller or equal than a given crisp number. In this context, three measures are used (e.g. Hwang 2006): possibility, necessity and credibility.

## Definition 11

Let $\tilde{A}$ be a fuzzy number and $r$ a crisp one.

a) the possibility of the realization of $\tilde{A}$ being not greater than $r$, denoted as $Pos(\tilde{A} \leq r)$, is defined as $\sup_{x \leq r}(\mu_A(x))$

b) the necessity of the realization of $\tilde{A}$ being not greater than $r$, denoted as $Nec(\tilde{A} \leq r)$, is defined as $1 - Pos(\tilde{A} > r) = 1 - \sup_{x > r}(\mu_A(x))$

c) the credibility of the realization of $\tilde{A}$ being not greater than $r$, denoted as $Cr(\tilde{A} \leq r)$, is defined as $\frac{1}{2}(Pos(\tilde{A} \leq r) + Nec(\tilde{A} \leq r))$

The possibility measure is very weak, i.e. it is very easy for a fuzzy number to be smaller than a given $r$ to a high degree. For example, $Pos(\tilde{A} \leq r) = 1$ for each $r \geq a_2$. The necessity measure is significantly stronger, and the credibility measure is the average of the two.

The same measures can also be used while comparing two fuzzy numbers. E.g.:

## Definition 12

Let $\tilde{A}$ and $\tilde{B}$ be two fuzzy numbers. Then the possibility of the realization of $\tilde{A}$ being not greater than the realization of $\tilde{B}$, denoted as $Pos(\tilde{A} \leq \tilde{B})$, equal to $Pos(\tilde{B} \leq \tilde{A})$, is defined as $\sup_{x \leq y}(\min\{\mu_A(x), \mu_B(y)\})$.

Fuzzy numbers are sometimes compared by means of a crisp characteristic which is calculated for each of them and then the respective crisp values are compared. For example in (Yao and Wu 2000) the following crisp characteristic is considered.

## Definition 13

Let $\tilde{A} = (a_1, a_2, a_3, a_4, f_1^A(\lambda), f_2^A(\lambda))$ be a fuzzy number. Then its crisp characteristic is defined as $\hat{A} = \frac{1}{4}(a_1 + a_2 + a_3 + a_4)$.

In (Campos and González 1989) another crisp characteristic is used:

## Definition 14

Let $\tilde{A} = (a_1, a_2, a_3, a_4, f_1^A(\lambda), f_2^A(\lambda))$ be a fuzzy number. Then its crisp characteristic is defined as $F_A^{\alpha, \lambda^*} = \int_{\lambda^*}^{1}(\alpha f_1^A(\lambda) + (1 - \alpha)f_2^A(\lambda))$, where $\lambda^*$ is any number from the interval $[0,1]$ and $\alpha$ represents the decision maker's degree of optimism. Both parameters $\alpha$ and $\lambda$ are determined by the decision maker.

In (Sheen 2005 a) the membership functions of a triangular or trapezoidal fuzzy number $\tilde{A}$ is converted into a probability distribution in the following way:

$$f_A(x) = h\mu_A(x) \tag{7}$$

54    D. Kuchta

where $f_A(x)$ is the density function and $h$ is chosen in such a way that $f_A(x)$ fulfills the necessary conditions of a density function. Thanks to this approach the author of (Sheen 2005 a) can use the Mellin transform, described for probability distributions in (Giffin 1975) and (Debnath 1995), to calculate another crisp characteristic of a positive triangular or trapezoidal fuzzy number:

**Definition 15**

Let $\tilde{A} = (a_1, a_2, a_3, a_4)$ or $\tilde{A} = (a_1, a_2, a_4)$ be a positive triangular or trapezoidal fuzzy number. Then its crisp characteristic is defined as $M_A(s) = \int_0^\infty x^s f_A(x)dx$, where the integer positive parameter $s$ is determined by the decision maker and $f_A$ is defined by (7).

Another method of comparing fuzzy umbers is presented in (Kaufmann and Gupta 1988). It is based on the ides of removals, which are defined as follows:

**Definition 16**

Let $\tilde{A}$ be a fuzzy numbers. Then

a) The left removal of $\tilde{A}$, denoted as $A_L$, is equal to $a_2 - \int_{a_1}^{a_2} f_1^A(x)dx$

b) The right removal of $\tilde{A}$, denoted as $A_R$, is equal to $a_3 + \int_{a_3}^{a_4} f_2^A(x)dx$

Then a crisp characteristic of $\tilde{A}$, denoted as $M(\tilde{A})$, is defined in the following way:

$$M(\tilde{A}) = \frac{1}{2}(A_L + A_R) \tag{8}$$

Other crisp characteristics which can be used to rank fuzzy numbers are defined e.g. in (Chang 1981, Chiu and Park 1994, Dubois and Prade 1983, Jain 1976, Kaufmann and Gupta 1988).

Sometimes fuzzy parameters are used as coefficients in optimization problems. For example, we can deal with the following linear programming problem with fuzzy coefficients:

$$\sum_{i=1}^{n} \tilde{c}_i x_i \to \max$$
$$\sum_{i=1}^{n} \tilde{a}_i x_i \le \tilde{b}_j \ (j = 1,...,m) \tag{9}$$

where $\tilde{a}_{ij}, \tilde{b}_j, \tilde{c}_i \ (i = 1,...,n; j = 1,...,m)$ are fuzzy numbers.

The definition of the optimal solution of problem (9) is not unequivocal, as it is not unequivocal how to compare fuzzy numbers. In the literature there exist several proposals of this definition. E.g. using the results from (Inuiguchi and Ichihashi 1990, Tanaka and Asai 1984), we can consider the following definition:

## Definition 17

Let $\tilde{b}_0$ be the fuzzy aspiration level for the objective, it should be defined by the decision maker. Then $(x_1^*,...,x_n^*)$ is the optimal solution of (9) if for this vector the following expression attains its maximum

$$\min_{j=0,1,...m} \{\mu_j(x)\}$$

where

$$\mu_0(x) = Pos\left(\sum_{i=1}^{n} \tilde{c}_i x_i \geq \tilde{b}_0\right)$$

$$\mu_j(x) = Pos\left(\sum_{i=1}^{n} \tilde{a}_i x_i \leq \tilde{b}_j\right) \quad (i = 1,...n; j = 1,...m)$$

The problem from the above definition can be reformulated as a classical programming problem, which can be solved using conventional methods. Also in (Słowinski 1986) a similar reformulation of problem (9) into classical programming problem is proposed.

In (Terceño et al. 2003) problem (9) is considered with only triangular fuzzy numbers as parameters (Definition 5). Thus we deal with the following problem:

$$\sum_{i=1}^{n} (c_1^i, c_2^i, c_4^i) x_i \rightarrow \max$$

$$\sum_{i=1}^{n} (a_1^{ij}, a_2^{ij}, a_4^{ij}) x_i \leq (b_1^j, b_2^j, b_4^j)(j = 1,...,m) \tag{10}$$

The problem is the reformulated as a one criterial or a multicriterial crisp linear problem in the following way:

1. the objective function is replaced with

   1.1. either one crisp objective function $\sum_{i=1}^{n} F_{C,i}^{\alpha,\lambda} x_i \rightarrow \max$ (Def. 14)

   1.2. or three crisp objective functions:

   $$\sum_{i=1}^{n} c_l^i x_i \rightarrow \max (l = 1,2,4))$$

The problem is treated like a crisp multiobjective problem - any method known for solving such problems can be used.

2. the constrains are replaced with the following crisp ones:

$$\sum_{i=1}^{n} a_l^{ij} x_i \leq b_l^j \quad (l = 1,2,4; j = 1,...,m)$$

Another method of solving (9) consists in replacing all the fuzzy coefficients with their crisp characteristics, e.g. those from Definition 13 and 14, which of course leads to a conventional optimization problem.

Fuzzy approach is also used in fuzzy logic (e.g. Ross 1995), where fuzzy numbers are used in their second role - to define fuzzy concepts (e.g. "high income", "low cost"

56    D. Kuchta

etc.). There, the starting point is an equivalent of the membership function from Definition 4, let us call it also a membership function and denote it also by $\mu_A$, where $\tilde{A}$ stands for the fuzzy concept. The value $\mu_A(x)$ represents the degree to which $x$ belongs to the fuzzy concept, i.e. to which degree it is a high income or a low cost. Here quite often left hand or right hand fuzzy numbers are used (although fuzzy numbers from Definition 4 can be used too, all depends on the concept being described). The left-hand and right-hand fuzzy numbers are defined as follows:

**Definition 18**
a) The triplet $\tilde{A} = (a_1, a_2, f_1^A)$ will be called a right-hand fuzzy number. Its membership function is defined as:

$$\mu_A(x) = \begin{cases} 0 \text{ for } x \le a_1 \\ f_1^A(x) \text{ for } x \in (a_1, a_2) \\ 1 \text{ for } x \ge a_2 \end{cases}$$

b) The triplet $\tilde{A} = (a_3, a_4, f_2^A)$ will be called a left-hand fuzzy number. Its membership function is defined as:

$$\mu_A(x) = \begin{cases} 1 \text{ for } x \le a_3 \\ f_1^A(x) \text{ for } x \in (a_3, a_4) \\ 0 \text{ for } x \ge a_4 \end{cases}$$

Once the fuzzy concepts are defined, fuzzy rules are applied, which assume the following form:

$$\text{If } x_1 \text{ is } \tilde{A}_1 \text{ and/or } x_2 \text{ is } \tilde{A}_2 \text{ and/or ... and/or } x_n \text{ is } \tilde{A}_n \text{ then } y \text{ is } \tilde{B} \tag{11}$$

or

$$\begin{aligned} &\text{If } x_1 \text{ is } \tilde{A}_1 \text{ and/or } x_2 \text{ is } \tilde{A}_2 \text{ and/or ... and/or } x_n \text{ is } \tilde{A}_n \\ &\text{then take a specified decision} \end{aligned} \tag{12}$$

where $x_i$ $(i = 1,...,n)$ - input values and $y$ - output values - are crisp numbers, $\tilde{A}_i$ $(i = 1,...,n)$ and $\tilde{B}$ are fuzzy concepts. While applying fuzzy rules of type (11) a defuzzification method has to be used (e.g. Ross 1995), so that a precise decision about the output variables can be met. Defuzzification makes a crisp number of a fuzzy set, while this crisp number has to be in some way representative for the fuzzy number $\tilde{B}$ being used. For example, if the input variables are the demand and the production cost and the output variable is the planned production quantity, the fuzzy rule may be: if demand is low and if the production cost high, then the planned production should be low. A defuzzification method would then choose a specific value (which would be to a high degree "a low production") for the production quantity.

In (Kwakernaak 1978, 1979) so called random fuzzy variables are introduced. Let us now define this generalization of fuzzy numbers, which is also used in capital budgeting.

**Definition 19**

Let $\tilde{A}$ be a fuzzy number. Then the triplet $\left( \Re, P(\Re), Pos \right)$ , where $P(\Re)$ is the power set of the set $\Re$ of real numbers is called a possibility space, where $Pos$ is defined on $P(\Re)$ in the following way: for each $X \subset \Re$ $Pos(X) = \sup_{x \in X} \mu_A(x)$.

For the sets from $P(\Re)$ the necessity and credibility measure can be defined:

**Definition 20**

Let $\left( \Re, P(\Re), Pos \right)$ be a possibility space. Then for each $X \subset \Re$ :

$$Nec(X) = 1 - Pos(\Re \setminus X)$$

$$Cr(X) = \frac{1}{2}\left( Pos(X) + Nec(X) \right)$$

Now we are in position to define fuzzy random variable:

**Definition 21**

Let $\left( \Re, P(\Re), Pos \right)$ be a possibility space. A fuzzy random variable is any function $\xi$ from the possibility space to the set of random variables.

For example, if $\tilde{A}$ is a fuzzy number and $N(a,1)$ stands for a normal distribution with parameters $a,1$, then $\xi = N(\tilde{A},1)$ is a fuzzy random variable. The sum and the multiplication of fuzzy random variables can be defined in natural way.

The following definition will be used further on:

**Definition 22 (Liu 2002)**

Let $\left( \Re, P(\Re), Pos \right)$ be a possibility space, $\left( \xi_1, \ldots, \xi_n \right)$ be an $n$-dimensional vector of fuzzy random variables and $f$ a function from $\Re^n$ to $\Re$ . Then the chance of a random fuzzy event $f\left( \left( \xi_1, \ldots \xi_n \right) \right) \leq 0$ on the credibility level $\alpha$ is a function from $(0,1]$ to $[0,1]$ defined as:

$$Ch\left( f\left( \left( \xi_1, \ldots \xi_n \right) \right) \leq 0 \right)(\alpha) = \sup_{Cr(X) \geq \alpha} \inf_{x \in X} \Pr\left( f\left( \left( \xi_1(x), \ldots \xi_n(x) \right) \right) \leq 0 \right)$$

where Pr stands for probability.

We have presented basic concepts and applications concerning the Present Value, the Net Present value in the crisp case and the fuzzy numbers (as well as their generalization). Now the two aspects will be combined.

## 58    D. Kuchta

## 3 Fuzzy Present Value - Concept and Applications

### 3.1 Fuzzy Data

If we consider fuzzy net present value, we admit fuzzy data as characteristics of the project. Also the duration of the project can be fuzzy, but here we assume it be crisp (for the case of fuzzy project duration see e.g. (Greenhut et al. 1995 and Kuchta 2000)). Thus each project will be characterized by:

- Crisp duration $n$
- $C\tilde{O}F_i$ $(\geq 0)$ – fuzzy cash outflow at the end of the $i$th year $(i=0,1,...,n)$, $C\tilde{I}F_i$ $(\geq 0)$ - cash inflow at the end of the $i$th year $(i=1,...,n)$, and fuzzy cash flow $C\tilde{F}_i$, occurring at the end of the $i$th year $(i=0,1,...,n)$, where $C\tilde{F}_0 = -C\tilde{O}F_0$, $C\tilde{F}_i = C\tilde{I}F_i - C\tilde{O}F_i$ $(i=1,...,n)$.
- $\tilde{r}$ $(>0)$ - the fuzzy required rate of return or cost of capital.

It is quite natural in many projects to consider fuzzy cash flows, fuzzy rate of return and fuzzy duration. The element of uncertainty is present in each investment project, but in a special way in those where social cost and benefits should be taken into account. These characteristics are usually non-measurable, and here the use of linguistic expressions, which can be then transformed to fuzzy numbers (Dompere 1995) is especially important. An example are of non-measurable data which should be taken into account while evaluating an investment project are social costs and benefits or the entrepreneur's energy (Greenhut et al. 1995).

The fuzzy cash outflows, inflows and flows are calculated applying fuzzy arithmetic operations to various (fuzzy) components, like quantities of items, prices, sales, fixed and variable cost, depreciation, wages etc. In (Sheen 2005) formula (1) is used in its fuzzy version, where both the cash flow in the first year and the inflation rate are given in the form of triangular or trapezoidal fuzzy numbers:

$$C\tilde{F}_i = C\tilde{F}_1\left(1+\tilde{d}\right)^{n-1} \qquad (i=1,...,n) \qquad (12)$$

However, as the author of (Sheen 2005) considers a very practical case of projects optimizing the delivery of electricity and the models presented there should be used by practitioners, he - what is done very often in practice - does not use Definition 9 but it simplification, Definition 10. He analyses the divergence between the results given by both definitions, coming to the conclusion that Definition 10 is sufficient.

If certain dependencies exist between individual components and Definition 8 (or 9) cannot be applied, because certain couples of crisp argument will never occur, the approach from (Lesage 2001) should be used. In the computational example in (Lesage 2001) the following dependencies in the calculation of cash flows are considered (each of them can occur in various degrees of strength - the wider the "1" region (see Fig. 1), the smaller the strength of the relation):

- relation price-quantity: the selling price of a product decreases if the quantity sold increases;
- relation revenues-charges: if the revenues increase, so do the charges, and the stronger proportion of the variable cost in the charges, the stronger the relation.

## 3.2 Calculation of Fuzzy Present Value and Fuzzy Net Present Value

As all arithmetic operations are on fuzzy numbers are defined (Definitions 8-10), the definitions of fuzzy present value and fuzzy net present value are straightforward (Buckley 1987, Goutiérez 1989, Buckley 1992, Chiu and Park 1994, Kuchta 2000, Kahraman et al. 2002), a slightly different definition has been proposed in (Ward 1985).

**Definition 23**

The present worth (or present value) $PV(C\tilde{F}_i)$ of a cash flow $C\tilde{F}_i$ (or cash outflow or inflow) $(i=0,1,...,n)$ is defined as

$$PV(C\tilde{F}_i) = \frac{C\tilde{F}_i}{(1+\tilde{r})^i}$$

Also here we can use the Definitions 8-10 only if there is no relationship between the discount rates and the cash flows. In (Lesage 2001) an increasing relation is considered between the two values: the higher the estimated cash inflow, the higher the risk linked to it, and the higher the discount rate.

**Definition 24**

Net Present value of a project is defined as:

$$NPV(P) = CF_0 + \sum_{i=1}^{n} PV(CF_i)$$

Again, we can use here Definitions 8-10 only when the cash flows from corresponding periods are independent on each other. Lesage in (Lesage 2001) considers other cases: a good year (high cash inflows) may imply that next year will also be good (an increasing relation), or in some cases we can assume that if one year is good, the next one will be bad (saturation of demand) or, on the contrary, if one year is bad, the next one will be better (a decreasing relation). In such cases we have to use the modification of Definition 8 from (Lesage 2001).

## 3.3 Interpretation of Fuzzy Present Value and Fuzzy Net Present Value

Analogously to the crisp case, the higher the Net Present Value, the better the project. The only problem is that, as mentioned above, fuzzy values cannot be usually compared to each other in an unequivocal way. That is why the decision maker has to decide how he wants to compare fuzzy values, i.e. when one fuzzy NPV will he preferred by him to another one. For example in (Terceño et al 2003) the crisp characteristic from Definition 14 is used (in case the fuzzy data used are triangular fuzzy numbers). The authors have chosen this method because, as compared to many other crisp characteristics of fuzzy numbers, it allows an intuitive introduction of the decision maker's risk aversion and the fuzziness of the triangular numbers used can be reduced by fixing $\lambda^*$.

60     D. Kuchta

### 3.4 Fuzzy and Probabilistic Approach to the Present Value

There has been a long discussion in the literature about the differences, advantages and disadvantages of the fuzzy and probabilistic approach to modeling of risk and uncertainty (e.g. Choobineh and Behren,1992), where in the probabilistic approach the magnitudes which are unknown at the decision moment are expressed in the form of probability distributions instead of fuzzy numbers. There are lot of arguments in favour and against each of the approaches, and maybe the best conclusion of the discussion would be to say that both approaches are different and do complete each other. There are even numerous proposals of switching from one approach to the other for the same type of problems and the same data (e.g. Yager 1982), and it turns out that both approaches can contribute something important to one and the same decision problem. In (Rebiasz 2007) both approaches are compared for the NPV. The fuzzy input data required for the calculation of the NPV is transformed, using the approach from (Yager 1982), into probabilistic data and than the NPV in the form of a probability distribution is calculated using simulation. Then the fuzzy NPV is calculated for the original fuzzy input data of the same problem. The two NPV's are compared. The author claims the results are in favour of the probabilistic approach: in the opinion of the author that information concerning the probability of various crisp numbers being the actual value of the NPV is more useful to the decision maker than the information about the corresponding possibility degree.

Another proposal combining fuzzy and probabilistic concepts with respect to the NPV can be found in (Tsao 2005). There for each project the following data is given:

a) Crisp duration $n$

b) set of possible scenarios $\{j: j = 1,...m\}$ with the corresponding probabilities

$p_j(j = 1,...m)$, where $\sum_{j=1}^{m} p_j = 1$ (it is assumed that for each project one scenario

can occur which will persist throughout its whole duration)

c) $\widehat{COF}_i^{\,j}$ $(\geq 0)$ – fuzzy cash outflow at the end of the $i$th year ($i=0,1,...,n$) for the $j$th

scenario ($j=1,...,m$), $\widehat{CIF}_i^{\,j}$ $(\geq 0)$ - cash inflow at the end of the $i$th year ($i=1,...,n$)

for the $j$th scenario and fuzzy cash flow $\widetilde{CF}_i^{\,j}$, occurring at the end of the $i$th year

($i=0,1,...,n$) for the $j$th scenario ($j=1,...,m$)

d) $\tilde{r}_j$ $(>0)$ - the fuzzy required rate of return or cost of capital for the $j$th scenario

($j=1,...,m$).

Then fuzzy $\widetilde{NPV}^{\,j}$ for each scenario is calculated according to Definition 24, and finally three characteristic of the global fuzzy $\widetilde{NPV}$ of the project are calculated

( $NPV^{\,j\lambda} = \left[ \underline{NPV}^{\,j\lambda}, \overline{NPV}^{\,j\lambda} \right]$ are the $\lambda$-cuts of $\widetilde{NPV}^{\,j}$).

a) the fuzzy expected value $\widetilde{ENPV} = \sum_{j=1}^{m} \widetilde{NPV}^{\,j} p_j$ , being a fuzzy number with $\lambda$-cuts

denoted as $ENPV^{\,\lambda} = \left[ \underline{ENPV}^{\,\lambda}, \overline{ENPV}^{\,\lambda} \right]$

b) the standard deviation of $\tilde{NPV}$, denoted as $\sigma_{NPV}$, being a crisp number defined as:

$$\sigma_{NPV} = \frac{1}{2}\left[ \begin{array}{l} \int\limits_0^1 \left\{ \sum\limits_{j=1}^m \left[ \underline{NPV}^{j\lambda} - \underline{ENPV}^\lambda \right]^2 p_j \right\} d\lambda + \\ \int\limits_0^1 \left\{ \sum\limits_{j=1}^m \left[ \overline{NPV}^{j\lambda} - \overline{ENPV}^\lambda \right]^2 p_j \right\} d\lambda \end{array} \right]^{\frac{1}{2}}$$

c) the fuzzy performance index $\tilde{FPI}$, defined as:

$$\tilde{FPI} = \tilde{ENPV} \Big/ \sigma_{NPV}$$

Further in (Tsao 2005) investment projects are ranked on the basis of $\tilde{ENPV}$ or $\tilde{FPI}$, where (8) is used to compare fuzzy numbers.

Another approach combining fuzzy and probabilistic concepts with respect to the NPV, based on the notion of random fuzzy variables (Definition 21), can be found in (Huang 2007). There for each project the following data is given:

a) Crisp duration $n$

b) $\hat{COF}_i$ ($\geq 0$) – a fuzzy random variable representing the cash outflow at the end of the $i$th year ($i=0,1,....,n$), $\hat{CIF}_i$ ($\geq 0$) - fuzzy random variable representing the cash inflow at the end of the $i$th year ($i=1,...,n$), and fuzzy random variable representing the cash flow $\hat{CF}_i$, occurring at the end of the $i$th year ($i=0,1,...,n$)

c) $r$ ($>0$) - a crisp required rate of return or cost of capital.

For such projects we get, applying formulae analogous to that from Definition 24, a random fuzzy variable $\hat{NPV}$ representing the NPV (Huang 2007).

Let us now go over to the applications of fuzzy present value and fuzzy net present value.

## 3.5 Fuzzy Net Present Value as Objective in Project Selection Problems

Exactly like the crisp NPV is used as objective in project selection problems (Example 1), fuzzy NPV can also play such a role. However, as in case of fuzzy numbers it is not clear which one is the biggest, special methods of fuzzy programming should be used. What is more, also the constraints will usually comprise fuzzy data, so that also they should be treated using special approaches. Usually the solution obtained will not be "optimal" and/or "feasible" in a crisp, decisive sense, but rather in a fuzzy one - it will be to some degree possible that the solution will be so once the realizations of the fuzzy parameters are known.

62     D. Kuchta

The authors of (Terceño et al. 2003) and of (Kahraman and Ulukan 1999) consider the following problem:

$$\sum_{i=1}^{k} N\tilde{P}V_i x_i \to \max$$

$$\sum_{i=1}^{k} C\tilde{F}_{ij} x_i \le \tilde{B}_j \; \left(j = 1,...,n\right)$$

(13)

where the notation is exactly like in Example 1. In (Terceño et al 2003) the fuzzy parameters are triangular fuzzy numbers and the solution method described for model (10) is used. In (Kahraman and Ulukan 1999) the fuzzy parameters do not have to be triangular and the problem is reformulated into a crisp classical one using the approach from (Slowinski 1986).

In (Hwang 2006) the chance constrained programming model is applied. The author considers the following two models, concerning a similar decision situation as in Example 1, but with in- and outflows, thus also NPV, being fuzzy (the discount rate remains crisp).

Let $N\tilde{P}V_i \; \left(i = 1,...k\right)$ denote the fuzzy Net Present value of the $i$th project (calculated according to Definition 24), $C\tilde{O}F_{ij} \; \left(i = 1,...,n; j = 1,...k\right)$ - cash outflows caused by the $i$th project in the $j$th year, and $B_j \; \left(j = 1,...,n\right)$ - the crisp cash budget of the company available for the $j$-th year. The problem formulated below - with a fuzzy objective and fuzzy constrains - will give us the information which selection of projects should be chosen to be realized so that the chances that the budget will not be exceeded are at least equal to a preset level. The objective being maximized is not directly the total NPV, because its values are here fuzzy and it is not clear which project should be considered to be the one with the maximal NPV. Instead, the objective will be:

- either the crisp limit that the total fuzzy NPV has to exceed with chances not smaller than a preset level
- or the credibility degree of the total fuzzy NPV being greater or equal than a preset level.

The chances in (Hwang 2006) are measured using the credibility measure (Definition 11). The models from (Hwang 2006) corresponding to cases a) and b) are given below:

**Case a).** ($\alpha$ and $\beta_j \; \left(j = 1,...n\right)$ from the interval (0, 1) should be preset by the decision maker)

$$z \to \max$$

$$Cr\left(\sum_{i=1}^{k} NPV_i x_i \ge z\right) \ge \alpha$$

$$Cr\left(\sum_{i=1}^{k} CF_{ij} x_i \le B_j\right) \ge \beta_j \; \left(j = 1,...,n\right)$$

$$x_i \in \{0,1\} \left(i = 1,...,k\right)$$

(14)

## Optimization with Fuzzy Present Worth Analysis and Applications 63

**Case b).** ( $\beta_j$ $(j = 1,...n)$ from the interval $(0, 1)$ and a real value $z$ should be preset by the decision maker)

$$Cr\left(\sum_{i=1}^{k} NPV_i x_i \geq z\right) \rightarrow \max$$

$$Cr\left(\sum_{i=1}^{k} CF_{ij} x_i \leq B_j\right) \geq \beta_j \ (j = 1,...,n) \tag{15}$$

$$x_i \in \{0,1\} (i = 1,...,k)$$

In (Hwang 2006) a fuzzy simulation based genetic algorithm - along with the results of some computational experiments - is proposed, which solves the above problems.

In (Hwang 2007) a generalization of problem (14) is considered: the cash flows are assumed to be fuzzy random variables (Definition 21), thus the NPV's are also fuzzy random variables. Then, using Definition 22, the above problem becomes (where $(\alpha, \delta, \beta_j$ and $\gamma_j$ $(j = 1,...n)$ from the interval $(0, 1)$ should be preset by the decision maker):

$$z \rightarrow \max$$

$$Ch\left(\sum_{i=1}^{k} N\hat{P}V_i x_i \geq z\right)(\delta) \geq \alpha$$

$$Ch\left(\sum_{i=1}^{k} C\hat{F}_{ij} x_i \leq B_j\right)(\gamma_j) \geq \beta_j \ (j = 1,...,n) \tag{16}$$

$$x_i \in \{0,1\} (i = 1,...,k)$$

Also in (Hwang 2007) a goal programming approach for problem (16) is presented. Let $g$ be the desired (crisp) goal value of $\sum_{i=1}^{k} N\hat{P}V_i x_i$ and $b_j$ $(j = 1,...n)$ the desired (crisp) goal values of $\sum_{i=1}^{k} C\hat{F}_{ij} x_i$ . The problem is then formulated in he following way:

$$(d_1,...d_n,d) \rightarrow \min$$

$$Ch\left(g - \sum_{i=1}^{k} N\hat{P}V_i x_i \leq d\right)(\delta) \geq \alpha$$

$$Ch\left(\sum_{i=1}^{k} C\hat{F}_{ij} x_i - b_j \leq d_j\right)(\gamma_j) \geq \beta_j \ (j = 1,...,n) \tag{17}$$

$$x_i \in \{0,1\} (i = 1,...,k)$$

where $(d_1,...d_n,d)$ is the vector of undesired deviations for the goals $(b_1,...b_n,g)$.

This is a problem with $n+1$ objectives, any method of multiobjective programming can be used here. In (Hwang 2007) a lexicographic approach is used. In the same paper genetic-kind algorithm for problems (16) and (17) can be found.

64    D. Kuchta

## 3.6 Applications to the Valuation of Projects with Future Options

The NPV approach is often criticized because of the fact that it looks at the project in question as "now or never" and the main question it answers is whether to go ahead with an investment or not (Balasubramanian et al. 2000). Sometimes the real question is more complex: we cannot say in advance what the cash flows will be, even if we apply the modeling by means of fuzzy numbers, because there can be certain investment options available later on, which may be taken on or not, and the decision has to be made only later, while the project is already going on. The Net Present value approach ignores the value of flexibility and that is why leads to the rejection of many interesting and innovative projects.

In such cases of flexible projects the real-option approach should be applied (Balasubramanian et al. 2000), which has a fuzzy equivalent (Carlsson and Fullér 2003). However, the Present value is not completely rejected when the real-option approach is used, but it constitutes an element of the new approach. The Present Values of various options are calculated and used in the formulae to calculate the options prices as well as the moment till which we can defer he decision to take on or not each individual investment option. In the fuzzy equivalent of the real-option approach (Carlsson and Fullér 2003) the fuzzy Present Values are calculated and - while looking for the best moment to take an investment decision - fuzzy numbers are compared among each other, so that the maximal one is selected. As mentioned earlier, the latter choice is not unequivocal and the decision maker has to decide upon the method of ranking fuzzy numbers.

Also in (Collan 2004) a dynamic approach to investment evaluation is used: it is admitted that the start of the project can be postponed, because there may be some value in waiting (for example, certain prices may increase in the future). On the other hand, waiting may also causes some losses. The two aspects of waiting and of its positive or negative impact on the cash in- and out flows are taken into account in the Fuzzy Real Investment Valuation Model from (Collan 2004). The fuzzy NPV is calculated for various project starting moments and the starting moment with the best (according to some assumed ranking method) fuzzy NPV is selected.

## 3.7 Fuzzy Net Present Value as Objective in Optimization Problems from the Industry

In (Liu and Sahinidis 1997) a fuzzy version of the problem presented in Example 2 is considered. The authors introduce fuzziness into the market demands and supplies (which constitute the right-hand sides of the constraints) and into the objective function coefficients $\alpha_{it}, \beta_{it}, \quad \gamma_{ilt}, \Gamma_{ilt}, \delta_{it}, \quad i = 1,..,NP; \; j = 1,...,NC$, $l = 1,...,NM; t = 1,...,NT$. In this way we get a problem of the (9) type, which is then solved using Definition 17. The authors develop an adequate algorithm for the resulting model, which is an integer and non-linear one.

In (Zhou and Gong 2000) a fuzzy version of Example 3 is presented. The net present values $NPV_{ik}$ are given as fuzzy numbers, thus fuzzy $NPV$ occurs in two problems: in the objective function (5) and in the constraints (6). In the problem with objective (5) also some coefficients of the constraints are fuzzy. What is more, $NPV_{TV}$ from (6) is also given as a fuzzy number. Both problems are solved by turning them into conventional ones: all the fuzzy coefficients are replaced with their crisp characteristics from Definition 12.

In (Chan et al 2005) a fuzzy version of Example 4 is presented. All the weightings (those of the criteria, those of the technologies under individual criteria and the final ratings of the technologies) are determined in the form of a fuzzy number, and the ratings concerning the NPV criterion are calculated on the basis of the fuzzy NPV's of individual technologies. And in (Wong et al 2005) the application of the same fuzzy multicriteria rating algorithm taking into account the fuzzy NPV's of the alternatives is considered for the choice of technology for intelligent building projects.

In (Sheen 2005) various projects optimizing the delivery of electricity are considered. The fuzzy cash flows are assumed to be constant if there is no inflation, and the inflation is assumed to be fuzzy. The cash flows are calculated according to formula (12). Fuzzy Net Present Values of the projects are ranked according to several criteria: using Definition 14 and the crisp characteristics defined in (Chang 1981, Chiu and Park 1994, Dubois and Prade 1983, Jain 1976, Kaufmann and Gupta 1988). As no ranking criterion for fuzzy numbers is ideal, the ranking gives different results for different ranking methods. Finally the project is selected which has been classified as lbelonging to the best ones by all (most) ranking methods.

### 3.8 Fuzzy Net Present Value as an Objective in Spatial Games

Example 5 presents the crisp example of spatial games. As mentioned there, the optimal strategy is searched for using *NPV* as the reward of the players. In (Bogataj and Usenik 2005) the fuzzy equivalent of the same problem is presented. Here the fuzziness is used in the sense of fuzzy logic, fuzzy concepts and fuzzy rules (11), as well as neutral nets. The input values (the set-up costs, prices, demands etc.) are given as fuzzy concepts: low, high, medium etc. The optimal NPV can also take on fuzzy values and it can be found to be bad, good, very good etc. according to predefined fuzzy concepts. After defuzzification crisp values of NPV are determined. The authors of (Bogataj and Usenik 2005) claim that their approach can also be extended to more competitors and other types of competing.

### 3.9 Fuzzy Classification Based on Net Present Value

Fuzzy rules of type (12) can be applied to make decisions about rejecting or accepting a project. The boundary between the projects that should be accepted and those to be rejected is never quite clear (also if crisp NPV is used), that is why fuzzy rules can be useful here. An example of the application of such rules is described in (Vranes et al. 2002). There various fuzzy concepts are defined, among them the fuzzy concepts concerning the NPV. All the fuzzy concepts in (Vranes et al. 2002) linked to NPV are defined by means of right-hand, left-hand (Definition 18) or trapezoidal fuzzy numbers, whose membership functions are defined for the ratios $NPV/COF_0$ (section 2.1), called relative NPV. Here are the fuzzy concepts for the NPV from (Vranes et al. 2002), together with the respective fuzzy numbers:

- a very good NPV (defined by means of $\tilde{A}_{vg} = \left(a_1, a_2, f_1^A\right)$ , where $a_1 = 0.5, a_2 = 1, f_1^A(x) = 2(x - 0.5))$

66     D. Kuchta

- a good NPV (defined by means of $\tilde{A}_g = \left(a_1, a_2, a_3, a_4, f_1^A, f_2^A\right)$,

where
$a_1 = -0.4, a_2 = 0, a_3 = 0.5, a_4 = 0.9, f_1^A(x) = 2.5(x + 0.4), f_1^B(x) = -2.5(x - 0.9)))$

- a bad NPV (defined by means of $\tilde{A}_b = \left(a_1, a_2, a_3, a_4, f_1^A, f_2^A\right)$,

where
$a_1 = -0.9, a_2 = -0.5, a_3 = 0, a_4 = 0.4,$

$f_1^A(x) = 2.5(x + 0.9), f_1^B(x) = -2.5(x - 0.4)))$

- a very bad NPV (defined by means of $\tilde{A}_{vb} = \left(a_3, a_4, f_2^A\right)$,

where $a_3 = -1, a_2 = -0.5, f_2^A(x) = -2(x + 0.5))$

In (Vranes et al. 2002) the criterion of NPV is combined with several other criteria, also qualified as "good" or "bad" by means of fuzzy numbers. Let us simplify the reasoning from (Vranes et al. 2002) and denote the other criteria by a single symbol $p$. Let us then use the symbols $\mu_{vg}^{NPV}, \dots, \mu_{vb}^{NPV}, \mu_{vg}^p, \dots, \mu_{vb}^p$ for the membership functions corresponding to the respective fuzzy sets (i.e. for an investment project $I$ $\mu_{vg}^{NPV}$ represents the degree to which $I$ has a very good NPV). Then sets like A= {projects whose NPV is very good and $p$ is very good}, B = {projects whose NPV is good and $p$ is very good}, C = {projects whose NPV is very bad and $p$ is bad} etc. are defined. For each project we calculate the degree of belonging to each of those sets as the minimum of the degree to which the project has the respective characteristics, e.g. an investment project $I$ belongs to A to the degree equal to $\min\left\{\mu_{vg}^{NPV}(I), \mu_{vg}^p(I)\right\}$. Finally each project is assumed to belong to the set for which it has the maximal membership degree. The decision maker should then apply some decision rules, e.g. "if a project belongs to A or B it should be accepted", "if it belongs to C it should be rejected" etc. Also in (Dimova et al 2006) NPV is used, in the fuzzy setting, as one of project assessment criteria.

(Haven 1998) contains a proposal of a fuzzy transitive relation between projects characterized by crisp NPV's. This relation allows to calculate for each couple of projects the measure of preference of one project to another one, while this measure is based on cash flows of the projects in consecutive years and is included in the interval [0,1] (that is why the relation is a fuzzy one). Thanks to this relation a set of projects with equal NPV's can be ranked and the best project can be identified or it can be stated that the projects are incomparable.

## 4 Conclusions

We presented the concept of fuzzy present value and fuzzy net present value together with selected applications. The present value is considered by many to be the most important criterion of investment project evaluation. It is thus extremely important to know how to apply it in the fuzzy case. We hope that this section will be a help in this respect.

# References

Balasubramanian, P., Kulatilaka, N., Storck, J.: Managing information technology investments using a real-options approach. The Journal of Strategic Information Systems 9(1), 39–62 (2000)

Bogataj, M., Bogataj, L.: Supply chain coordination in spatial games. International Journal of Production Economics 71, 277–285 (2001)

Bogataj, M., Usenik, J.: Fuzzy approach to the spatial games in the total market area. International Journal of Production Economics 93(94), 493–503 (2005)

Buckley, J.J.: The Fuzzy Mathematics of Finance. Fuzzy Sets and Systems 21, 257–273 (1987)

Buckley, J.J.: Solving Fuzzy Equations in Economics and Finance. Fuzzy Sets and Systems 48, 289–296 (1992)

Carlsson, C., Fullér, R.: A Fuzzy Approach to Real Options Valuation. Fuzzy Sets and Systems 139, 297–312 (2003)

Campos, L.M., Gonzáles, A.: A subjective approach for ranking fuzzy numbers. Fuzzy Sets and Systems 29, 145–153 (1989)

Chan, F.T.S., Chan, M.H., Tang, N.K.H.: Evaluation methodologies for technology selection. Journal of Materials Processing Technology 107, 330–337 (2000)

Chang, W.: Ranking of fuzzy utilities with triangular membership function. In: Proc. Int. Conf. of Policy Anal. and Inf. Systems, pp. 263–272 (1981)

Choobineh, F., Behrens, A.: Use of interval and possibility distributions in economic analysis. Journal of Operations Research Society 43(9), 907–918 (1992)

Chiu, C.Y., Park, C.S.: Fuzzy Cash Flow Analysis Using Present Worth Criterion. Eng. Econom. 39(2), 113–138 (1994)

Collan, M.: Fuzzy real investment valuation model for giga-investments, and a note on giga-investment lifecycle and valuation. TUCS Technical Report 617, Abo Akademi University, Institute for Advanced Management Systems Research (2004)

Debnath, L.: Integral transforms and their application. CRC Press, New York (1995)

Dimova, L., Sevastianov, P., Sevastianov, D.: MCDM in a fuzzy setting: Invetsment projects assessment application. International Journal of Production Economics 100, 10–26 (2006)

Dubois, D., Prade, H.: Ranking fuzzy numbers in the setting of possibility theory. Information Sciences 30, 183–224 (1983)

Dompere, K.K.: The theory of social cost and costing for cost-benefit analysis in a fuzzy-decision space. Fuzzy Sets and Systems 76, 1–24 (1995)

Dubois, D., Prade, H.: Fuzzy Numbers: An overview, analysis of fuzzy information. In: Mathematics and Logic (1), pp. 3–39. CRC Press, Boca Raton (1985)

Giffin, W.C.: Transform techniques for probability modeling. Academic Press, New York (1975)

Gutiérrez, I.: Fuzzy Numbers and Net Present Value. Scand. J. Mgmt. 5(2), 149–159 (1989)

Greenhut, J.G., Norman, G., Temponi, C.: Toward a fuzzy theory of oligopolistic competition. In: IEEE Procedings of ISUMA-NAFIPS 1995, pp. 286–291 (1995)

Haven, E.E.: The Fuzzy Multicrteria Analysis Method: An Application on NPV Ranking. Int. Journal of Intelligent Systems in Accounting, Finance and Management 7, 243–252 (1998)

Horngren, T.H., Foster, G., Datar, S.M.: Cost accounting: A Managerial Emphasis. Prentice Hall, Englewood Cliffs (1994)

Huang, X.: Chance-constrained programming problems for capital budgeting with NPV as fuzzy parameters. Journal of Computational and Applied Mathematics 198(1), 149–159 (2006)

Huang, X.: Optimal project selection with random fuzzy parameters. International Journal of Production Economics 106, 513–522 (2007)

Inuiguchi, M., Ichihashi, H.: Relative modalities and their use in possibilistic linear programming. Fuzzy Sets and Systems 35(3), 303–323 (1990)

Jain, R.: Decision-making in the presence of fuzzy variables. IEEE Trans. Syst. Man Cybern. 6, 698–703 (1976)

Kahraman, C.: Fuzzy Versus Probabilistic Benefit/Cost Ratio Analysis for Public Work Projects. Int. J. of Applied Math. and Comp. Sc. 11(3), 705–718 (2001)

Kahraman, C., Ulukan, Z.: Multi-Criteria Capital Budgeting Using FLIP. In: Third International Conference on Computational Intelligence and Multimedia Applications (ICCIMA 1999), pp. 426–431 (1999)

Kahraman, C., Ruan, D., Tolga, E.: Capital Budgeting Techniques Using Discounted Fuzzy Versus Probabilistic Cash flows. Information Sciences 142, 57–76 (2002)

Kaufmann, A., Gupta, M.M.: Fuzzy mathematical models in engineering and management science. Elsevier Scinece Publishers BV, Amsterdam (1988)

Kuchta, D.: Fuzzy Capital Budgeting. Fuzzy Sets and Systems 111, 367–385 (2000)

Lesage, C.: Discounted Cash-flows Analysis: An Interactive fuzzy arithmetic approach. European Journal of Economic and Social Systems 15(2), 49–68 (2001)

Kwakernaak, H.K.: Fuzzy random variables-I. Information Sciences 15, 1–29 (1978)

Kwakernaak, H.K.: Fuzzy random variables-II. Information Sciences 17, 153–1768 (1979)

Liu, B.: Theory and Practice of Uncertain Programming. Phisica-Verlag, Heidelberg (2002)

Liu, L.M., Sahinidis, N.V.: Process planning in a fuzzy environment. European Journal of Operational Research 100, 142–169 (1997)

Rebiasz, B.: Fuzziness and randomness in investment project risk appraisal. Computers and Operations Research 34, 199–210 (2007)

Ross, T.J.: Fuzzy logic with engineering applications. McGraw Hill Inc., New York (1995)

Sheen, J.N.: Fuzzy financial analyses of demand side management alternatives. IEE Proc. Gener. Transm. Distrib. 152(2), 215–226 (2005a)

Sáez, C.A., Requena, J.C.: Reconciling sustainability and discounting in Cost-Benefit Analysis: A methodological approach. Ecological Economics 60, 712–725 (2007)

Sheen, J.N.: Fuzzy financial profitability analyses of demand side management alternatives from participant perspective. Information Sciences 169, 329–364 (2005b)

Sheen, J.N.: Fuzzy evaluation of cogeneration alternatives in a petrochemical industry. Computers and Mathematics with applications 49, 741–755 (2005c)

Słowiński, R.: A multicriteria fuzzy linear programming method for water supply system development planning. Fuzzy Sets and Systems 19(3), 217–237 (1986)

Tanaka, H., Asai, K.: Fuzzy linear programming with fuzzy numbers. Fuzzy Sets and Systems 13, 110 (1984)

Terceño, A., de Andrés, J., Barberà, G.: Using fuzzy set theory to analyse investments and select portfolios of tangible investments in uncertain environments. International Journal of Uncertainty, Fuzziness and Knowledge-Based Systems 3, 263–281 (2003)

Tsao, C.-T.: Assessing the Probabilistic Fuzzy Net Present Value for a Capital. Investment Choice Using Fuzzy Arithmetic, J. of Chin. Inst. of Industrial Engineers 22(2), 106–118 (2005)

Vranes, S., Stanojevic, M., Stevanovic, V.: Financial decision aid based on heuristics and fuzzy logic. In: Damiani, E., et al. (eds.) Knowledge based Intelligent Information Engineering systems and allied technologies, Proceedings of KES 2003 conference, Crema, Itally, September 16-18, pp. 381–387 (2002)

Wong, J.K.W., Li, H., Wang, S.W.: Intelligent building research: a review. Automation in Construction 14, 143–159 (2005)

Ward, T.L.: Discounted fuzzy Cash flow Analysis. In: 1985 Fall Industrial Engineering Conference Proceedings, pp. 476–481 (1985)

Yager, R.R.: Level sets for membership evaluations of fuzzy subsets. In: Yager, R.R. (ed.) Fuzzy sets and possibility theory: recent developments, pp. 90–97. Pergamon Press, Oxford (1982)

Yao, J., Wu, K.: Ranking fuzzy numbers based on decomposition principle and signed distance. Fuzzy Sets and Systems 116, 275–288 (2000)

Zadeh, L.A.: Fuzzy Sets. Inf. and Control 8, 338–353 (1965)

Zhou, W., Gong, P.: Economic effects of environmental concerns in forest management: an analysis of the cost of achieving environmental goals. Journal of Forest Economics 10, 97–113 (2004)

# Fuzzy Equivalent Annual-Worth Analysis and Applications

Cengiz Kahraman and İhsan Kaya

Istanbul Technical University, Department of Industrial Engineering,
34367, Macka, İstanbul, Turkey

**Abstract.** Equivalent annual worth analysis is one of the most used analysis techniques for the evaluation of investments. The main process is to convert any cash flow to an equivalent uniform cash flow. Later, the decision is given taking the economic category of the problem into account. If it is a fixed input problem, you should maximize equivalent uniform annual benefit. If it is a fixed output problem, minimize equivalent uniform annual cost, and if it is neither output nor input fixed problem, maximize equivalent uniform annual profit. In this chapter, it is shown that how you can apply this technique when the parameters are all fuzzy. Numerical examples are given for different analysis periods.

## 1 Introduction

Decision makers who are facing different investment situations when selecting alternatives use several well developed engineering economy techniques to analyze those alternatives. The common economic analysis techniques that take the time value of money into account and are used in selecting the alternatives are the present worth method the future worth method and the annual worth method. When the benefits are not considered, decision makers often think in terms of annual expenses, annual receipts and annual budgets. They are not calculated as present worth. The solution is to state equivalence in yearly terms that is equivalent annual worth (EAW) or equivalent annual cost. EAW is the value per period. More rigorously: EAW represents N identical cash flow equivalents occurring at the ends of the periods 1 through n. a desirable project or alternative has an EAW>0, just as a desirable project or alternative has a PW>0. In fact EAW equals $PW(A/P,i,n)$. (Eschenbach 2002). The "equivalent annual" approach is the most convenient method for comparing project evaluations with annual budget because saying that this project will cost or save \$500 per year make intuitive sense even to people who have never heard of engineering economy.

The formula for EAW is (Young 1993);

$$A = P \frac{i}{1 - \beta^n} \qquad for \qquad i \neq 0 \qquad (1)$$

where $\beta$ is equal to $(1+i)^{-1}$.

For a zero interest rate, $A = \dfrac{P}{n}$.

The factor after $P$ in Eq.1 is called *"the equal payment series capital recovery factor"*:

$$(A/P,i,n) = \frac{i}{1-\beta^n} = \frac{1}{(P/A,i,n)} \qquad (2)$$

As $n \to \infty$, the infinite equal-payment series $\{A_t = A : t = 1,2,...\}$ approaches an amount $A = iP$.

The equivalent uniform annual worth method has the advantage of comparing alternatives with different lifetimes, compared to the equivalent present and final worth methods (Blank and Tarquin 1998). The conventional economic analysis methods such as EAW analysis based on the exact numbers (Chan et al. 2000). In reality, however, it is not practical to expect that certain data, such as the future cash flows and discount rates, are known exactly in advance. Usually, based on past experience or educated guesses, decision makers modify vague data to fit certain conventional decision-making models. The modified data consist of vagueness such as approximately between 3% to 5 % or around $15,000—in linguistic terms. Fuzzy set theory, first introduced by Zadeh (1965) can be used in the uncertain economic decision environment to deal with the vagueness. Fuzzy numbers can be equal for these inputs, such as cash amounts and interest rates in the future, for conventional decision-making models (Chan et al. 2000). Fuzzy set theory was applied extensively to solve the problem of alternative selection when it was introduced in engineering economy. Ward (1985) develops fuzzy present worth analysis by introducing trapezoidal cash flow amounts. Buckley (1987) proposes a fuzzy capital budgeting theory in the mathematics of finance. Kaufmann and Gupta (1988) are using the fuzzy number to the discount rate. They derive a fuzzy present worth method for investment alternative selection. Karsak and Tolga (2001) propose a fuzzy present worth model for financial evaluation of advanced manufacturing system (AMS) investments. Kuchta (2001) considers net present value as a quadratic 0-1 programming problem in a fuzzy form. Kahraman et al. (2002) derivates the formulas for the analyses of fuzzy present value, fuzzy equivalent uniform annual value, fuzzy final value, fuzzy benefit-cost ratio and fuzzy payback period in capital budgeting. Liou and Chen (2007) propose a fuzzy equivalent uniform annual worth (FEUAW) method to assist practitioners in evaluating investment alternatives utilizing the fuzzy set theory.

The rest of this chapter is organized as follows: In Section 2, the fuzzy equivalent uniform annual worth method is explained and two approaches to analyze FEUAWM are presented. Two different outranking methods are used to determine the best alternative via FEUAWM and they are summarized in Section 3. In the next section, an application on selection among CNC machines is illustrated. In section 5, the situation where the analysis period is infinite is analyzed. Section 6 presents conclusions.

## 2 Fuzzy Equivalent Uniform Annual Value Method

The equivalent uniform annual value (EUAV) means that all incomes and disbursements (irregular and uniform) must be converted into an equivalent uniform annual amount, which is the same each period. The major advantage of this method over all the other methods is that it does not require making the comparison over the least common multiple of years when the alternatives have different lives (Chiu and Park, 1994). The general equation for this method is

$$EUAV = A = NPV\gamma^{-1}(n,r) = NPV \frac{(1+r)^n r}{(1+r)^n - 1} \qquad (3)$$

where NPV is the net present value. In the case of fuzziness, $N\tilde{P}V$ will be calculated and then the fuzzy $EU\tilde{A}V\left(\tilde{A}_n\right)$ will be found. The membership function $\mu\left(x|\tilde{A}_n\right)$ for $\left(\tilde{A}_n\right)$ is determined by

$$f_{ni}\left(y|\tilde{A}_n\right) = f_i\left(y|N\tilde{P}V\right)\gamma^{-1}\left(n, f_i\left(y|\tilde{r}\right)\right) \qquad (4)$$

and $TFN(y)$ for fuzzy EUAV is

$$\tilde{A}_n(y) = \left( \frac{NPV^{l(y)}}{\gamma\left(n, r^{l(y)}\right)}, \frac{NPV^{r(y)}}{\gamma\left(n, r^{r(y)}\right)} \right) \qquad (5)$$

**Example**

Assume that, $N\tilde{P}V = (-\$3525.57, \$24.47, +3786.34)$ and $\tilde{r} = (3\%, 5\%, 7\%)$. Calculate the fuzzy EUAV:

$$f_{6,1}\left(y|\tilde{A}_6\right) = (3501.1y - 3525.57)\frac{(1.03 + 0.02y)^6 (0.02y + 0.03)}{(1.03 + 0.02y)^6 - 1}$$

$$f_{6,2}\left(y|\tilde{A}_6\right) = (-3810.81y + 3786.34)\frac{(1.07 - 0.02y)^6 (0.07 - 0.02y)}{(1.07 - 0.02y)^6 - 1}$$

For $\quad y = 0, \quad f_{6,1}\left(y|\tilde{A}_6\right) = -\$650.96$

For $\quad y = 1, \quad f_{6,1}\left(y|\tilde{A}_6\right) = f_{6,2}\left(y|\tilde{A}_6\right) = -\$4.82$

For $\quad y = 0, \quad f_{6,2}\left(y|\tilde{A}_6\right) = +\$795.13$

The literature has the different methods to calculate the fuzzy annual worth of alternatives. Liou and Chen (2007) proposed an approach to this aim. The detail of their approach is follow:

They used triangular fuzzy number to determine the best alternative. The experts decide the lower bound ($l$), upper bound ($u$) and the most likely number ($m$) of the relevant amounts concerning the cash flows. They used them to define the following fuzzy triangular numbers:

$\tilde{P} = (P_l, P_m, P_u)$: The fuzzy present worth,

$\tilde{F} = (F_l, F_m, F_u)$: The fuzzy final worth,

$\tilde{A} = (A_l, A_m, A_u)$: The fuzzy equivalent uniform annual worth,

$\tilde{i} = (i_l, i_m, i_u)$: The fuzzy discount worth.

According to the arithmetic operations on fuzzy numbers;

$$1 \oplus \tilde{i} = (1,1,1) \oplus (i_l, i_m, i_u) = (1 + i_l, 1 + i_m, 1 + i_u),$$

$$\left(1 \oplus \tilde{i}\right)^n = \left((1 + i_l)^n, (1 + i_m)^n, (1 + i_u)^n\right),$$

$$\tilde{F} / \left(1 \oplus \tilde{i}\right)^n = \left(\frac{F_l}{(1 \oplus i_u)^n}, \frac{F_m}{(1 \oplus i_m)^n}, \frac{F_u}{(1 \oplus i_l)^n}\right),$$

$$\tilde{P} \otimes \left(1 \oplus \tilde{i}\right)^n = \left(P_l(1 \oplus i_l)^n, P_m(1 \oplus i_m)^n, P_u(1 \oplus i_u)^n\right).$$

$$(6)$$

The relation between $\tilde{A}$ and $\tilde{P}$ expresses in Eq. (7):

$$\tilde{A} = \tilde{P} \otimes \left(\tilde{A} / \tilde{P}, \tilde{i}, N\right) \tag{7}$$

where

$$\left(\tilde{P} / \tilde{A}, \tilde{i}, N\right) = \left(\frac{i_l(1 + i_l)^N}{(1 + i_l)^N - 1}, \frac{i_m(1 + i_m)^N}{(1 + i_m)^N - 1}, \frac{i_u(1 + i_u)^N}{(1 + i_u)^N - 1}\right).$$

The relation between $\tilde{A}$ and $\tilde{F}$ expresses in Eq. (6):

$$\tilde{A} = \tilde{F} \otimes \left(\tilde{A} / \tilde{F}, \tilde{i}, N\right) \tag{8}$$

where

$$\left(\tilde{F} / \tilde{A}, \tilde{i}, N\right) = \left(\frac{i_u}{(1 + i_u)^N - 1}, \frac{i_m}{(1 + i_m)^N - 1}, \frac{i_l}{(1 + i_l)^N - 1}\right).$$

If the decision makers have to determine the best alternatives between "$m$" alternatives, they can use this approach. One of the "$m$" alternatives have different economic situation, fuzzy equivalent uniform annual worth of the $k$th alternative $\left(A\tilde{W}_k\right)$ is shown in Eq. (7).

$$A\tilde{W}_k = \left(\sum_{j=1}^{N_k} \frac{\tilde{A}_{kj}}{\left(1 \oplus \tilde{i}\right)^j} - \tilde{C}_k\right) \otimes \left(\tilde{A} / \tilde{P}, \tilde{i}, N\right) \oplus S_k\left(\tilde{A} / \tilde{F}, \tilde{i}, N\right) \tag{9}$$

where

$\tilde{i}$ : The discounted rate,

$\tilde{C}_k = (C_{kl}, C_{km}, C_{ku})$: The initial investment of the $k$th alternative,

$N_k$ : The duration of the $k$th alternative,

$\tilde{A}_{kj} = (C_{kl}, C_{km}, C_{ku})$: The cash flow of the $j$th period for the $k$th alternative,

$\tilde{S}_k = (S_{kl}, S_{km}, S_{ku})$: The salvage value of the $k$th alternative

After all the $A\tilde{W}_k$ ($k=1, 2,..., m$) of the alternatives are calculated, we need to ranking methods to compare the alternatives. The outranking method which is used in this chapter is explained in Section 3.

## 3 Ranking Methods

Due to the fact that fuzzy number is frequently not linearly ordered, some fuzzy numbers are not directly comparable, and there is need in converting a fuzzy number to a crisp number for ranking. There are a number of methods that are devised to rank mutually exclusive projects such as Chang's method (1981), Jain's method (1976), Dubois and Prade's method (1983), Yager's method (1980), Baas and Kwakernaak's method (1977). In this chapter, the method of ranking fuzzy numbers with integral value by Liou and Wang (1992) and Yuan's (1991) approaches are used.

### 3.1 The First Ranking Method

In this paper, one of the two ranking methods is proposed by Liou and Wang (1992). Let,

$$u_{\tilde{A}}^L(x) = \begin{cases} \dfrac{(x-a)}{(b-a)}, & a \le x \le b, \quad a \ne b \\ 1, & a = b \end{cases} \tag{10}$$

$$u_{\tilde{A}}^R(x) = \begin{cases} \dfrac{(x-c)}{(b-c)}, & b \le x \le c, \quad b \ne c \\ 1, & b = c \end{cases} \tag{11}$$

In this equation, $u_{\tilde{A}}^L$ and $u_{\tilde{A}}^R$ are continuous and increasing functions. Since, they have inverse function, denoted as $g_{\tilde{A}}^L$ and $g_{\tilde{A}}^R$ as follows:

$$g_{\tilde{A}}^L(x) = \begin{cases} a + (b-a)u, & a \ne b, \quad u \in [0,1] \\ a, & a = b \end{cases} \tag{12}$$

$$g_{\tilde{A}}^R(x) = \begin{cases} c + (b-c)u, & b \ne c, \quad u \in [0,1] \\ c, & b = c \end{cases} \tag{13}$$

The definition of integral values for triangular fuzzy number $\tilde{A}$ is proposed in Eq. (14):

$$I(\tilde{A}) = (1-\alpha)\int_0^1 g_{\tilde{A}}^L(u)du + \int_0^1 g_{\tilde{A}}^R(u)du, \qquad where \quad 0 \le \alpha \le 1$$

$$I(\tilde{A}) = \frac{1-\alpha}{2}a + \frac{1}{2}b + \frac{\alpha}{2}c \qquad\qquad (14)$$

In this formulation, $\alpha$ is called *"the index of optimism"* (Liou and Chen, 2007). It is representing the degree of optimism for an expert.

### 3.2 The Second Ranking Method

The second ranking method which is used in this chapter is proposed by Yuan (1991). In this approach, a binary relation on F(R) is a fuzzy set such as $\mu : F(R) \times F(R) \to [0,1]$, where $\mu(A,B)$ represents the truth level or the strength of the relation between A and B, where $\mu(A,B) = 1$ means that the relation between A and B is true or the strongest, and $\mu(A,B) = 0$ means the relation is false or the weakest. Let $C_i, C_j \in F(R)$ are normal and convex. A fuzzy relation which compares the right spread of $C_i$ with the left spread of $C_j$, defined in Eq. (15).

$$\Delta_{ij} = \int_{c_{i\alpha}^+ > c_{j\alpha}^-}(c_{i\alpha}^+ - c_{j\alpha}^-)d\alpha + \int_{c_{i\alpha}^- > c_{j\alpha}^+}(c_{i\alpha}^- - c_{j\alpha}^+)d\alpha,$$

$$c_{i\alpha}^+ = \sup\{x : x \in C_{i\alpha}\}, \qquad c_{i\alpha}^- = \inf\{x : x \in C_{i\alpha}\} \qquad\qquad 15)$$

Then $\mu(C_i, C_j)$ is point that the degree of bigness of $C_i$ relative to $C_j$ and it can be calculated from Eq. (16).

$$\mu(C_i, C_j) = \frac{\Delta_{ij}}{\Delta_{ij} + \Delta_{ji}} \qquad\qquad (16)$$

Based on this formula, we have two situations:

- $C_i$ is bigger than $C_j$ if and only if $\mu(C_i, C_j) > 0.5$,
- $C_i$ and $C_j$ are equal if and only if $\mu(C_i, C_j) = 0.5$.

## 4 Application

An automotive company wants to buy a CNC machines. They have four alternatives which have different economic criterion. We can assume that fuzzy discount rate is $\tilde{i} = (11, 13, 15\%)$ for each alternative. The main economic characterization of the CNC machine is summarizing in Table 1.

According to the information which is presented in Table 1, fuzzy annual worth for each alternative is calculated and they are presented in Table 2.

## Fuzzy Equivalent Annual-Worth Analysis and Applications

**Table 1.** Information for Different CNC Alternatives

| Alternative (k) | I | II | III | IV |
|---|---|---|---|---|
| Duration($N_k$) | 15 | 15 | 14 | 13 |
| Initial Cost $\left(\widetilde{C}_k\right)$ | (10,000; 12,000; 14,000) | (9,000; 10,000; 11,000) | (10,000; 11,000; 11,500) | (7,800; 8,500; 9,000) |
| Salvage Value $\left(\widetilde{S}_k\right)$ | (2,000; 2,200; 2,400) | (1,000; 1,400; 1,600) | (1,980; 2,000; 2,450) | (850; 900; 990) |
| Cash Flow $\left(\widetilde{A}_{kj}\right)$ | The First Five Years: (2,000; 2,400; 2,500) The Next Two Years: (2,100; 2,400; 2,600) The Last Eight Years: (2,100; 2,200; 2,300) | The First Five Years: (1,500; 1,700; 2,000) The Next Seven Years: (1100; 1400; 1600) The Last Three Year: (2,100; 2,200; 2,300) | The First Five Years: (1,300; 1,400; 1,500) The Next Five Years: (2,100; 2,700; 3,000) The Last Four Years: (2,000; 2,400; 2,800) | The First Five Years: (1,000; 1,200; 1,700) The Next Four Year: (1,100; 1,400; 1,600) The Last Four Years: (1,100; 1,400; 1,800) |

The fuzzy annual worth of alternative-IV is determined as follow:

$$A\widetilde{W}_{IV} = \left(\sum_{j=1}^{N_k} \frac{\widetilde{A}_{kj}}{\left(1 \oplus \widetilde{i}\right)^j} - \widetilde{C}_k\right) \otimes \left(\widetilde{A}/\widetilde{P}, \widetilde{i}, 13\right) \oplus S_k \left(\widetilde{A}/\widetilde{F}, \widetilde{i}, 13\right)$$

$$= \left[ \begin{array}{l} \left( \dfrac{(1000,1200,1700)}{(1.11,1.13,1.14)} \oplus \dfrac{(1000,1200,1700)}{(1.11,1.13,1.14)^2} \oplus \ldots \oplus \right. \\[2mm] \dfrac{(1000,1200,1700)}{(1.11,1.13,1.14)^5} \oplus \dfrac{(1100,1400,1600)}{(1.11,1.13,1.14)^6} \oplus \ldots \oplus \\[2mm] \dfrac{(1100,1400,1600)}{(1.11,1.13,1.14)^9} \oplus \dfrac{(1100,1400,1800)}{(1.11,1.13,1.14)^{10}} \oplus \ldots \oplus \\[2mm] \left. \dfrac{(1100,1400,1800)}{(1.11,1.13,1.14)^{13}} \right) \end{array} \right] - (7800, 8500, 9000)$$

$$\otimes \left( \frac{0.11 \times 1.11^{13}}{1.11^{13} - 1}, \frac{0.13 \times 1.13^{13}}{1.13^{13} - 1}, \frac{0.14 \times 1.14^{13}}{1.14^{13} - 1} \right) \oplus (850,900,990)$$

$$\otimes \left( \frac{0.14}{1.14^{13} - 1}, \frac{0.13}{1.13^{13} - 1}, \frac{0.11}{1.11^{13} - 1} \right)$$

$$= (5806.25, 7867.09, 11411.95) - (7800, 8500, 9000) \otimes (0.15, 0.16, 0.18)$$

$$\oplus (850,900,990) \otimes (0.003, 0.03, 0.04)$$

$$= (-473.16, -103.39, 646.94) \oplus (24.74, 30.02, 37.77)$$
$$= (-448.71, -73.37, 684.71)$$

The other results are summarized in Table 2.

**Table 2.** FAW for Each Alternative

| Alternative (k) | I | II | III | IV |
|---|---|---|---|---|
| Fuzzy AW | (-243.85, 534.40, 1,366.59) | (-368.40, 117.93, 811.41) | (-270.64, 244.79, 986.86) | (-448.71, -73.37, 684.71) |

Based on the Table 2, each alternative is ordered by Liou and Wang's method. The results of this outranking are summarized in Table 3.

**Table 3.** Ranking Order according to Liou and Wang's Method

| Alternative (k) | $\alpha$ | I | II | III | IV | Ranking Order |
|---|---|---|---|---|---|---|
| | 0.5 (Objective) | 547.89 | 169.72 | 301.45 | 95.69 | I>III>II>IV |
| Ranking Results | 0.1 (Pessimistic) | 225.80 | -66.24 | 49.95 | -131.00 | I>III>II>IV |
| | 0.9 (Optimistic) | 869.97 | 405.68 | 552.95 | 322.37 | I>III>II>IV |

According to Table 3, we can say that Alternative-I is the best also Alternative-IV is the worst for all situation (objective, pessimistic and optimistic).

In this section, the four alternatives whose fuzzy values are proposed in Table 2 are outranking by Yuan's (1991) approach. The first stage, the graph of fuzzy number must be created. The graph of the fuzzy AW for each alternative is created by Ms Excel and it is proposed in Figure 1.

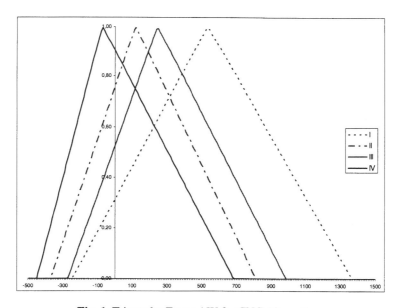

**Fig. 1.** Triangular Fuzzy AW for CNC Alternatives

The degree of bigness for each alternative is calculated and proposed in Table 4.

**Table 4.** Ranking Order According to Yuan's Method

| Alternatives | $\Delta_{ij}$ | $\Delta_{ji}$ | Decision | Degree of Bigness |
|---|---|---|---|---|
| I-II | 1134.040 | 379.890 | I is better than II | 0.749 |
| I-III | 990.930 | 498.440 | I is better than III | 0.665 |
| I-IV | 1333.090 | 278.570 | I is better than IV | 0.827 |
| II-III | 481.512 | 753.110 | III is better than II | 0.390 |
| III-IV | 916.640 | 358.260 | III is better than IV | 0.719 |
| II-IV | 740,060 | 447,570 | II is better than IV | 0,623 |

According to the Yuan's ranking approach, we have the following order of the alternatives: $\{I - III - II - IV\}$ and this result is same as results of Liou and Wang's Method.

## 5 Analysis Period is Infinite

In the crisp case, equivalent annual worth of a present value is calculated by the following equation:

$$A = P \times i \tag{17}$$

In the fuzzy case, Eq. (17) will change to be Eq. (18):

$$\tilde{A} = \tilde{P} \otimes \tilde{i} = \left(P_1 \times i_1, P_2 \times i_2, P_3 \times i_3\right) \tag{18}$$

Lets compare the following two mutually exclusive investment alternatives using an annual interest rate of (10%, 12%, 14%).

**Table 5.** Investment Alternatives

| Alternatives | I | II |
|---|---|---|
| First costs, US$ | (120,000; 125,000; 130,000) | (180,000; 185,000; 190,000) |
| Uniform Annual Operating Cost, US$ | (13,000; 15,000; 17,000) | (22,000; 24,000; 26,000) |
| Uniform Annual Benefit, US$ | (32,000; 34,000; 36,000) | (43,000; 45,000;47,000) |
| Salvage Value, US$ | (85,000; 90,000; 95,000) | (90,000; 95,000; 95,000) |
| Useful life, years | (40, 42, 44) | $\infty$ |

## For Alternative I

$$\tilde{A} = A\tilde{B}\Theta AO\tilde{C}\Theta\left[F\tilde{C}\otimes\left(A/P,\tilde{i},\tilde{n}\right)\right]\oplus\left[S\tilde{V}\otimes\left(A/F,\tilde{i},\tilde{n}\right)\right] \tag{19}$$

$$f_1\left(y\middle|\tilde{A}\right) = (32,000+2,000\,y)-(17,000-2,000\,y)$$

$$-\left[(130,000-5,000\,y)\times\left(\frac{(1.14-0.02\,y)^{40+2y}(0.14-0.02\,y)}{(1.14-0.02\,y)^{40+2y}-1}\right)\right]$$

$$+\left[(85,000+5,000\,y)\times\frac{(0.14-0.02\,y)}{(1.14-0.02\,y)^{40+2y}-1}\right]$$

$$f_2\left(y\middle|\tilde{A}\right) = (36,000-2,000\,y)-(13,000+2,000\,y)$$

$$-\left[(120,000+5,000\,y)\times\left(\frac{(1.10+0.02\,y)^{44-2y}(0.10+0.02\,y)}{(1.10+0.02\,y)^{44-2y}-1}\right)\right]$$

$$+\left[(95,000-5,000\,y)\times\frac{(0.10+0.02\,y)}{(1.10+0.02\,y)^{44-2y}-1}\right]$$

When y=0, $f_1\left(y\middle|\tilde{A}\right)=$ US$ -3,240 and when y=1, $f_1\left(y\middle|\tilde{A}\right)=f_2\left(y\middle|\tilde{A}\right)=$ US$ 4,963 and finally when y=0, $f_2\left(y\middle|\tilde{A}\right)=$ US$ 10,962.

## For Alternative II

$$f_1\left(y\middle|\tilde{A}\right) = (43,000+2,000\,y)-(26,000-2,000\,y)-(190,000-5,000\,y)(0.14-0.02\,y)$$

$$f_2\left(y\middle|\tilde{A}\right) = (47,000-2,000\,y)-(22,000+2,000\,y)-(180,000+5,000\,y)(0.10+0.02\,y)$$

When y=0, $f_1\left(y\middle|\tilde{A}\right)=$ US$ -9,600 and when y=1, $f_1\left(y\middle|\tilde{A}\right)=f_2\left(y\middle|\tilde{A}\right)=$ US$ -1,200 and finally when y=0, $f_2\left(y\middle|\tilde{A}\right)=$ US$ 7,000.

Without using a ranking method, it is obvious that Alternative I is superior to Alternative II.

## 6 Conclusions

In equivalent uniform cash flow analysis, instead of computing equivalent present sums as it is made in present worth analysis, we could compare alternatives based on their equivalent annual cash flows. Depending on the particular situation we may wish to compute the equivalent uniform annual cost (EUAC), the equivalent uniform annual benefit (EUAB), or their difference (EUAB-EUAC). The equivalent annual cash flow (EACF) is an annuity having the same term, rate of return and net present value as the project that it represents. This analysis is advantageous when the analysis period is different from the alternative lives.

Fuzzy parameters have been used in the classical equivalent uniform annual worth analysis in this chapter. The obtained results have a nonlinear membership function, but for simplicity the least possible, the most possible, and the largest possible values have been calculated. The membership degree of any point between these values can be approximately calculated assuming a linear function.

# References

Baas, S.M., Kwakernaak, H.: Rating and ranking multiple aspect alternatives using fuzzy sets. Automatica 13, 47–58 (1977)

Blank, L.T., Tarquin, A.J.: Engineering economy. McGraw-Hill, New York (1998)

Buckley, J.J.: The fuzzy mathematics of finance. Fuzzy Sets and Systems 21, 257–273 (1987)

Chan, F.T.S., Chan, M.H., Tang, N.K.H.: Evaluation methodologies for technology selection. Journal of Material Technology 107, 330–337 (2000)

Chang, W.: Ranking of fuzzy utilities with triangular membership functions. In: Proceedings of the International Conference of Policy Anal. And Inf. Systems, pp. 263–272 (1981)

Chiu, C.Y., Park, C.S.: Fuzzy cash flow analysis using present worth criterion. The Engineering Economist 39(2), 113–138 (1994)

Dubois, D., Prade, H.: Ranking fuzzy numbers in the setting of possibility theory. Information Sciences 30, 183–224 (1983)

Eschenbach, G.T.: Engineering economy applying theory to practice. Oxford University Press, Oxford (2002)

Jain, R.: Decision-making in the presence of fuzzy variables. IEEE Transactions on Systems, Man, and Cybernetics 6, 693–703 (1976)

Kahraman, C., Ruan, D., Tolga, E.: Capital budgeting techniques using discounted fuzzy versus probability cash flows. Information Sciences 142, 57–76 (2002)

Karsak, E.E., Tolga, E.: Fuzzy multi-criteria decision-making procedure for evaluating advanced manufacturing system investments. International Journal of Production Economics 69, 49–64 (2001)

Kaufmann, A., Gupta, M.M.: Fuzzy mathematical models in engineering and management science, pp. 19–35, 159–165. Elsevier Science Publishers BV, Amsterdam (1988)

Kuchta, D.: A fuzzy model for R&D project selection with benefit, outcome and resource interactions. The Engineering Economist 46(3), 164–180 (2001)

Liou, T.S., Wang, M.J.: Ranking fuzzy numbers with integral value. Fuzzy Sets and Systems 50, 247–255 (1992)

Liou, T.S., Chen, C.W.: Fuzzy decision analysis for alternative selection using a fuzzy annual worth criterion. The Engineering Economist 51, 19–34 (2007)

Ward, T.L.: Discounted fuzzy cash flow analysis. In: Proceedings of 1985 Fall Industrial Engineering Conference Institute of Industrial Engineers, pp. 476–481 (1985)

Yager, R.R.: On choosing between fuzzy subsets. Cybernetics 9, 151–154 (1980)

Yuan, Y.: Criteria for evaluating fuzzy ranking methods. Fuzzy Sets and Systems 43, 139–157 (1991)

Young, D.: Modern Engineering Economy. John Wiley & Sons, Chichester (1993)

Zadeh, L.A.: Fuzzy Sets. Information and Control 8, 338–353 (1965)

# Case Studies Using Fuzzy Equivalent Annual Worth Analysis

Manuel Matos[1] and Aleksandar Dimitrovski[2]

[1] INESC Porto & Universidade do Porto, Faculdade de Engenharia , Rua Dr Roberto Frias 378
[2] Washington State University, School of EE&CS, P.O. Box 642752, Pullman, WA 99164,
USA, 4200-465 Porto, Portugal

**Abstract.** This chapter presents some case studies using fuzzy equivalent uniform annual worth analysis. It includes three case studies and each case is studied for both crisp and fuzzy cases. Trapezoidal fuzzy numbers and correlated, uncorrelated, and partially correlated cash flows are handled in the cases.

## 1 Use of the A/P Factor

One of the ways of obtaining the Annualized Value (Worth), AW, in the crisp case, departs from a previous calculation of the Present Value (Worth), PW, using Equation (4) in Chapter 3.2:

$$AW = (A/P, i, n) \cdot PW = \frac{i(1+i)^n}{(1+i)^n - 1} \cdot PW \tag{1}$$

At first, one could think that extending this procedure to the fuzzy case would consist merely in multiplying (following the rules of fuzzy arithmetic) the fuzzy $A/P$ factor by the fuzzy PW. Using the reasoning described in Chapter 3.2 about the $P/A$ factor, we would get, for each $\alpha$-level of the $A/P$ factor[1]:

$$(A/P, i, n) = \left[ \frac{i_{min} \cdot (1+i_{min})^n}{(1+i_{min})^n - 1} \quad \frac{i_{max} \cdot (1+i_{max})^n}{(1+i_{max})^n - 1} \right] \tag{2}$$

where $i = [i_{min} \ i_{max}]$ is the $\alpha$-level interval of the discount rate.

Now, if the $\alpha$-level interval of the PW is $[PW_{min} \ PW_{max}]$, our simplified approach that assumes independence between $A/P$ and PW, would lead to:

$$AW_{indep} = \left[ PW_{min} \cdot \frac{i_{min} \cdot (1+i_{min})^n}{(1+i_{min})^n - 1} \quad PW_{max} \cdot \frac{i_{max} \cdot (1+i_{max})^n}{(1+i_{max})^n - 1} \right] \tag{3}$$

---

[1] We dropped the index $\alpha$ whenever possible, in order to alleviate the equations.

C. Kahraman (Ed.): Fuzzy Engineering Economics with Appl., STUDFUZZ 233, pp. 83–95, 2008.
springerlink.com © Springer-Verlag Berlin Heidelberg 2008

Unfortunately, the assumption leads to inaccurate results when the cash flow consists of an initial investment followed a series of future revenues (positive), which is by far the most common situation in project assessment. In that case, the discount rate associated to $PW_{min}$ is $i_{max}$ (the same with $PW_{max}$ and $i_{min}$), so the use of the preceding formula would lead to inconsistency, by considering simultaneously the two extreme values of the discount rate. Note, however, that different cash-flow structures, with mixed positive and negative payments during the project horizon, may lead to the reverse situation ($i_{min}$ associated to $PW_{min}$, etc.) or even to cases where the rate leading to $PW_{min}$ is neither the minimum nor the maximum, but rather an intermediate value.

Coming to the most frequent situation, where $PW_{min} = PW(i_{max})$ and $PW_{max} = PW(i_{min})$, we find an additional difficulty, due to the fact that AW results from a multiplication of two quantities (PW and $A/P$), one that increases with the rate $i$ ($A/P$), the other decreasing with the rate $i$ (PW). In this situation, the minimum and maximum values of AW may occur for intermediate values of the rate $i$, so just multiplying the extremes of $A/P$ and PW in reverse order may not be the correct operation.

Therefore, the correct procedure to obtain $AW_{min}$ and $AW_{max}$ relies on finding the discount rate $i$ that minimizes (maximizes) the product of PW and $A/P$:

$$AW_{min} = \min_{i_{min} \le i \le i_{max}} \left\{ PW(i) \cdot \frac{i \cdot (1+i)^n}{(1+i)^n - 1} \right\}$$

$$AW_{max} = \max_{i_{min} \le i \le i_{max}} \left\{ PW(i) \cdot \frac{i \cdot (1+i)^n}{(1+i)^n - 1} \right\}$$

(4)

Operationally, this requires establishing first the expression of $PW(i)$, but a numerical approach is also possible, of course.

## 2 EAW Example 1

Let us consider again Example 1 in Chapter 3.2, but this time we will apply the EAW method to asses economic efficiency and compare alternatives. All the data in both crisp and fuzzy case are assumed to be the same as before. Consequently, the results from the analyses will also be the same, but presented in a different way.

### 2.1 Crisp Case

Table 1 shows equivalent annual worth of cumulative costs for both alternatives as they evolve during their lifetime. This is also shown on Figure 1. The conclusion, of course, is the same as before. The break-even year occurs within the lifetime (year 16) after which alternative A yields smaller final equivalent annuity than alternative B and, thus, is preferable.

**Table 1.** Equivalent annuities ($) – crisp case

| Year | Alternative A | Alternative B |
|---|---|---|
| 1  | 698,000 | 120,000 |
| 2  | 386,462 | 120,000 |
| 3  | 282,820 | 120,000 |
| 4  | 231,152 | 120,000 |
| 5  | 200,274 | 120,000 |
| 6  | 179,789 | 120,000 |
| 7  | 165,243 | 120,000 |
| 8  | 154,409 | 120,000 |
| 9  | 146,048 | 120,000 |
| 10 | 139,418 | 120,000 |
| 11 | 134,046 | 120,000 |
| 12 | 129,617 | 120,000 |
| 13 | 125,913 | 120,000 |
| 14 | 122,778 | 120,000 |
| 15 | 120,098 | 120,000 |
| 16 | 117,786 | 120,000 |
| 17 | 115,778 | 120,000 |
| 18 | 114,021 | 120,000 |
| 19 | 112,477 | 120,000 |
| 20 | 111,111 | 120,000 |

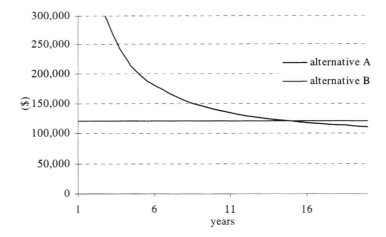

**Fig. 1.** Equivalent annuities of cumulative costs for alternatives A and B – crisp case

## 2.2 Fuzzy Case

Table 2 shows the fuzzy equivalent annuities for alternatives A and B in this case. Again, the table shows the lower and upper bounds of both 0-cuts and 1-cuts, which are the characteristic points of the trapezoidal approximations of the fuzzy numbers.

The data for the final year, from the last row in Table 2, are also shown on Figure 2 as trapezoidal fuzzy numbers.

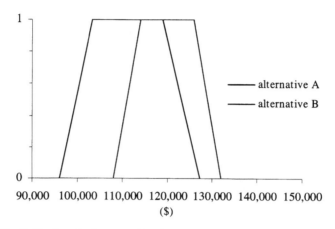

**Fig. 2.** Final equivalent annuities for alternatives A and B – fuzzy case

The analysis of Figure 2 is analogous to that of Figure 10 in Chapter 3.2. At first sight, it may seem that alternative A is always preferred over alternative B. However, this will be true only if the alternatives are totally correlated. To clearly see how they stand with respect to each other in a fuzzy environment it is best to find the difference between the two. This is shown on Figure 3, for three cases of correlated, partially correlated and uncorrelated costs. This figure corresponds to Figure 14 in Chapter 3.2 from the NPW analysis of this problem and the conclusions that follow are the same.

Alternative A is always preferable when the costs are correlated since the fuzzy difference between the alternatives is always a negative number. When the costs are uncorrelated, the decision is not straightforward and additional analysis is required. Partially correlated cases lie between these two marginal cases. The one shown on Figure 3 is the same case shown on Figure 14 in Chapter 3.2 with $k = 0.2$ which is the threshold that still produces a clear outcome.

A careful reader may notice that, although both NPW and EAW methods lead to same outcomes, application of the latter is slightly simpler in this particular example. The annual costs are taken directly and only one equivalence calculation is needed. It turns the initial investment into equivalent annuity by applying Equation 3 in Chapter 3.2. As a result, the calculation of difference between the fuzzy annuities is less prone to errors than the calculation of difference between the fuzzy present values of total costs, as was shown on Figure 11 and Figure 12 in Chapter 3.2. The chances to introduce artificial uncertainty by inaccurate arithmetic are much reduced when only one calculation is involved, in comparison to the two required for the annual costs in the NPW case.

**Table 2.** Equivalent annuities ($) – fuzzy case

| | Alternative A | | | | Alternative B | | | |
|---|---|---|---|---|---|---|---|---|
| Year | $\alpha=0$ | $\alpha=1$ | $\alpha=1$ | $\alpha=0$ | $\alpha=0$ | $\alpha=1$ | $\alpha=1$ | $\alpha=0$ |
| 1  | 622,800 | 660,250 | 736,050 | 774,400 | 108,000 | 114,000 | 126,000 | 132,000 |
| 2  | 343,670 | 364,949 | 408,208 | 430,189 | 108,000 | 114,000 | 126,000 | 132,000 |
| 3  | 250,768 | 266,686 | 299,170 | 315,736 | 108,000 | 114,000 | 126,000 | 132,000 |
| 4  | 204,423 | 217,683 | 244,831 | 258,721 | 108,000 | 114,000 | 126,000 | 132,000 |
| 5  | 176,701 | 188,384 | 212,372 | 224,681 | 108,000 | 114,000 | 126,000 | 132,000 |
| 6  | 158,290 | 168,936 | 190,852 | 202,127 | 108,000 | 114,000 | 126,000 | 132,000 |
| 7  | 145,199 | 155,116 | 175,583 | 186,136 | 108,000 | 114,000 | 126,000 | 132,000 |
| 8  | 135,433 | 144,814 | 164,218 | 174,245 | 108,000 | 114,000 | 126,000 | 132,000 |
| 9  | 127,883 | 136,857 | 155,457 | 165,087 | 108,000 | 114,000 | 126,000 | 132,000 |
| 10 | 121,884 | 130,541 | 148,517 | 157,841 | 108,000 | 114,000 | 126,000 | 132,000 |
| 11 | 117,013 | 125,418 | 142,901 | 151,985 | 108,000 | 114,000 | 126,000 | 132,000 |
| 12 | 112,987 | 121,188 | 138,276 | 147,169 | 108,000 | 114,000 | 126,000 | 132,000 |
| 13 | 109,611 | 117,647 | 134,414 | 143,154 | 108,000 | 114,000 | 126,000 | 132,000 |
| 14 | 106,746 | 114,645 | 131,151 | 139,766 | 108,000 | 114,000 | 126,000 | 132,000 |
| 15 | 104,289 | 112,074 | 128,365 | 136,879 | 108,000 | 114,000 | 126,000 | 132,000 |
| 16 | 102,163 | 109,853 | 125,967 | 134,398 | 108,000 | 114,000 | 126,000 | 132,000 |
| 17 | 100,310 | 107,920 | 123,887 | 132,251 | 108,000 | 114,000 | 126,000 | 132,000 |
| 18 | 98,683  | 106,227 | 122,071 | 130,380 | 108,000 | 114,000 | 126,000 | 132,000 |
| 19 | 97,247  | 104,734 | 120,478 | 128,742 | 108,000 | 114,000 | 126,000 | 132,000 |
| 20 | 95,972  | 103,413 | 119,073 | 127,301 | 108,000 | 114,000 | 126,000 | 132,000 |

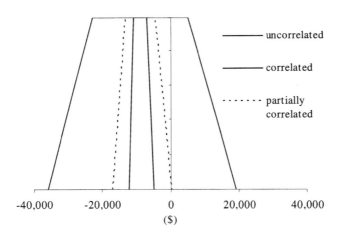

**Fig. 3.** Final difference between annuities for alternatives A and B

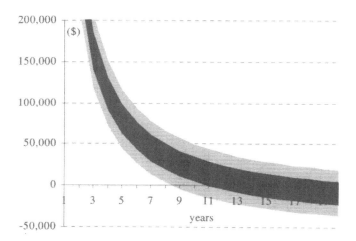

**Fig. 4.** Difference between annuities for alternatives A and B – uncorrelated case

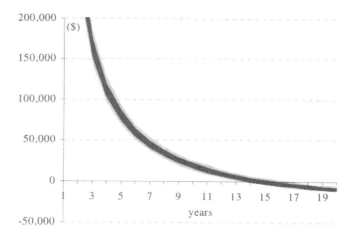

**Fig. 5.** Difference between annuities for alternatives A and B – correlated case

Figure 4 and Figure 5 show how the cost differences between annuities develop over time for the correlated and uncorrelated case, respectively. These figures correspond to Figure 15 and Figure 16 in Chapter 3.2 from the NPW analysis of the same problem. Again, the same conclusions regarding the growth of uncertainty apply, but the differences may not appear so dramatic this time because the values involved in the EAW method are smaller.

## 3 EAW Example 2

Example 2 in Chapter 3.2 is revisited using the EAW method. As we already know, the results will be consistent with those obtained with the NPW method, just presented in a different way.

## 3.1 Crisp Case

The equivalent annuities for the three alternatives are shown on Table 3 and Figure 6.

**Table 3.** Equivalent annuities ($) – crisp case

| Year | Alternative A | Alternative B | Alternative C |
|---|---|---|---|
| 1 | -48500 | -98000 | -195000 |
| 2 | -22423 | -45847 | -90694 |
| 3 | -13753 | -28505 | -56011 |
| 4 | -9433 | -19867 | -38734 |
| 5 | -6855 | -14709 | -28418 |
| 6 | -5146 | -11292 | -21584 |
| 7 | -3935 | -8869 | -16738 |
| 8 | -3034 | -7067 | -13135 |
| 9 | -2340 | -5680 | -10360 |
| 10 | -1791 | -4582 | -8164 |
| 11 | -1347 | -3695 | -6389 |
| 12 | -983 | -2965 | -4930 |
| 13 | -678 | -2357 | -3713 |
| 14 | -422 | -1843 | -2687 |
| 15 | -203 | -1406 | -1812 |
| 16 | -15 | -1030 | -1060 |
| 17 | 148 | -705 | -409 |
| 18 | 289 | -421 | 158 |
| 19 | 413 | -173 | 654 |
| 20 | 523 | 45 | 1091 |

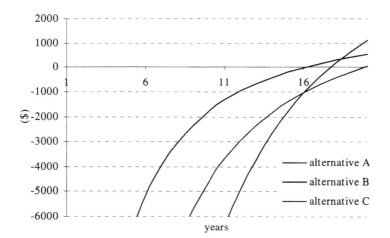

**Fig. 6.** Equivalent annuities – crisp case

In summary, all alternatives have positive final equivalent annuity and, thus, generate a better return rate than the assumed discount rate of 9%/year. Among them, alternative C has the largest equivalent annuity and is the preferable alternative.

### 3.2 Fuzzy Case

Table 4 shows the equivalent annuities in this case and Figure 7 summarizes the final results in the form of approximate trapezoidal FNs.

The results shown on Figure 7, Figure 8, and Figure 9 correspond to those on Figures 21, 22 and 23 in Chapter 3.2, respectively, only the metrics is different. They show again that there is no single best solution in every possible future. The choice will depend on the acceptable level of risk to the decision maker.

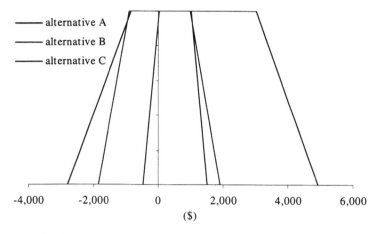

**Fig. 7.** Final equivalent annuities of the alternatives – fuzzy case

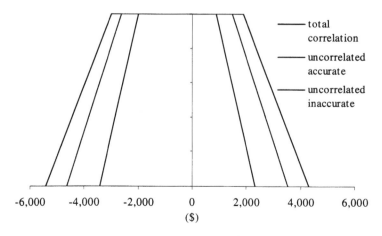

**Fig. 8.** Final difference between annuities for alternatives A and C – correlated and uncorrelated cases

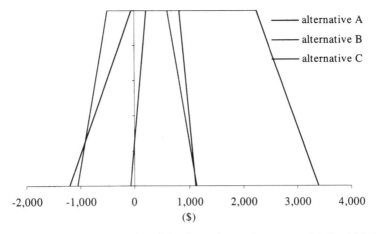

**Fig. 9.** Final equivalent annuities of the alternatives – fuzzy case with fixed MARR

**Table 4.** Equivalent annuities ($) – fuzzy case

|      | Alternative A |   |   |   | Alternative B |   |   |   | Alternative C |   |   |   |
|---|---|---|---|---|---|---|---|---|---|---|---|---|
| Year | $\alpha=0$ | $\alpha=1$ | $\alpha=1$ | $\alpha=0$ | $\alpha=0$ | $\alpha=1$ | $\alpha=1$ | $\alpha=0$ | $\alpha=0$ | $\alpha=1$ | $\alpha=1$ | $\alpha=0$ |
| 1  | -49,600 | -49,050 | -47,950 | -47,400 | -100,100 | -99,050 | -96,950 | -95,900 | -199,300 | -197,150 | -192,850 | -190,700 |
| 2  | -23,410 | -22,916 | -21,931 | -21,438 | -47,719 | -46,783 | -44,912 | -43,977 | -94,538 | -92,615 | -88,773 | -86,854 |
| 3  | -14,706 | -14,229 | -13,277 | -12,802 | -30,311 | -29,408 | -27,604 | -26,703 | -59,723 | -57,866 | -54,158 | -52,307 |
| 4  | -10,374 | -9,903 | -8,964 | -8,496 | -21,647 | -20,756 | -18,979 | -18,092 | -42,394 | -40,563 | -36,908 | -35,084 |
| 5  | -7,790 | -7,322 | -6,388 | -5,923 | -16,480 | -15,594 | -13,827 | -12,946 | -32,059 | -30,237 | -26,603 | -24,791 |
| 6  | -6,080 | -5,613 | -4,680 | -4,216 | -13,061 | -12,175 | -10,411 | -9,532 | -25,221 | -23,401 | -19,771 | -17,963 |
| 7  | -4,870 | -4,402 | -3,468 | -3,004 | -10,641 | -9,754 | -7,987 | -7,107 | -20,381 | -18,557 | -14,924 | -13,114 |
| 8  | -3,972 | -3,502 | -2,567 | -2,101 | -8,844 | -7,955 | -6,183 | -5,301 | -16,789 | -14,959 | -11,316 | -9,503 |
| 9  | -3,282 | -2,810 | -1,871 | -1,404 | -7,464 | -6,570 | -4,792 | -3,908 | -14,028 | -12,191 | -8,535 | -6,716 |
| 10 | -2,737 | -2,263 | -1,320 | -851 | -6,375 | -5,477 | -3,691 | -2,803 | -11,849 | -10,003 | -6,332 | -4,506 |
| 11 | -2,298 | -1,822 | -875 | -404 | -5,496 | -4,594 | -2,799 | -1,908 | -10,093 | -8,237 | -4,549 | -2,715 |
| 12 | -1,938 | -1,459 | -508 | -35 | -4,776 | -3,869 | -2,065 | -1,170 | -8,653 | -6,788 | -3,081 | -1,239 |
| 13 | -1,639 | -1,158 | -201 | 274 | -4,178 | -3,265 | -1,452 | -552 | -7,456 | -5,580 | -1,855 | -4 |
| 14 | -1,387 | -903 | 58 | 535 | -3,675 | -2,757 | -934 | -30 | -6,449 | -4,564 | -818 | 1,041 |
| 15 | -1,174 | -687 | 279 | 759 | -3,247 | -2,324 | -492 | 417 | -5,595 | -3,699 | 66 | 1,934 |
| 16 | -991 | -502 | 469 | 951 | -2,882 | -1,953 | -111 | 802 | -4,863 | -2,957 | 827 | 2,705 |
| 17 | -833 | -342 | 634 | 1,119 | -2,566 | -1,633 | 219 | 1,137 | -4,233 | -2,316 | 1,488 | 3,374 |
| 18 | -697 | -202 | 778 | 1,265 | -2,293 | -1,355 | 507 | 1,430 | -3,686 | -1,759 | 2,064 | 3,960 |
| 19 | -577 | -81 | 905 | 1,394 | -2,055 | -1,111 | 760 | 1,687 | -3,209 | -1,273 | 2,570 | 4,474 |
| 20 | -473 | 26 | 1,016 | 1,507 | -1,846 | -898 | 983 | 1,915 | -2,792 | -845 | 3,016 | 4,930 |

## 4 EAW Example 3

In all the previous examples the alternatives considered had equal book lives of 20 years. This is also known as *cotermination*. In reality, different alternatives usually have different lifetimes. In such case, the alternatives have to be compared on equal time basis. Otherwise, the comparison is invalid and it would be like comparing "potatoes and tomatoes".

There are several approaches to deal with this problem. A commonly used one defines the study period as the least common multiple of the given lifetimes and assumes continual substitution of the alternatives. Other approaches define different study periods and then deal with the residual effects in a different way. Some estimate the salvage value of the assets at the end of the period, others distribute the "unused" costs/benefits back over the study period, and yet others, especially in the fast changing industries, completely neglect the residual effects.

If a common multiple of lifetimes is used, the NPW technique can be cumbersome. It requires finding the present value equivalence of the complete cash flow over the extended period. Thus, the computation effort is multiplied by the number of lifetimes for each alternative in the extended period. On the other hand, the EAW technique does not require any additional effort. Because each alternative is substituted by exactly the same one, the equivalent annuity stays constant over time. Hence, for each alternative, only a single lifetime analysis is sufficient.

To illustrate the concept, let's assume that the alternatives from Example 2 have different lifetimes. Alternative A and C last 15 years, while alternative B remains at 20 years.

In order to use the NPW technique with the least common multiple of lifetimes, we need to use an extended period of 60 years. Thus, alternatives A and C are repeated four times during this period and alternative B three times. The computational effort will increase proportionally. However, the EAW analysis stays the same.

In the crisp case, it is easy to see from rows 15 and 20 in Table 3 that only alternative B has a positive EAW and is the preferred one. In the fuzzy case, all the alternatives have fuzzy EAWs around zero. They are shown on Figure 10.

Lifetimes are variables that were assumed certain so far, but in reality that is not the case by any means. Therefore, they can be modeled also as FNs. In general, these FNs can take any real number value, but it is a firmly established practice in engineering economics to work only with integer values. Thus, integer FNs are used that have discrete possibility membership functions.

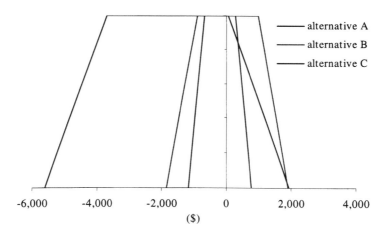

**Fig. 10.** Final equivalent annuities of the alternatives – fuzzy case with different lifetimes

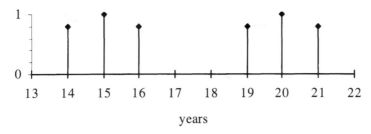

**Fig. 11.** Fuzzy lifetimes of the alternatives

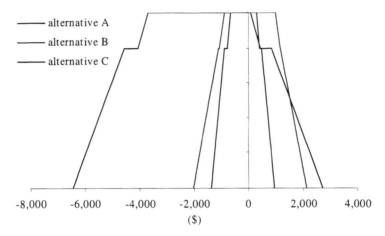

**Fig. 12.** Final equivalent annuities of the alternatives – fuzzy case with different fuzzy lifetimes

Let us assume that the lifetimes of the alternatives from Example 2 are modeled with the following integer FNs: [0.8/14 + 1/15 + 0.8/16] for alternatives A and C, and [0.8/19 + 1/20 + 0.8/21] for alternative B, as shown on Figure 11. Then, the final equivalent annuities in this case will be as shown on Figure 12. Note that the resultant FNs are composed of more piecewise linear parts than the usual three (or two in case of triangular FNs), because of the discontinuities introduced by the discrete membership functions of the assumed fuzzy lifetimes.

Cases like this are especially prone to improper application of the Extension Principle and the resultant FNs must be constructed carefully. Otherwise, besides the more complex shapes, the analysis is the same as in a case with regularly shaped FNs.

# References

Beskese, A., Kahraman, C., Irani, Z.: Quantification of flexibility in advanced manufacturing systems using fuzzy concept. International Journal Of Production Economics 89(1), 45–56 (2004)

Botterud, A., Ilic, M.D., Wangensteen, I.: Optimal investments in power generation under centralized and decentralized decision making. IEEE Transactions on Power Systems 20(1), 254–263 (2005)

Buckley, J.J.: The Fuzzy Mathematics of Finance. Fuzzy Sets and Systems 21, 257–273 (1987)

Carlsson, C., Fuller, R.: A fuzzy approach to real option valuation. Fuzzy Sets And Systems 139(2), 297–312 (2003)

Carlsson, C., Fuller, R., Heikkila, M., et al.: A fuzzy approach to R&D project portfolio selection. International Journal Of Approximate Reasoning 44(2), 93–105 (2007)

Dimitrovski, A.D., Matos, M.A.: Fuzzy Engineering Economic Analysis. IEEE Transactions on Power Systems 15(1), 283–289 (2000)

Dubois, D., Prade, H.: Fuzzy Sets and Systems - Theory and Applications. Academic Press Inc., Orlando (1980)

Huang, X.X.: Chance-constrained programming models for capital budgeting with NPV as fuzzy parameters. Journal Of Computational And Applied Mathematics 198(1), 149–159 (2007)

Huang, X.X.: Optimal project selection with random fuzzy parameters. International Journal of Production Economics 106(2), 513–522 (2007)

Kahraman, C., Ruan, D., Tolga, E.: Capital budgeting techniques using discounted fuzzy versus probabilistic cash flows. Information Sciences 142(1-4), 57–76 (2002)

Kaufmann, A., Gupta, M.M.: Fuzzy Mathematical Models in Engineering and Management Science. Elsevier, Amsterdam (1988)

Matthews, J., Broadwater, R., Long, L.: The Application of Interval Mathematics to Utility Economic Analysis. IEEE Transactions On Power Systems 5(1), 177–181 (1990)

Merrill, H.M., Wood, A.J.: Risk and Uncertainty in Power System Planning. In: Proceedings of the 10th Power Systems Computation Conference, Graz (1990)

Miranda, V.: Fuzzy Reliability Analysis of Power Systems. In: Proceedings of the 12th Power Systems Computation Conference, Dresden (1996)

Mohanty, R.P., Agarwal, R., Choudhury, A.K., et al.: A fuzzy ANP-based approach to R&D project selection: a case study. International Journal Of Production Research 43(24), 5199–5216 (2005)

Rebiasz, B.: Fuzziness and randomness in investment project risk appraisal. Computers & Operations Research 34(1), 199–210 (2007)

Riggs, J.L., West, T.M.: Essentials of Engineering Economics. McGraw-Hill Book Co., New York (1986)

Rubinson, T.: Multi-period risk management using fuzzy logic. International Journal Of Uncertainty Fuzziness And Knowledge-Based Systems 4(5), 449–466 (1996)

Saade, J.J.: A Unifying Approach to Defuzzification and Comparison of the Outputs of Fuzzy Controllers. IEEE Transactions On Fuzzy Systems 4(3), 227–237 (1996)

Sheen, J.N.: Fuzzy financial profitability analyses of demand side management alternatives from participant perspective. Information Sciences 169(3-4), 329–364 (2005a)

Sheen, J.N.: Fuzzy evaluation of cogeneration alternatives in a petrochemical industry. Computers & Mathematics With Applications 49(5-6), 741–755 (2005b)

Sheen, J.N.: Fuzzy financial analyses of demand-side management alternatives. IEE Proceedings Generation Transmission And Distribution 152(2), 215–226 (2005c)

Sheen, J.N.: Fuzzy Financial Decision-Making: Load Management Programs Case Study. IEEE Transactions On Power Systems 20(4), 1808–1817 (2005d)

Terceno, A., De Andres, J., Barbera, G.: Using fuzzy set theory to analyse investments and select portfolios of tangible investments in uncertain environments. International Journal Of Uncertainty Fuzziness And Knowledge-Based Systems 11(3), 263–281 (2003)

Tolga, E., Demircan, M.L., Kahraman, C.: Operating system selection using fuzzy replacement analysis and analytic hierarchy process. International Journal Of Production Economics 97(1), 89–117 (2005)

Zadeh, L.A.: Fuzzy Sets. Information and Control 8, 338–353 (1965)

Zimmermann, H.-J.: Fuzzy Set Theory - and Its Applications, 2nd edn. Kluwer Academic Publishers, Boston (1991)

# Fuzzy Rate of Return Analysis and Applications

Dorota Kuchta

Wroclaw University of Technology, Institute of Industrial Engineering and Management, Smoluchowskiego, 25, 50-371 Wroclaw, Poland

**Abstract.** The definition of the fuzzy internal rate of return is not unequivocal and requires some discussion. In this chapter, crisp rate of return analysis is first introduced. Then fuzzy rate of return analysis is handled. The problem of defining fuzzy internal rate of return together with certain applications is discussed.

## 1 Introduction

In this section we will present some information concerning the definition and calculation of the fuzzy internal rate of return (IRR) as well as the way this notion is applied in practice. The definition of the fuzzy internal rate of return is not unequivocal and requires some discussion. We will also show how fuzzy numbers can be used when projects are classified (ranked) on the basis of their IRR value. The notation used in this section is based on that used in Chapter 3.1, also there basic notions are discussed.

## 2 Crisp Internal Rate of Return

We consider a project $P$ or another type of investment that a company might be interested to undertake. For this project the following data has to be given:

a) $n$ - the duration of the project in assumed time units (we assume $n$ to be integer)
b) $COF_0$ $(\geq 0)$ - cash outflow at the moment when the project is starting $(i=0,1,...,n)$ and $CIF_i$ $(\geq 0)$ - cash inflow at the end of the $i$th year $(i=1,...,n)$
c) $r>0$ – the cost of acquiring capital $COF_0$ .
d) $rr>0$ – the rate off return with which we can invest $CIF_i$ $(\geq 0)$ $(i=1,...,n)$ in some other projects to withdraw them together with interests in the moment $n$.

In cases where the cash flows have different signs and are not only non-negative, as defined in point b) above, the Internal Rate of Return does not exist in an unequivocal manner and therefore is usually not calculated (e.g. Horngen et al. 1994). Some details about how the cash flows are determined and about $r$ can be found in Chapter 3.1. The estimation of $rr$ is usually difficult. It requires a thorough analysis of investment possibilities in the future and is always biased by a significant risk of error.

---

C. Kahraman (Ed.): Fuzzy Engineering Economics with Appl., STUDFUZZ 233, pp. 97–104, 2008.
springerlink.com © Springer-Verlag Berlin Heidelberg 2008

98  D. Kuchta

And now let us introduce the definition of IRR:

**Definition 1**

The internal rate of return (IRR) a project $P$ is such a number $E$ for which the following equality is fulfilled:

$$NPV(P) = CF_0 + \sum_{i=1}^{n} PV(CIF_i) = 0,$$

where

$$PV(CIF_i) = \frac{CIF_i}{(1+E)^i} \quad \text{for } i = 1,\dots,n$$

The IRR is thus the cost of capital at which the project in question has NPV equal to zero.

A modification of the above notion is also considered:

**Definition 2**

The Modified Internal Rate of Return (MIRR) a project is such a number $E$ for which the following equality is fulfilled:

$$NPV(P) = CF_0 + \sum_{i=1}^{n} PV(CIF_i)(1+rr)^{n-i} = 0,$$

where

$$PV(CIF_i) = \frac{CIF_i}{(1+E)^i} \quad \text{for } i = 1,\dots,n.$$

The MIRR is thus the rate of return of the project, calculated under the assumption that the cash generated by the project can be reinvested (in some other projects) with the rate of return $rr$ and will be available (together with the interests) at the end of project in question.

The IRR is the maximal cost of capital at which the project in question does not have a negative NPV. Thus, it is the maximal cost of capital that can be accepted when money is borrowed for project realisation. Consequently, the higher the IRR, the better – in the respect that we have more manoeuvre possibilities while looking for project financing sources. However, we have to remember that is only in this respect that a project with a higher IRR can be considered to be better than a project with a lower IRR. If we are not interested in the possibility of being able to finance the project at a high cost, but simply in getting more cash flow (or advantages in the more general sense of the cash flows, see Chapter 3.1.) out of the project than we put into it, we have to use e.g. the NPV as the evaluation criterion (Chapter 3.1). The two criteria can give contradictory answers while evaluating two or more projects (e.g. Horngen et al 1994). While evaluating a single project, it generally accepted to reject it if its IRR is lower than $r$ – because it means that the cost at which we have to acquire the capital is higher than the maximal accepted one. But this acceptance criterion is equivalent to NPV>0.

Generally, the IRR is not considered to be a basic project evaluation criterion. Apart from some special uses, it is generally seen as one which does not more advantages than NPV (Hajdasinski 2003, Hazen 2003, Tang and Tang 2003), apart maybe from the fact that it is calculated in percentages (and NPV in absolute values), which may be an advantage to some decision makers.

Sometimes the IRR is understood – erroneously – as the rate of return of the project. This would mean that if you put an amount $COF_0$ in the project which takes $n$ years and $E$ stands for its IRR, you get $COF_0(1+E)^n$ once the project is terminated. This a wrong interpretation unless the flows $CIF_i$ $(\geq 0)$ $(i=1,...,n)$ can be reinvested in some other project(s) till the end of the project in question with the rate of return $E$, which is usually wrong. The rate of return of the project is calculated in MIRR, where the actual reinvestment rate is taken into account. Of course, the greater the MIRR, the better.

# 3 Fuzzy Internal Rate of Return

Some of the input information concerning a project to be evaluated can be fuzzy numbers. For computational reasons (it is evident already from Definitions 2 and 3 that the IRR and MIRR is computationally more difficult to obtain than e.g. the NPV) we will use here the following definition of a fuzzy number:

**Definition 3**

A fuzzy number, denoted by $\tilde{A} = \left(a_1, a_2, a_3, a_4, f_1^A, f_2^A\right)$, where $f_1^A(x)$ is defined and increasing for $x \in (a_1, a_2)$, $f_2^A(x)$ is defined and decreasing for $x \in (a_3, a_4)$, both function are continuous and transform their domain intervals into the interval $[0,1]$, is defined as the family of closed intervals for $\lambda \in [0,1]$:

$$A^\lambda = \left[\left(f_1^A\right)^{-1}(\lambda), \left(f_2^A\right)^{-1}(\lambda)\right] = \left[\underline{A}^\lambda, \overline{A}^\lambda\right]$$

In numerous practical applications it is not necessary to know all the $\lambda$-cuts, the knowledge of $\lambda$-cuts for selected values of $\lambda$ is usually sufficient.

There is a one to one dependency between the $\lambda$-cuts and the membership function (see Chapter 3.1-Definition 7) of a fuzzy number:

$$\mu_A(x) = \sup_{\lambda \in [0,1]} x \in A^\lambda$$

The membership function $\mu_A(x)$ of a fuzzy number $\tilde{A}$ can represent (for $x \in \mathfrak{R}$)

- either the degree to which it is possible that the crisp $x$ will be the yet unknown value of a magnitude modelled by $\tilde{A}$
- or the degree to which $x$ possesses a certain characteristic modelled by $\tilde{A}$.

For the latter use we need the explicit form of the membership function of fuzzy number $\tilde{A} = \left(a_1, a_2, a_3, a_4, f_1^A, f_2^A\right)$. It is as follows:

$$\mu_A(x) = \begin{cases} 0 & \text{for } x \le a_1 \text{ i } x \ge a_4 \\ f_1^A(x) & \text{for } x \in (a_1, a_2) \\ 1 & \text{for } x \in [a_2, a_3] \\ f_2^A(x) & \text{for } x \in (a_3, a_4) \end{cases}$$

We will also need right-hand and left-hand fuzzy numbers. They will be defined "the other way round", using the membership function as the starting point (and not the $\lambda$-cuts):

**Definition 4**

- a) The triplet $\tilde{A} = \left(a_1, a_2, f_1^A\right)$ will be called a right-hand fuzzy number. Its membership function is defined as:

$$\mu_A(x) = \begin{cases} 0 & \text{for } x \le a_1 \\ f_1^A(x) & \text{for } x \in (a_1, a_2) \\ 1 & \text{for } x \ge a_2 \end{cases}$$

- b) The triplet $\tilde{A} = \left(a_3, a_4, f_2^A\right)$ will be called a left-hand fuzzy number. Its membership function is defined as:

$$\mu_A(x) = \begin{cases} 1 & \text{for } x \le a_3 \\ f_1^A(x) & \text{for } x \in (a_3, a_4) \\ 0 & \text{for } x \ge a_4 \end{cases}$$

We consider the case where the information about the project is given in the form of a fuzzy number, in the sense of Definition 3:

a) $n$ - the duration of the project in assumed time units

b) $C\tilde{O}F_0 = \left[\underline{COF}_0^\lambda, \overline{COF}_0^\lambda\right]$, $\underline{COF}_0^\lambda \ge 0$ $(\lambda \in [0,1])$ - cash outflow at the moment when the project is starting and $C\tilde{I}F_i = \left[\underline{CIF}_i^\lambda, \overline{CIF}_i^\lambda\right]$, $\underline{CIF}_i^\lambda \ge 0$ $(\lambda \in [0,1])$- cash inflow at the end of the $i$th year $(i=1,...,n)$.

c) $\tilde{r} = \left[\underline{r}^\lambda, \overline{r}^\lambda\right] > 0$, $\underline{r}^\lambda > 0$ $(\lambda \in [0,1])$– the cost of acquiring capital $COF_0$.

d) $\tilde{r}r = \left[ \underline{rr}^{\lambda}, \overline{rr}^{\lambda} \right] > 0$, $\underline{rr}^{\lambda} > 0$ $(\lambda \in [0,1])$– – the rate off return with which we can invest $CIF_i$ $(\geq 0)$ $(i=1,...,n)$ to withdraw them together with interests in the moment $n$.

Using Definition 8 in Chapter 3.1, we can now define the fuzzy IRR and MIRR for a project with fuzzy parameters (Buckley 1985, Pohjola and Turunen 1990, Buckley 1992, (Kuchta 2000, Terceño et al 2003):

### Definition 5

Let $P$ be a project with fuzzy parameters. Then its fuzzy IRR is the family of intervals $\tilde{E} = \left\{ \left[ \underline{E}^{\lambda}, \overline{E}^{\lambda} \right] : \lambda \in [0,1] \right\}$ such that:

- $\underline{CF}_0^{\lambda} + \sum_{i=1}^{n} PV\left( \underline{CIF}_i^{\lambda} \right) = 0$, where $PV\left( \underline{CIF}_i^{\lambda} \right) = \dfrac{\underline{CIF}_i^{\lambda}}{\left( 1 + \underline{E}^{\lambda} \right)^i}$

- $\overline{CF}_0^{\lambda} + \sum_{i=1}^{n} PV\left( \overline{CIF}_i^{\lambda} \right) = 0$, where $PV\left( \overline{CIF}_i^{\lambda} \right) = \dfrac{\overline{CIF}_i^{\lambda}}{\left( 1 + \overline{E}^{\lambda} \right)^i}$

### Definition 6

Let $P$ be a project with fuzzy parameters. Then its fuzzy MIRR is the family of intervals $\tilde{E} = \left\{ \left[ \underline{E}^{\lambda}, \overline{E}^{\lambda} \right] : \lambda \in [0,1] \right\}$ such that:

- $\underline{CF}_0^{\lambda} + \sum_{i=1}^{n} PV\left( \underline{CIF}_i^{\lambda} \left( 1 + \underline{rr}^{\lambda} \right)^{n-i} \right) = 0$, where $PV\left( \underline{CIF}_i^{\lambda} \right) = \dfrac{\underline{CIF}_i^{\lambda}}{\left( 1 + \underline{E}^{\lambda} \right)^i}$

- $\overline{CF}_0^{\lambda} + \sum_{i=1}^{n} PV\left( \overline{CIF}_i^{\lambda} \right)\left( 1 + \overline{rr}^{\lambda} \right)^{n-i} = 0$, where $PV\left( \overline{CIF}_i^{\lambda} \right) = \dfrac{\overline{CIF}_i^{\lambda}}{\left( 1 + \overline{E}^{\lambda} \right)^i}$

Let us remark that the fuzzy IRR and MIRR do not have to fulfill Definition 3 in the sense that the corresponding functions $f_1^A$, $f_2^A$ for the fuzzy IRR and MIRR from Definitions 5 and 6 do not have to be continuous. However, in (Buckley 1992) it is shown that if we persist on e.g. the IRR to be a fuzzy number according to Definition 3 and define it like in Definition 7, we will have to deal with a significant practical problem:

### Definition 7

Let $P$ be a project with fuzzy parameters. Then its fuzzy IRR is a fuzzy number $\tilde{E}$ according to Definition 3 such that:

$$- C\tilde{F}_0 = \sum_{i=1}^{n} PV\left( \tilde{CIF}_i \right)$$

102    D. Kuchta

(in the sense of the identity of fuzzy numbers), where

$$PV\left(\underline{CIF}_i^\lambda\right) = \frac{CIF_i^\lambda}{\left(1 + \underline{E}^\lambda\right)^i}$$

The IRR defined in Definition 7 is very difficult to calculate (Buckley 1992, Carlsson and Fuller 1999) and if we assume Definition 5, in (Buckley 1992) it is shown that the $\lambda$-cuts $\left[\underline{E}^\lambda, \overline{E}^\lambda\right]$ include those of the fuzzy IRR according to Definition 7. Taking into account that the application Definition 5 is easy from the practical point of view and no information is lost with respect to Definition 7 (in the worst case the fuzzy IRR according to Definition 5 will present to the decision maker some crisp values of IRR as possible to a higher degree than shown by the IRR from Definition 7), in the following we assume Definition 5 and 6. It is obvious that, like for any fuzzy number according to Definition 3, also for the IRR and the MIRR from Definition 5 and 6 the following property is fulfilled:

$$\left[\underline{E}^{\lambda 1}, \overline{E}^{\lambda 1}\right] \subset \left[\underline{E}^{\lambda 2}, \overline{E}^{\lambda 2}\right] \quad \text{for } \lambda 1 > \lambda 2 \tag{1}$$

The interpretation of the fuzzy IRR and MIRR reduces itself to comparing families of intervals

$$\tilde{E}(1) = \left\{\left[\underline{E}^\lambda(1), \overline{E}^\lambda(1)\right] : \lambda \in [0,1]\right\}$$

and

$$\tilde{E}(2) = \left\{\left[\underline{E}^\lambda(2), \overline{E}^\lambda(2)\right] : \lambda \in [0,1]\right\}$$

corresponding to two projects.

Of course, it is not unequivocal to decide which family of intervals can be considered to be larger. One of many possibilities (taking into account Eq. (1)) is to choose a confidence level $\lambda 0$ and compare the corresponding intervals $\left[\underline{E}^{\lambda 0}(1), \overline{E}^{\lambda 0}(1)\right]$ and $\left[\underline{E}^{\lambda 0}(2), \overline{E}^{\lambda 0}(2)\right]$, taking into account the pessimistic values (the lower ends of the intervals), the optimistic values (the upper ends of the intervals) or the average values of the intervals. More possibilities of comparing intervals, including some fuzzy relations between intervals, which can be defined by the decision maker himself, can be found e.g. in (Kuchta 2003). Otherwise it is also possible to use (modifying them a bit to adopt to intervals) the many ways of comparing fuzzy numbers, described e.g. in Chapter 3.1.

Applications of fuzzy IRR are in principle limited to the calculation of fuzzy IRR for various alternatives and ranking those alternatives, using various fuzzy numbers ranking methods (see e.g. Chapter 3.1). The project with the fuzzy IRR considered to be the highest one is then selected. Thus, in (Terceño et al 2003) the Campos and González index is used ((Campos and González 1989), see also 3.1), the authors of (Greenhut et al. 1995) use the index from (Yao and Wu 2000) (see also Definition 13 in Chapter 3.1)

Also if crisp IRR is used, fuzzy approach can be useful. IRR can be used as one of the criteria of acceptance/rejection of a project. Sometimes the boundary between projects to be accepted or those to be rejected is fuzzy. In such cases right-hand, left-hand (Definition 1) or "normal" fuzzy numbers can be used to define the notions "good IRR", "bad IRR" etc. This approach is used in (Vranes et al. 2002), where the IRR (crisp) is one of the several criteria used to evaluate a project.

The authors define six notions – types of IRR, out of which we quote 3 examples here:

- a very good IRR (defined by means of $\tilde{A}_{vg} = (a_1, a_2, f_1^A)$, where

$$a_1 = r + 5, a_2 = r + 15, f_1^A(x) = 0.1(x - r - 5))$$

- a good IRR (defined by means of $\tilde{A}_g = (a_1, a_2, a_3, a_4, f_1^A, f_2^A)$, where

$$a_1 = r - 6, a_2 = r - 4, a_3 = r + 15, a_4 = r + 20,$$
$$f_1^A(x) = 0.5(x - r + 6), f_1^B(x) = -0.2(x - r - 20)))$$

- a very bad IRR (defined by means of $\tilde{A}_{vb} = (a_3, a_4, f_2^A)$, where

$$a_3 = r - 6, a_2 = r - 4, f_2^A(x) = -0.5(x - r + 4))$$

In (Vranes et al. 2002) the criterion of IRR is combined with NPV and several other criteria, also qualified as "good" or "bad" by means of fuzzy numbers. Sets like $A$={projects whose NPV is very good, IRR good and other criteria are good}, $B$={projects whose NPV is good, IRR fairly good and other criteria are good}, $C$={projects whose NPV is very bad and $p$ is bad} etc. are defined. For each project we calculate the membership degree to each of those sets as the minimum of the degree to which the project has the respective characteristics (details can be found in section 3.1). Finally each project is assumed to belong to the set for which it has the maximal membership degree. The decision maker should then apply some decision rules, e.g. "if a project belongs to $A$ or $B$ it should be accepted", "if it belongs to C it should be rejected" etc.

# 4  Conclusions

In this section the problem of defining fuzzy Internal Rate of Return, together with certain applications, was discussed. In the literature also other ways of defining fuzzy IRR can be found, however, we presented the concept which seemed to us most natural.

# References

Buckley, J.J.: The fuzzy mathematics of finance. Fuzzy Sets and Systems 21, 257–273 (1987)

Buckley, J.J.: Solving fuzzy equations in economics and finance. Fuzzy Sets and Systems 48, 289–296 (1992)

Campos, L.M., Gonzáles, A.: A subjective approach for ranking fuzzy numbers. Fuzzy Sets and Systems 29, 145–153 (1989)

Carlsson, C., Fuller, R.: On fuzzy capital budgeting problem. In: Proceedings of the ICSC Congress on Computational Intelligence: Methods and Applications, Rochester, NY, vol. 5 (1999)

Greenhut, J.G., Norman, G., Temponi, C.: Toward a fuzzy theory of oligopolistic competition. In: IEEE Procedings of ISUMA-NAFIPS 1995, pp. 286–291 (1995)

Hajdasinski, M.M.: Technical note – the internal rate of return (IRR) as a financial indicator. The Eng. Econ. 49, 185–197 (2004)

Hazen, G.B.: A new perspective on multiple internal rates of return. The Eng. Econ. 48(1), 31–51 (2003)

Horngren, T.H., Foster, G., Datar, S.M.: Cost accounting: A Managerial Emphasis. Prentice Hall, Englewood Cliffs (1994)

Kaufmann, A., Gupta, M.M.: Fuzzy mathematical models in engineering and management science. Elsevier Science Publishers BV, Amsterdam (1988)

Kuchta, D.: Fuzzy capital budgeting. Fuzzy Sets and Systems 111, 367–385 (2000)

Kuchta, D.: User-tailored fuzzy relations between intervals. In: Conference of the European Society for Fuzzy Logic and Technology, Proceedings of EUSFLAT 2003, Zittau, Germany, September 10-12, pp. 437–441 (2003)

Pohloja, V.J., Turunen, I.: Estimating the internal rate of return form the fuzzy data. Eng. Costs Prod. Econ. 18, 215–221 (1990)

Tang, S.L., Tang, H.J.: Technical note – the variable financial indicator IRR and the constant economic indicator NPV. The Eng. Econ. 48(1), 69–78 (2003)

Terceño, A., de Andrés, J., Barberà, G.: Using fuzzy set theory to analyse investments and select portfolios of tangible investments in uncertain environments. International Journal of Uncertainty, Fuzziness and Knowledge-Based Systems 3, 263–281 (2003)

Vranes, S., Stanojevic, M., Stevanovic, V.: Financial decision aid based on heuristics and fuzzy logic. In: Damiani, E., et al. (eds.) Knowledge based Intelligent Informatiopn Engineering systems and allied technologies, Proceedings of KES 2003 conference, Crema, Itally, September 16-18, pp. 381–387 (2002)

# On the Fuzzy Internal Rate of Return

P. Sewastjanow and L. Dymowa

Institute of Comp. & Information Sci., Technical University of Czestochowa, Dabrowskiego 73, 42-200 Czestochowa, Poland

**Abstract.** In a capital investment we usually deal with projects taking a long time - as a rule some years - for its realization. In such cases, a description of uncertainty within a framework of traditional probability methods usually is impossible due to the absence of objective information about the probabilities of future events. This is the reason for the growing interest in the application of interval and fuzzy methods in budgeting, which has been observed for the last two decades. There are many financial parameters proposed in literature for the project quality assessment, but the two primary among them --- net present value, $NPV$ , and internal rate of return, $IRR$ - are necessarily used in a financial analysis. Whereas the problem of $NPV$ fuzzy estimation is now well studied and many authors have contributed to its solution, obtaining of fuzzy $IRR$ seems to be rather an open problem. This problem is a consequence of inherent properties of fuzzy and interval mathematics, but it seems unnatural to have crisp $IRR$ in a fuzzy environment when all other financial parameters are fuzzy. In this paper, the problem of $IRR$ estimation in fuzzy setting is considered in the framework of more general problem of fuzzy equations solving. Finally, the concept of restricted fuzzy $IRR$ as the solution of the corresponding non-linear fuzzy equation is proposed and analyzed.

## 1 Introduction

There are a lot of financial parameters proposed in the literature (Belletante and Arnaud 1989, Brigham 1992, Chansa-ngavej and Mount-Cambell 1991, Liang and Song 1994) for budgeting. The main ones being: Net Present Value ( $NPV$ ), Internal Rate of Return ( $IRR$ ), Payback period ( $PB$ ), Profitability Index ( $PI$ ). It is shown in (Bogle and Jehenck 1985) that the most important parameters are $NPV$ and $IRR$ . A good review of other useful financial parameters may be found in (Babusiaux and Pierru 2001).

Net Present Value is usually calculated as follows:

$$NPV = \sum_{t=1}^{T} \frac{P_t}{(1+d)^t} - KV \tag{1}$$

where $d$ is a discount rate, $KV$ is a starting capital investment, $P_t$ is the total income (cash flow) in a year $t$ , $T$ is a duration of an investment project in years. Usually the discount rate is equal to the average bank interest rate in a country of an investment or other value corresponding to a profit rate of alternative capital investments. The value of $IRR$ is a solution of a non-linear equation with respect to $d$ :

$$\sum_{t=1}^{T} \frac{P_t}{(1+d)^t} - KV = 0 \tag{2}$$

C. Kahraman (Ed.): Fuzzy Engineering Economics with Appl., STUDFUZZ 233, pp. 105–128, 2008.
springerlink.com © Springer-Verlag Berlin Heidelberg 2008

If $P_t$, $KV$ (or at least one of them) are fuzzy numbers with use of fuzzy extension of Eq. (2), i.e., by replacement of its parameters and variable with fuzzy ones and all arithmetic operations with relevant fuzzy operations, Eq. (2) can be transformed to fuzzy equation. The problem is to find a fuzzy solution of such fuzzy equation, i.e., to obtain fuzzy $IRR$. An economic nature of the $IRR$ can be explained as follows. If an actual bank discount rate or return of any other alternative investment under consideration is less than $IRR$ of considered project, then investment in this project is more preferable. Only the cases when Eq.(2) has single root will be analyzed. The reasons behind this are as follows. The problem of multiple roots of Eq. (2) rises when the negative cash flows take place after starting investment. In practice, an appearance of some negative cash flow after initial investment is usually treated as a local "force majeur" or even a total project's failure. That is why, on the stage of planning, investors try to avoid situations when such negative cash flows are possible, except the cases when they dealing with long-term projects consist of some phases. Let us see to the Fig. 1.

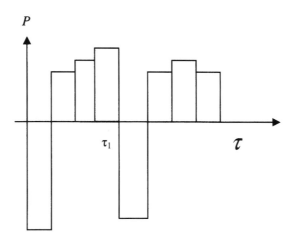

**Fig. 1.** Two stage investment project

This is a typical two-phase project: after initial investment the project brings considerable profits and at the time $\tau_1$ a part of accumulated earnings and, perhaps, an additional banking credit is invested once again. Factually, an investor buys new production equipment and buildings (in fact, creating a new enterprise) and from his/her point of view a quite new project is started. It is easy to see that investor's creditors which are interested in repayment of a credit always analyze phases $\tau < \tau_1$ and $\tau > \tau_1$ separately. It worth noting that what we describe is only an investment planning routine, not some theoretical considerations we can find in financial textbooks. On the other hand, a separate assessment of different projects' phases reflects economic sense of capital investment better. Indeed, if we consider a two phase project as a whole, we often get the $IRR$ s performed by two roots so different that it is impossible to make any decision. For example, we can obtain $IRR_1$ =4% and $IRR_2$ =120%. It is clear that average $IRR$ = (4+120)/2=62% seems as rather fantastic estimation, whereas when considering the two phases of project separately we usually get quite acceptable values, e.g., for the first phase $IRR_1$ =20% and for the second phase $IRR_2$ =25%. So we can say that the problem of "multiple $IRR$ values" exists only in some financial textbooks, not in the practice of capital investment. An estimation of $IRR$ is frequently used as a first step of the financial analysis. Only projects with $IRR$ above some accepted threshold value (usually 15--20%) can be chosen for further consideration. Nowadays, traditional approaches to the evaluation of $NPV$ , $IRR$ and other financial parameters is subjected to quite deserved criticism, since the future incomes $P_t$ and rates $d$ are rather uncertain parameters. Uncertainties which one meets in capital budgeting cannot be adequately described in terms of probability. Really, in budgeting we usually deal with a business-plan that takes a long time - as a rule some years - for its realization. In such cases, the description of uncertainty via probability representation of $P_t$ , $KV$ and $d$ usually is impossible due to a lack of information about probabilities of future events. Thus, what we really have in such cases, are some subjective expert judgments. In real-world situations, investors or experts involved are able to estimate only intervals of possible values $P_t$ and $d$ and the expected (more probable) values inside these intervals. That is why during last two decades the growing interest to applications of the interval arithmetic (Moore 1966) and fuzzy sets theory methods (Zadeh 1965) in budgeting has being observed. After pioneer works of Ward (1985) and Buckley (1987), some other authors contributed to the development of fuzzy capital budgeting theory (Chen 1995, Chiu and Park 1994, Choobineh and Behrens 1992, Calzi 1990, Perrone 1994, Kahraman et al. 2000, Kahraman et al. 2002, Kahraman et al. 1997a, Kahraman et al. 1997b, Sevastianov and Sevastianov 1997, Kuchta 2000, Dimovaet al. 2000, Kahraman 2001). It is safe to say now that almost all problems of the fuzzy $NPV$ estimation are solved. An unresolved problem is the fuzzy estimation of the $IRR$ . Ward (1985) considers Eq. (2) and states that such an expression cannot be applied to the fuzzy case because the left side of Eq. (2) is fuzzy, 0 in the right hand side is a crisp value and an equality is impossible. Hence, the Eq. (2) is senseless from the fuzzy viewpoint. Kuchta (2000) proposed a method for fuzzy $IRR$ estimation where $\alpha$ -cut representation of fuzzy numbers (Kaufmann and Gupta 1985) was used. The method is based on an assumption (Kuchta 2000, p. 380) that a set of equations for $IRR$ determination on each $\alpha$ -level may be presented as (in our notation)

$$
(KVV^\alpha)_1 + \sum_{t=1}^{T} \frac{(P_t^\alpha)_1}{(1 + IRR_1^\alpha)^t} = 0
$$

$$
(KVV^\alpha)_2 + \sum_{t=1}^{T} \frac{(P_t^\alpha)_2}{(1 + IRR_2^\alpha)^t} = 0
$$

$$(3)$$

where $KVV = -KV$ , indexes "1","2" stand for the left and right bounds of corresponding intervals respectively, $P_t^\alpha = [(P_t^\alpha)_1, (P_t^\alpha)_2]$ are crisp interval representations of fuzzy cash flows at time $\tau$ on the $\alpha$ -levels. Of course, from the equations (3) all crisp intervals $d^\alpha = [d_1^\alpha, d_2^\alpha]$, expressing the fuzzy valued $IRR$ may be obtained. On the other hand, Eqs.(3) are not a direct cosequence of conventional fuzzy and interval arithmetics rules. Eqs.(3) were obtained in (Kuchta

2000) using fuzzy extension of (1) assuming that $P_t$, $KV$ (or at least one of them) are fuzzy numbers and representing them by the sets of $\alpha$-levels. Since Eqs.(3) should be verified on each $\alpha$-level, it is quite enough to consider only crisp interval extension of (2), which is the particular case of more general equation

$$F(d) - B = 0 \qquad (4)$$

where $B$ is an interval ($B = KV$ in our case) and $F(d)$ is an interval valued function of interval argument $d$ (in our case $F(d) = \sum_{t=1}^{T} \frac{P_t}{(1+d)^t}$). Using regular interval arithmetic (Jaulin et al. 2001), this equation can be transformed to $[F_1(d) - B_2, F_2(d) - B_1] = 0$, and finally we get two equations $F_1(d) - B_2 = 0, F_2(d) - B_1 = 0$. Of course, if we deal with a linear interval function $F(d) = Ad$ ($A$ is an interval), then $F_1(d) = A_1 d_1$ and $F_2(d) = A_2 d_2$, but if $F(d) = \frac{A}{d}$ we have $F_1(d) = \frac{A_1}{d_2}$, $F_2(d) = \frac{A_2}{d_1}$ since $F_1$ is the left bound ($min$ value in interval) and $F_2$ is the right bound ($max$ value in interval) of interval value $F(d)$.

Hence, using the regular interval arithmetic rules leads to the following equations:

$$(KVV^{\alpha})_1 + \sum_{t=1}^{T} \frac{(P_t^{\alpha})_1}{(1 + IRR_2^{\alpha})^t} = 0$$
$$(KVV^{\alpha})_2 + \sum_{t=1}^{T} \frac{(P_t^{\alpha})_2}{(1 + IRR_1^{\alpha})^t} = 0$$
$$(5)$$

There is no way to get a correct not inverted interval solution of (5). Only inverted intervals $IRR$, i.e., such that $IRR_1^{\alpha} > IRR_2^{\alpha}$ can be obtained from (5). Since it is hard or even impossible to interpret reasonable such results, they can not be used in practice. It is shown in (Dimova et al. 2000) that only approximate real valued $IRR$ (represented by usual non interval numbers) may be obtained from (5). In Section 2, we show that main problem is that conventional interval extension (and the fuzzy as well) of usual equation, which leads to the interval or fuzzy equation such (5) is not a correct procedure. Less problems we meet when dealing with interval or fuzzy equation in form

$$F(d) = B \qquad (6)$$

An important feature of interval and fuzzy mathematics is that Eq.(6) not equivalent to Eq. (4). This fact deserves more detailed analysis in context of fuzzy $IRR$ problem and we study it thoroughly in Section 2. Generally, the problem of interval or fuzzy equations solving is not trivial even for linear ones such as

$$AX = C \qquad (7)$$

As it is stated in (Buckley and Qu 1990), "...for certain values of $A$ and $C$, Eq.(7) has no solution for $X$. That is, for some triangular fuzzy numbers $A$ and $C$ there is no fuzzy set $X$ so that, using regular fuzzy arithmetic, $AX$ is exactly equal to $C$ ". Although many different numerical methods were proposed for solving interval and fuzzy equations including such complicated as Neural Net solutions (Buckley and Eslami 1997, Buckley et al. 1997) and fuzzy extension of Newton's method (Abbasbandy and Asady 2004) only particular solutions valid in specific conditions were obtained. Summarizing, we can say that the problem of $IRR$ estimation in a fuzzy setting should be considered in the framework of more general problem of fuzzy equations solving.

The rest of the chapter is set out as follows. In Section 2, some basic statements of interval and fuzzy mathematics needed for the analysis are recalled. The $\alpha$-cut representation of fuzzy numbers is used as a base of the analysis. Further, a new method for solving of linear interval equations is presented. In Section 3, this method is used to solve Eq. (2) in the simple case of one-year project and for interval cash flows. It is shown that restricted solution of Eq. (2) may be obtained as a fuzzy interval even for a crisp interval representation of cash flows. In Section 4, a numerical method for restricted solution of non-linear Eq. (2) in fuzzy setting is presented. The possible applications of fuzzy $IRR$ are described in Section 5. Finally, Section 6 concludes with some remarks.

## 2   A New Method for Solving of Interval Equations

The technique is based on a fuzzy extension principle (Zadeh 1965). So, the values of uncertain parameters $P_t$ and $d$ are substituted with corresponding fuzzy intervals. In practice it means that an expert sets lower - $P_{t1}$ (pessimistic value) and upper - $P_{t4}$ (optimistic value) boundaries of intervals and internal intervals consist of most probable values [ $P_{t2}$, $P_{t3}$ ] for the analyzed parameters (see Fig. 2). The function $\mu(P_t)$ is usually interpreted as a membership function, i.e., a degree to which values of a parameter belong to an interval (in our case [ $P_{t1}$, $P_{t4}$ ]). The membership function changes continuously from 0 (an area out of the interval) up to maximum value 1 in an area of the most probable values. It is obvious that the membership function is a generalization of the characteristic function of usual set, which equals 1 for all values inside a set and 0 in all other cases.

The linear character of function is not obligatory, but such a mode is most used and allows to represent fuzzy intervals in a convenient form of a quadruple $P_t = \{P_{t1}, P_{t2}, P_{t3}, P_{t4}\}$ . Then all necessary calculations are carried out using special fuzzy-interval arithmetic rules. Consider some basic principles of the fuzzy arithmetic (Kaufmann and Gupta 1985). In general, for the arbitrary form of the membership function the technique of fuzzy-interval calculations is based on the representation of initial fuzzy intervals in a form of so-called $\alpha$-cuts (Fig. 2), which are, in fact, crisp intervals associated with corresponding degrees of the membership. All further calculations are made with those $\alpha$-cuts according to the well known crisp interval-arithmetic rules and resulting fuzzy intervals are obtained as the conjunction of corresponding final $\alpha$-cuts. Thus, if $A$ is a fuzzy number then $A = \bigcup_{\alpha} \alpha A_\alpha$, where

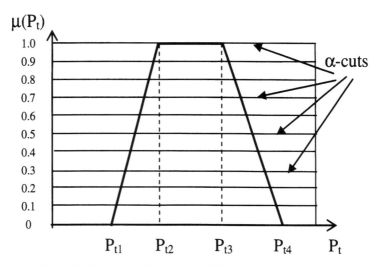

**Fig. 2.** Fuzzy interval of an uncertain parameter $P_t$ and its membership function $\mu(P_t)$

$A_\alpha$ is a crisp interval $\{x : \mu_A(x) \geq \alpha\}$, $\alpha A_\alpha$ is a fuzzy interval $\{(x,\alpha) : x \in A_\alpha\}$. Thus, if $A, B, Z$ are fuzzy numbers (intervals) and @ is an operation from $\{+,-,*,/\}$ then

$$Z = A @ B = \bigcup_\alpha (A @ B)_\alpha = \bigcup_\alpha A_\alpha @ B_\alpha. \tag{8}$$

Since in a case of $\alpha$-cut presentation the fuzzy arithmetic is based on crisp interval arithmetic rules, basic definitions of the applied interval analysis is also presented. One of the most negative features of interval arithmetic is the fast increasing of width of intervals obtained as the results of interval calculations. To reduce this undesirable effect, several different modifications of interval arithmetic were proposed. The most known are : Non- standard interval arithmetic (Markov 1977) based on the special form of interval subtraction and division, Generalized interval arithmetic (Hansen 1975), Segment interval analysis (Sendov 1977), Centralized interval arithmetic (Wang and Kerre 2001), MV-form (Caprani and Madsen 1980). All of these approaches provide good results only in specific conditions. On the other hand, in practice so-called "naive" form proposed by Moore (1966) is proved to be the best one (Jaulin et al. 2001). According to it, if $[x] = [\underline{x}, \overline{x}]$ and $[y] = [\underline{y}, \overline{y}]$ are crisp intervals and $@ = \{+,-,*,/\}$, then

$$[x] @ [y] = \{x @ y, \forall x \in A, \forall y \in B\} \tag{9}$$

As a direct outcome of the basic definition ((9))the following expressions were obtained:

$$[x] + [y] = [\underline{x} + \underline{y}, \overline{x} + \overline{y}] \tag{10}$$

$$[x]-[y]=[\underline{x}-\overline{y},\overline{x}-\underline{y}] \tag{11}$$

$$[x]*[y]=[\min(\underline{x}\underline{y},\overline{x}\underline{y},\underline{x}\overline{y},\overline{x}\overline{y}),\max(\underline{x}\underline{y},\overline{x}\underline{y},\underline{x}\overline{y},\overline{x}\overline{y})] \tag{12}$$

$$[x]/[y]=[\underline{x},\overline{x}]*[1/\overline{y},1/\underline{y}],\ 0\notin[y] \tag{13}$$

Of course, there are many inherent problems within the applied interval analysis, for example, a division by zero-containing interval, but in general, it can be considered as a good mathematical tool for modeling under conditions of uncertainty. In essence, the problem of fuzzy *IRR* evaluation is a fuzzy solution of the Eq. (2) with respect to $d$. It is proved that a solution of equations with fuzzy parameters is possible through an expression of these parameters in a form of sets of corresponding $\alpha$-cuts (Kaufmann and Gupta 1985). As a result, for the problem of evaluating *IRR* we obtain a system of the non-linear crisp-interval equations. Let us look at this problem from more general point of view. There is an important methodological problem of interval equations solution which is not widely discussed in scientific literature we name as "interval equation's right hand side problem". Suppose there exists some basic, non-interval algebraic equation $f(x)=0$. Its natural interval extension can be obtained by replacement of its variables with interval ones and all arithmetic operations with relevant interval operations. As a result we get an interval equation $[f]([x])=0$. Observe that this equation is senseless because its left part represents interval value whereas the right part is a non-interval degenerated zero. Obviously, if $[f](x)=[\underline{f},\overline{f}]$ then equation $[f]([x])=0$ is true only when $\underline{f}=\overline{f}=0$. It is easy to show that equation $[f]([x])=0$ in general can be verified only for inverted interval $[x]$, i.e., when $\overline{x}<\underline{x}$. Inverted intervals are analyzed in the framework of Modal Interval Arithmetic (Gardnes et al. 1985), but it is very hard and perhaps even impossible to encounter real-life situation when notation $\overline{x}<\underline{x}$ is meaningful. It is known (Moore 1966) that if mathematical expression can be presented in the different, but algebraically equivalent forms , they can bring the different interval results after interval extension. The same is true for the equations. Let us consider interval extensions of simplest linear equation

$$ax=b \tag{14}$$

and its algebraically equivalent forms

$$x=\frac{b}{a} \tag{15}$$

$$ax-b=0 \tag{16}$$

for $a$, $b$ being crisp intervals ($0\notin a$ ).

Since there are no strong rules in interval analysis for choosing the best form of equation among its algebraically equivalent representations to be extended, it is natural to compare the results we get from interval extensions of Eqs.(14)-(16). Let

$[a]=[\underline{a},\overline{a}]$ and $[b]=[\underline{b},\overline{b}]$ be intervals. Then interval extension of Eq. (14) is $[\underline{a},\overline{a}][\underline{x},\overline{x}]=[\underline{b},\overline{b}]$. Using conventional interval arithmetic rule (10) from this equation we obtain $[\underline{a}\underline{x},\overline{a}\overline{x}]=[\underline{b},\overline{b}]$. It is clear that equality of the right and left hand sides of this equation is possible only if $\underline{a}\underline{x}=\underline{b}$ and $\overline{a}\overline{x}=\overline{b}$ and finally we have

$$\underline{x}=\frac{\underline{b}}{\underline{a}},\overline{x}=\frac{\overline{b}}{\overline{a}} \tag{17}$$

As a consequence of the rule (13), interval extension of Eq. (15) results in the expressions

$$\underline{x}=\frac{\underline{b}}{\overline{a}},\overline{x}=\frac{\overline{b}}{\underline{a}} \tag{18}$$

Consider some examples.

**Example 1.** Let $a=[3,4],[b]=[1,2]$. Then from Eq. (17) we get $\underline{x}=0.333,\overline{x}=0.5$ and from Eq. (18) $\underline{x}=0.25,\overline{x}=0.666$.

**Example 2.** Let $a=[1,2],[b]=[3,4]$. Then from Eq. (17) we get $\underline{x}=3,\overline{x}=2$ and from Eq. (18) $\underline{x}=1.5,\overline{x}=4$.

**Example 3.** Let $a=[0,3],[b]=[1,1]$ (i.e., $b$ is a real number). Then from Eq. (17) we obtain $\underline{x}=10,\overline{x}=3.333$ and from Eq. (18) $\underline{x}=3.333,\overline{x}=10$.

We can see that interval extension of Eq. (14) may result in the inverse intervals $[x]$, i.e., such that $\overline{x}<\underline{x}$ (see Examples 2 and 3), whereas extension of Eq. (15) gives us the correct intervals ($\underline{x}<\overline{x}$). It is worth noting that interval extension of Eq. (15) will always deliver the correct resulting intervals because Eqs. (18) are inferred directly from the basic definition (10). For our purposes it is quite enough to state that interval extension of Eq. (15) guarantees the resulting intervals be correct ones in all cases, whereas interval extension of Eq. (14) can result in practically senseless inverse intervals. It is worth noting that in Example 3 the formal interval extension of Eq. (14) leads to contradictory interval equation since on the right hand side of extended Eq. (14) we have degenerated interval $b$ (real value), whereas the left hand side is an interval. In all such cases the solution of interval extension of Eq. (14) will be an inverse interval. This, at first glance, strange result is easy to explain from common methodological positions. Really, the rules of interval mathematics are constructed in such a way that any arithmetical operation with intervals results in interval as well. These rules are in a full correspondence with a well known common viewpoint, according to which any arithmetical operation with uncertainties must increase total uncertainty (and entropy) of system. Therefore, placing the degenerated intervals in the right hand side of (14) would be equivalent to the request of reducing uncertainty of the left hand sides down to zero, which is possible only in the case of inverse character of the interval $[x]$, which in turn can be interpreted as a request to introduce negative entropy into the system.

The standard interval extension of Eq. (16) is $[\underline{a}x,\overline{a}x]-[\underline{b},\overline{b}]=0$. With use of interval arithmetic rules (10) and (11) from this equation we obtain $[\underline{a}x-\overline{b},\overline{a}x-\underline{b}]=0$ and finally

$$\underline{x}=\frac{\overline{b}}{\underline{a}},\overline{x}=\frac{\underline{b}}{\overline{a}}$$

It is easy to see that in any case $\underline{x}>\overline{x}$, i.e., we obtain an invert interval. Obviously, such solution may be considered only as absurd one. We can say that this fact is the direct consequence of that conventional interval extension of Eq. (16) is in contradiction with the basic assumptions of interval analysis since the right hand side of this equation always is degenerated zero, whereas left hand side is represented by interval. Summarizing, we can say that only the Eq. (15) can be considered as reasonable base for interval extension. On the other hand, Eqs. (18) often results in drastic extension of output interval in comparison with the input interval (see Example 3, where width of the resulting interval $[x]$ is almost 3 times greater than width of initial interval $[a]$). It is easy to see that there is no way to improve interval solution we obtain from interval extensions of Eq. (14) and Eq. (15). That is why let us turn to the consideration of equation Eq. (16) and look at it from another point of view. Formally, when extending equation Eq. (16) one obtains not only interval on its left hand side, but interval zero on the right hand side. In general case this interval zero cannot be degenerated interval $[0,0]$. In conventional interval analysis (Jaulin et al. 2001), it is assumed that any interval containing zero may be considered as "interval zero".This is a satisfactory definition to suppress the division by zero in conventional interval arithmetic, but for our purposes a more restrictive definition is needed. It is easy to see that for any interval $[a]$ from basic definitions (9) and (11) we get $[\underline{a},\overline{a}]-[\underline{a},\overline{a}]=[\underline{a}-\overline{a},\overline{a}-\underline{a}]=[-(\overline{a}-\underline{a}),\overline{a}-\underline{a}]$. Therefore, in any case the result of interval subtraction $[a]-[a]$ is an interval symmetrical in respect to 0 . Thus, if we want to treat a result of subtraction of two identical intervals as "interval zero", the most general definition of such "zero" will be "interval zero is an interval symmetrical with respect to $0$". It must be emphasized that introduced definition says nothing about the width of "interval zero". Hence as the result of interval extension of Eq. (16) in general case we get

$$[\underline{a},\overline{a}][\underline{x},\overline{x}]-[\underline{b},\overline{b}]=[-y,y]. \tag{19}$$

In fact, the right hand side of Eq. (19) is some interval centered around zero which can be treated as interval extension of right hand side of Eq. (16) or, in other words, as interval extension of 0 . This is the reason for us to call our approach "interval extended zero" method. Of course, the value of $y$ in Eq. (19) is not yet defined. This seems to be quite natural since the values of $\underline{x},\overline{x}$ are also not defined. For the sake of

114    P. Sewastjanow and L. Dymowa

simplicity let us consider only a case of positive interval numbers $[a]$ and $[b]$, i.e. $\underline{a}, \overline{a}, \underline{b}, \overline{b} > 0$.

Then from Eq. (19) we get

$$\begin{cases} \underline{a}x - \overline{b} = -y \\ \overline{a}x - \underline{b} = y. \end{cases} \qquad (20)$$

Finally, summing the left hand sides and the right ones in Eqs. (20) we obtain only one linear equation with two unknown variables $\underline{x}$ and $\overline{x}$:

$$\underline{a}x + \overline{a}x - \underline{b} - \overline{b} = 0 \qquad (21)$$

If there are some restrictions on the values of unknown variables, then Eq. (21) with these restrictions may be considered as a so called Constraint Satisfaction Problem (Cleary 1987) and an interval solution may be obtained. The first restriction on the variables $\underline{x}$ and $\overline{x}$ is a solution of Eq. (21) assuming $\underline{x} = \overline{x}$.

In this degenerated case we get the solution of Eq. (21) as $x_m = \dfrac{\underline{b} + \overline{b}}{\underline{a} + \overline{a}}$.

It is easy to see that $x_m$ is an upper bound for $\underline{x}$ and a lower bound for $\overline{x}$: if $\underline{x} > x_m$ or $\overline{x} < x_m$ we get a degenerated solution of Eq. (21), i.e. $\underline{x} > \overline{x}$.

The natural lower bound for $\underline{x}$ and upper bond for $\overline{x}$ may be derived from basic definition of interval arithmetic (9) as $\underline{x} > \dfrac{\underline{b}}{\overline{a}}, \overline{x} < \dfrac{\overline{b}}{\underline{a}}$. Eq. (21) with all restrictions on $\underline{x}$ and $\overline{x}$ defined above is a typical constraint satisfaction problem (Cleary 1987) and its interval solution is

$$\underline{x} = \frac{\underline{b} + \overline{b} - \overline{a} \cdot \overline{x}}{\underline{a}} \cap \left[ \frac{\underline{b}}{\overline{a}}, \frac{\underline{b} + \overline{b}}{\underline{a} + \overline{a}} \right] \mid \overline{x} \in \left[ \frac{\underline{b} + \overline{b}}{\underline{a} + \overline{a}}, \frac{\overline{b}}{\underline{a}} \right]$$

$$\overline{x} = \frac{\underline{b} + \overline{b} - \underline{a} \cdot \underline{x}}{\overline{a}} \cap \left[ \frac{\underline{b} + \overline{b}}{\underline{a} + \overline{a}}, \frac{\overline{b}}{\underline{a}} \right] \mid \underline{x} \in \left[ \frac{\underline{b}}{\overline{a}}, \frac{\underline{b} + \overline{b}}{\underline{a} + \overline{a}} \right]. \qquad (22)$$

Eqs. (22) result in following interval solutions:

$$[\underline{x}] = \left[ \frac{\underline{b}}{\overline{a}}, \frac{\underline{b} + \overline{b}}{\underline{a} + \overline{a}} \right] \qquad (23)$$

$$[\overline{x}] = \left[ \frac{\underline{b} + \overline{b}}{\underline{a} + \overline{a}}, \frac{\underline{b} + \overline{b}}{\overline{a}} - \frac{\underline{a} \cdot \overline{b}}{\overline{a}^2} \right] \qquad (24)$$

On the Fuzzy Internal Rate of Return   115

Eqs. (23) and (24) define all possible solutions of Eq. (21). The values $\underline{x}_{min} = \dfrac{\underline{b}}{\underline{a}}, \overline{x}_{max} = \dfrac{\overline{b+b}}{\overline{a}} - \dfrac{\overline{a \cdot b}}{\overline{a}^{-2}}$ constitute interval $[\underline{x}, \overline{x}] = \left[ \dfrac{\underline{b}}{\underline{a}}, \dfrac{\overline{b+b}}{\overline{a}} - \dfrac{\overline{a \cdot b}}{\overline{a}^{-2}} \right]$ which derives the widest interval zero after its substitution into Eqs. (20).

In other words, the maximum interval solution width $w_{max} = \dfrac{\overline{b}}{\overline{a}} - \dfrac{\overline{a \cdot b}}{\overline{a}^{-2}}$ corresponds the maximum value of $y$: $y_{max} = \overline{b} - \dfrac{\overline{a \cdot b}}{\overline{a}}$.

Substitution of degenerated solution $\underline{x} = \overline{x} = x^*$ in Eq. (17) produces the minimum value of $y$: $y_{min} = \dfrac{\overline{a} \cdot \overline{b} - \underline{a} \cdot \underline{b}}{\underline{a} + \overline{a}}$.

It is clear that for any permissible solution $\underline{x}' > \underline{x}_{min}$ we have corresponding $\overline{x}' < \overline{x}_{max}$, for each $\underline{x}'' > \underline{x}'$ the inequalities $\overline{x}'' < \overline{x}'$ and $y'' < y'$ take place. We can see that values of $y$ characterize the closeness of right hand side of Eq. (19) to degenerated zero and that minimum value $y_{min}$ is defined exclusively by interval parameters $[a]$ and $[b]$. Hence, the values of $y$ may be considered, in a certain sense, as a measure of interval solution uncertainty caused by the initial uncertainty of Eq. (19).

Therefore we introduce

$$\alpha = 1 - \frac{y - y_{min}}{y_{max} - y_{min}} \qquad (25)$$

which may be connected with each permissible solution of Eq. (19), $[\underline{x}, \overline{x}]$. We can see that $\alpha$ rises from 0 to 1 with decreasing of interval's width from maximum value to 0, i.e. with increasing of solution's certainty. Consequently, the values of $\alpha$ may be treated as the labels of $\alpha$-cuts representing some fuzzy solution of interval Eq. (19). Finally, we obtain a solution as triangular fuzzy number

$$\tilde{x} = \left\{ \frac{\underline{b}}{\underline{a}}, \frac{\overline{b+b}}{\underline{a}+\overline{a}}, \frac{\overline{b+b}}{\overline{a}} - \frac{\overline{a \cdot b}}{\overline{a}^{-2}} \right\} \qquad (26)$$

illustrated in Fig. 3.

This result needs some comments.Using proposed approach to the solution of interval linear equation based on the Eq. (19) we obtain a triangular fuzzy number with a support which in all cases is included in a crisp interval obtainable from conventional interval extension of Eq. (15).

At first glance, it seems somewhat surprising to have a fuzzy solution of the crisp interval equation. But, on the other hand, the "interval extended zero" method is

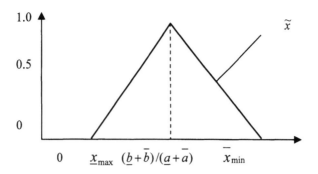

**Fig. 3.** Fuzzy interval solution, $\tilde{x}$, of Eq. (16)

based on some restricting assumptions. The main assumption is an introduction of an "interval zero" as the interval symmetrical with respect to 0. As a consequence we obtain a solution in a form of triangular fuzzy number, which is, in any case, more certain result than crisp interval representing its support, since it inherits more information about possible real valued solutions. Obviously, we can assume the support of obtained fuzzy number to be a solution of analyzed problem. This solution may be called as the "pessimistic" one since it corresponds to the lowest $\alpha$-level of resulting fuzzy value. We use here the word "pessimistic" to emphasize that this solution is charged with the largest imprecision as it is obtained in the most uncertain conditions possible on the set of considered $\alpha$-levels. On the other hand, it seems natural to utilize all additional information available in the fuzzy solution.

We can reduce the resulting fuzzy solution to the interval solution using well known defuzzification procedures. In our case defuzzified left and right solution's boundaries can be represented as

$$\underline{x}_{def} = \frac{\int_0^1 \underline{x}(\alpha) d\alpha}{\int_0^1 d\alpha}$$
$$\overline{x}_{def} = \frac{\int_0^1 \overline{x}(\alpha) d\alpha}{\int_0^1 d\alpha}. \tag{27}$$

From (20) and (25) we get the expressions for $\underline{x}(\alpha)$ and $\overline{x}(\alpha)$. Substituting them into Eqs. (27) we have

$$\underline{x}_{def} = \frac{\overline{b}}{\underline{a}} - \frac{y_{max} - y_{min}}{2\underline{a}}, \quad \overline{x}_{def} = \frac{\underline{b}}{\overline{a}} + \frac{y_{max} + y_{min}}{2\overline{a}}. \tag{28}$$

It is easy to prove that obtained interval $[\underline{x}_{def}, \overline{x}_{def}]$ is always included into support interval of initial fuzzy solution, i.e., $[\underline{x}_{def}, \overline{x}_{def}] \subset [\underline{x}_{min}, \overline{x}_{max}]$.

Useful for our further analysis characteristic of fuzzy solution $\tilde{x}$ is its mode $x_m = \dfrac{\overline{b}+\underline{b}}{\underline{a}+\overline{a}}$ (see Fig. 3 and expression (26)). It was shown above that $x_m$ is a degenerated solution of Eq. (19), i.e., $\underline{x} = \overline{x} = x_m$. On the other hand, $x_m$ is an asymptotic solution when the intervals $[a]$ and $[b]$ contract to the points. Thus, $x_m$ plays in fuzzy solution a role similar to a middle point in usual crisp interval. To illustrate, let us consider simple example: $[a] = [14,60], [b] = [25,99]$. In conventional interval arithmetic, Eq. (18), gives us $[\underline{x}, \overline{x}] = [0.42, 7.07]$. Using new method we get $[\underline{x}_{min}, \overline{x}_{max}] = [0.42, 1.97]$ and $[\underline{x}_{def}, \overline{x}_{def}] = [1.04, 1.82]$. It is easy to see that $[\underline{x}_{def}, \overline{x}_{def}] \subset [\underline{x}_{min}, \overline{x}_{max}] \subset [\underline{x}, \overline{x}]$. Moreover, the width of $[\underline{x}, \overline{x}]$ is in 4.3 times greater than that of $[\underline{x}_{min}, \overline{x}_{max}]$ and in 8.5 times greater than that of $[\underline{x}_{def}, \overline{x}_{def}]$. Thus, the new method provides the considerable reducing of resulting interval's width in comparison with that obtained using conventional interval arithmetic rules. The method presented in this Section may be considered as a general framework to solve not only linear interval equations and sets of such equations, but interval and fuzzy non-linear equations as well. That is why, it will be used for solving the Interval and Fuzzy *IRR* estimation problems considered in following Sections.

## 3  Fuzzy *IRR* for the Crisp Interval Cash Flows-Basics

To make the main idea more transparent, let us consider a simplified example of a one-year project, when a real valued investment $KV$ is made at the beginning of the first year and production starts right away. An interval profit $P = [\underline{P}, \overline{P}]$ is earned in the end of the first year and then the project is finished. In this case Eq. (2) transforms into following form:

$$\frac{P}{1+d} - KV = 0. \tag{29}$$

Rewriting Eq. (29) as:

$$d = \frac{P}{KV} - 1 \tag{30}$$

We get the simplest, but the widest (see Section 2) interval solution for $d$ :

$$[d] = [\overline{d}, \underline{d}] = \left[ \frac{\underline{P}}{KV} - 1, \frac{\overline{P}}{KV} - 1 \right]. \tag{31}$$

Using "interval extended zero" method described in previous Section we get an interval extension of Eq. (29) as follows

$$\frac{[\underline{P}, \overline{P}]}{1+[\underline{d}, \overline{d}]} - KV = [-y, y] \tag{32}$$

With a help of interval analysis rules (10)-(13), interval equation (32) may be represented in form

$$\left[\frac{\underline{P}}{1+\overline{d}}-KV,\frac{\overline{P}}{1+\underline{d}}-KV\right]=[-y,y] \tag{33}$$

and finally as

$$\begin{cases}\dfrac{\underline{P}}{1+\overline{d}}-KV=-y\\[2mm]\dfrac{\overline{P}}{1+\underline{d}}-KV=y\end{cases} \tag{34}$$

From equations (34) we obtain an explicit dependence between bounds of interval $[d]$:

$$\overline{d}=\frac{\underline{P}}{2\cdot KV-\dfrac{\overline{P}}{1+\underline{d}}}-1 \tag{35}$$

On the other hand, the original Eq. (29) can be rewritten in algebraically equivalent form

$$P-KV(1+d)=0 \tag{36}$$

With a help of "interval extended zero" method we get from Eq. (36):

$$[\underline{P},\overline{P}]-KV(1+[\underline{d},\overline{d}])=[-y,y] \tag{37}$$

and after some simple transformations

$$\overline{d}=\frac{\underline{P}+\overline{P}}{KV}-2-\underline{d} \tag{38}$$

It is easy to see that in both cases when using representation (32) or (37) for the degenerated solution we have $\underline{d}=\overline{d}=\dfrac{\underline{P}+\overline{P}}{2\cdot KV}-1$. Also taking into account the widest possible interval solution (31) we conclude that in both cases the next restrictions are verifying

$$\underline{d}\in\left[\frac{\underline{P}}{KV}-1,\frac{\underline{P}+\overline{P}}{2\cdot KV}-1\right],\overline{d}\in\left[\frac{\underline{P}+\overline{P}}{2\cdot KV}-1,\frac{\overline{P}}{KV}-1\right] \tag{39}$$

Taking into account the natural restriction $0\le\underline{d}\le\overline{d}$ (the values of *IRR* should be positive, but generally this restriction is not obligatory) we can solve Eq. (32) in interval form for $\underline{d}$ and $\overline{d}$ in a framework of Constraint Satisfaction Problem (Cleary 1987) as follows:

$$[\underline{d}] = \left( \frac{\overline{P}}{2 \cdot KV - \dfrac{P}{1+\overline{d}}} - 1 \right) \cap \left[ \max\left(0, \frac{P}{KV} - 1\right), \frac{P + \overline{P}}{2 \cdot KV} - 1 \right] \, |$$

$$| \, \overline{d} \in \left[ \frac{P + \overline{P}}{2 \cdot KV} - 1, \frac{\overline{P}}{KV} - 1 \right]$$

$$= \left( \frac{P}{2 \cdot KV - \dfrac{\overline{P}}{1+\underline{d}}} - 1 \right) \cap \left[ \frac{P + \overline{P}}{2 \cdot KV} - 1, \frac{\overline{P}}{KV} - 1 \right] \, |$$  (40)

$$| \, \underline{d} \in \left[ \max\left(0, \frac{P}{KV} - 1\right), \frac{P + \overline{P}}{2 \cdot KV} - 1 \right].$$

As a result we get from (40) following interval solutions

$$[\underline{d}] = \left[ \frac{\overline{P}}{KV \cdot \left( 2 - \dfrac{P}{\overline{P}} \right)} - 1, \frac{P + \overline{P}}{2 \cdot KV} - 1 \right], [\overline{d}] = \left[ \frac{P + \overline{P}}{2 \cdot KV} - 1, \frac{\overline{P}}{KV} - 1 \right]$$  (41)

Finally, using the approach described in Section 2 we get a fuzzy solution of Eq. (32) based on solution (41), i.e., the fuzzy $IRR$ as

$$\tilde{d} = \left\{ \frac{\overline{P}}{KV \cdot \left( 2 - \dfrac{P}{\overline{P}} \right)} - 1, \frac{P + \overline{P}}{2 \cdot KV} - 1, \frac{\overline{P}}{KV} - 1 \right\}$$  (42)

In a similar way from equations (37)-(39) and restriction $0 \leq \underline{d} \leq \overline{d}$ we obtain

$$[\underline{d}] = \left[ \max\left\{ 0, \frac{P}{KV} - 1 \right\}, \frac{P + \overline{P}}{2 \cdot KV} - 1 \right]$$

$$= \left[ \frac{P + \overline{P}}{2 \cdot KV} - 1, \min\left\{ \frac{P + \overline{P}}{KV} - 2, \frac{\overline{P}}{KV} - 1 \right\} \right]$$  (43)

$$\tilde{d} = \left\{ \max\left\{ 0, \frac{P}{KV} - 1 \right\}, \frac{P + \overline{P}}{2 \cdot KV} - 1, \min\left\{ \frac{P + \overline{P}}{KV} - 2, \frac{\overline{P}}{KV} - 1 \right\} \right\}$$  (44)

Some illustrative examples are presented in Fig. 4.

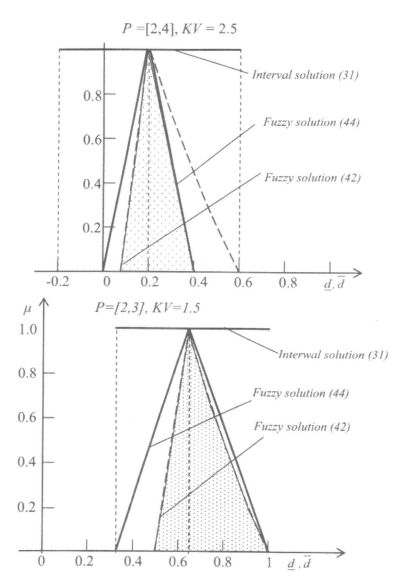

**Fig. 4.** Examples of fuzzy-interval solutions of interval equations (32) and (37): $\mu$ is a membership function of fuzzy value

Fig.4. Examples of fuzzy-interval solutions of interval equations (32) and (37): $\mu$ is a membership function of fuzzy value.

Thus, different but algebraically equivalent representations (29) and (37) of an equation for *IRR* give us after interval extension, different fuzzy solutions (42) and

(44). Basically, this result is not new, since it is well known (Moore 1966) that different representations of some original algebraic expression often derive different numerical results after interval extension. Now this circumstance is not considered as the drastic drawback, but rather as a specific feature inherent in applied interval analysis (Jaulin et al. 2001). That is why it seems quite natural in situations when we have some different interval or fuzzy solutions of some original problem to use an intersection of obtained solutions as the most reliable solution of considered problem. Such intersections are performed in Fig. 4 by the shaded regions.

## 4 A Numerical Solution of the Non-linear Fuzzy Problem of *IRR* Calculation

As it was shown in previous Section, the fuzzy *IRR* problem can be formulated with use of fuzzy extension of two different, but algebraically equivalent representations of initial non-fuzzy Eq.(2):

$$\sum_{t=1}^{T} \frac{P_t}{(1+d)^t} - KV = 0 \tag{45}$$

$$\sum_{t=1}^{T} P_t (1+d)^{T-t} - KV(1+d)^T = 0 \tag{46}$$

It is shown above that generally the solution of fuzzy *IRR* problem can be obtained in three phases:

- Fuzzy extension of above equations and $\alpha$-level representation of obtained fuzzy equations with use of "interval extended zero" method.
- Numerical solution of obtained two sets of interval equations. As the result, the different fuzzy values of *IRR* corresponding to the fuzzy extensions of equations (45) and (46) must be obtained in this phase.
- Calculating the resulting fuzzy *IRR* as the intersection of the fuzzy valued *IRR*s obtained in previous phase from the fuzzy extension of equations (45) and (46).

Proposed numerical algorithm for solving the fuzzy *IRR* problem is not too complicated, but its detailed description seems as rather cumbersome one. On the other hand, in previous Section we have shown that there may be only a bit difference between algorithms for solving the fuzzy extensions of equations (45) and (46). Therefore, to illustrate the proposed approach we restrict themselves only with the consideration of the algorithm for solving the fuzzy extensions of Eq.(46), as its description is shorter than that for fuzzy extensions of Eq. (45). In further analysis, we assume that *KV* is usual non-fuzzy value. As it is shown in Section 2, this assumption is not a simplification: quite contrary. In addition, it corresponds to the reality as usually the start investment *KV* is well known. Since the phase 1 is obvious, we start from phase 2. In this phase, on each $\alpha$-level in the framework of "interval extended zero" method we deal with interval equation (index $\alpha$ is omitted for the simplicity)

$$\sum_{t=1}^{T}\left[\underline{P}_t,\overline{P}_t\right]\left[1+\underline{d},1+\overline{d}\right]^{T-t} - KV\left[1+\underline{d},1+\overline{d}\right]^T = [-y,y] \tag{47}$$

As a result of transformation of the Eq. (47) according to rules of interval analysis we have

$$\left[\sum_{t=1}^{T}\underline{P}_t(1+\underline{d})^{T-t} - KV(1+\overline{d})^T, \sum_{t=1}^{T}\overline{P}_t(1+\overline{d})^{T-t} - KV(1+\underline{d})^T\right] \tag{48}$$
$$= [-y,y]$$

and

$$\sum_{t=1}^{T}\underline{P}_t(1+\underline{d})^{T-t} - KV(1+\overline{d})^T = -y, \sum_{t=1}^{T}\overline{P}_t(1+\overline{d})^{T-t} - KV(1+\underline{d})^T = y \tag{49}$$

The sum of the equations (49) gives an expression connecting the unknown left $\underline{d}$ and right $\overline{d}$ boundaries of $IRR$:

$$\sum_{t=1}^{T}\underline{P}_t(1+\underline{d})^{T-t} - KV(1+\overline{d})^T + \sum_{t=1}^{T}\overline{P}_t(1+\overline{d})^{T-t} - KV(1+\underline{d})^T = 0 \tag{50}$$

To obtain a fuzzy solution of Eq. (50), according to the results of Section 3 the natural restriction $0 \le \underline{d} \le \overline{d}$ should be used. The restriction allows to get a simple nonlinear equation with respect to $\overline{d}$ for each fixed value of $\underline{d}$. This equation can be solved using well-known numerical methods.

Finally, as in the linear case (see Section 3) the resulting fuzzy interval representation of $IRR$ with corresponding membership function is obtained.

When dealing with fuzzy cash flows, the Eq. (46) is represented as a set of $\alpha$-cuts. For each $\alpha$-cut an interval equation in a form of the Eq. (50) is obtained. Since the set of $\alpha$-levels is only an approximate representation of fuzzy number (as we deal with the method of discretization), the precision of final result depends on the number of $\alpha$-levels. In practice, this number is usually choosing when analyzing the shapes of used fuzzy values: if we deal with complicated shapes the number of $\alpha$-levels is rising.

Thus,the fuzzy problem is reduced to the crisp interval one. The algorithm for its solving is following:

- By partition into $n$ $\alpha$-cuts the fuzzy interval problem (Eq. (46)) is transformed into a set of crisp interval equations.
- On each $\alpha$-cut the crisp intervals $[\underline{P}_1,\overline{P}_1]\alpha,[\underline{P}_2,\overline{P}_2]\alpha,[\underline{P}_3,\overline{P}_3]\alpha,...,[\underline{P}_T,\overline{P}_T]\alpha$ are calculated.
- For each $\alpha$-cut the maximum value $\underline{d}_\alpha$ is calculated by solving the non-linear equation (50) in the assumption $\underline{d}_\alpha = \overline{d}\alpha$. Substituting obtained values of

On the Fuzzy Internal Rate of Return   123

$\underline{d}_\alpha = \overline{d}\alpha$ into one of the expressions (49) on observed $\alpha$ -cut the corresponding $y_{\alpha\min}$ is obtained.

- Substituting the $\underline{d}_{\alpha min} = 0$ in (50) the corresponding $\overline{d}_{\alpha max}$ are obtained. Substituting the $\underline{d}_{\alpha min}$, $\overline{d}_{\alpha max}$ in (49) we get the values of $y_{\alpha \max}$ .

- For each $\alpha$ -cut the interval of possible values $\underline{d}_\alpha = [0, \overline{d}\alpha]$ is divided into $m$ equal parts and $\underline{d}_{\alpha i}, i = 1, \ldots, m$ ($d_{1\alpha 0} = 0$, $\underline{d}_{\alpha m} = \overline{d}\alpha$) are calculated. Using $\underline{d}_{\alpha i}$ in (46) corresponding values $\overline{d}_{\alpha i}$ are obtained. Then using $\underline{d}_{\alpha i}$ and $\overline{d}_{\alpha i}$ in (45) the values $y_{\alpha i}$ are obtained. Finally, using expression (22) the values $\eta_{\alpha i}$ characterizing the degrees of membership of the crisp intervals $[\underline{d}_{\alpha i}, \overline{d}_{\alpha i}]$ in resulting fuzzy-interval solution of (45) on $\alpha$ -cuts are calculated.

To illustrate, let us consider a project with a crisp initial investment $KV = 1$ (one million of some monetary units) and trapezoidal fuzzy positive cash flows expected for the next three years and presented by their bottom and top $\alpha$ -cuts (see Table 1). The boundaries of intermediate $\alpha$ -cuts are obtained using simple interpolation. The result obtained with use of numerical algorithm for our example is presented in Fig. 5.

**Table 1.** Sample project

| Year | Bottom $\alpha$ -cuts of fuzzy Cash Flows | Top $\alpha$ -cuts of fuzzy Cash Flows |
|------|------|------|
| 1 | $[\underline{P}_1, \overline{P}_1] = [1,2]$ | $[\underline{P}_1, \overline{P}_1] = [1.1, 1.5]$ |
| 2 | $[\underline{P}_2, \overline{P}_2] = [1.5, 3]$ | $[\underline{P}_2, \overline{P}_2] = [1.6, 2]$ |
| 3 | $[\underline{P}_3, \overline{P}_3] = [1,2]$ | $[\underline{P}_3, \overline{P}_3] = [1.2, 1.5]$ |

Fig.5 shows that on each $\eta_{\alpha i}$ -cut we have $IRR = [\underline{d}_{\alpha i}, \overline{d}_{\alpha i}]$ being in turn the intervals on $\alpha$ -cuts. It seems to be natural for intervals $IRR = [\underline{d}_{\alpha i}, \overline{d}_{\alpha i}]$ with upper values of $\alpha$ to contribute more to the final fuzzy interval solution.

Hence, for defuzzification of obtained results it is possible to use, for example, following expressions:

$$\underline{d}_i = \frac{\sum\limits_{\alpha=0}^{n} \underline{d}_{\alpha i}\alpha}{\sum\limits_{\alpha=0}^{n} \alpha}, \overline{d}_i = \frac{\sum\limits_{\alpha=0}^{n} \overline{d}_{\alpha i}\alpha}{\sum\limits_{\alpha=0}^{n} \alpha}. \tag{51}$$

The result of defuzzification of $IRR$ using expressions (51) for considered example is shown in Fig. 6, where the result obtained for considered example with

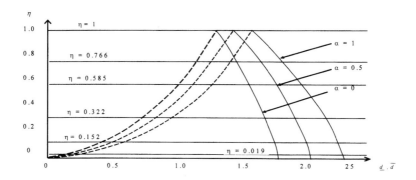

**Fig. 5.** Fuzzy *IRR* for the three-year investment project

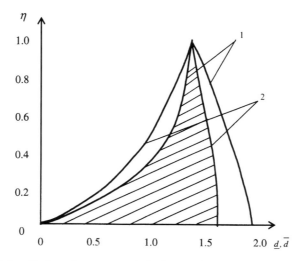

**Fig. 6.** *IRR* defuzzified on the $\alpha$-cuts: 1, 2 -the results obtained from fuzzy extensions (45) and (46) respectively

use of fuzzy extension of Eq. (45) is presented as well. It is shown in Section 3 that final solution may be obtained as the intersection of these fuzzy solutions. It is seen that such intersection is a narrowed final solution of fuzzy *IRR* problem.

## 5 Possible Applications

There may be different possible applications of fuzzy *IRR* in budgeting, but here we briefly describe the three more obvious ones.

- If several investment projects must be compared in regard to their fuzzy *IRR* s, the problem of fuzzy values comparison arises. There exist numerous definitions of ordering relation over fuzzy values (as well as crisp intervals) in literature.Usually the authors use some quantitative indices. The values of such indices present degree to which one interval or fuzzy number is greater/less than another one. The widest review of the problem of fuzzy values ordering based on more than 35 literature indices has been presented in (Wang and Kerre 2001), where the authors proposed a new interesting classification of methods for fuzzy number comparison. Nevertheless, using the real valued indices (usual numbers) for comparison instead of original fuzzy values leads inevitable to the loss of important information. That is why, we advise to use the methods based on so called probabilistic approach to fuzzy and interval value comparison (Wadman et al 1994, Sevastianov and Rog 2003, Yager et al. 2001) which makes it possible to compare fuzzy and interval values as a whole without any preliminary defuzzification. Recently the complete set of probabilistic interval and fuzzy relations involving separated equality and inequality relations and comparisons of real numbers with intervals and fuzzy values have been presented in (Sevastianov and Rog 2003, Sevastianov and Rog 2006). So if $IRR_1$ and $IRR_2$ are fuzzy valued $IRR$ s of two comparing projects, the next probabilities may be calculated $P(IRR_1 > IRR_2)$, $P(IRR_1 = IRR_2)$, $P(IRR_1 < IRR_2)$. Since in practice, the routine task is the comparison of $IRR$ (fuzzy value in our case) with known (non fuzzy) bank discount rate, it is important that these relations allow for comparing the fuzzy and interval values with usual numbers.

- Often the decision makers prefer to analyze instead of fuzzy $IRR$ its interval representation. This can be justified psychologically since the left and right bounds of interval valued $IRR$ are naturally treated as the pessimistic and optimistic assessments of $IRR$ respectively. An interval representation of fuzzy $IRR$, i.e., $[IRR_1, IRR_2]$ can be calculated using expressions

$$IRR_1 = \frac{\sum\limits_{\alpha=0}^{n} \alpha IRR_{\alpha 1}}{\sum\limits_{\alpha=0}^{n} \alpha}, IRR_2 = \frac{\sum\limits_{\alpha=0}^{n} \alpha IRR_{\alpha 2}}{\sum\limits_{\alpha=0}^{n} \alpha}$$

Of course, the problems of interval comparison and comparison of intervals with usual values arise, but they can be solved using above mentioned probabilistic methods.

- It is well known that fuzzy numbers can be characterized by the measure of its fuzziness (Yager 1979). The measure of fuzziness is directly connected with uncertainty measure, which in financial applications is usually treated as the measure of risk. Hence we can adopt an uncertainty measure of fuzzy $IRR$ as the assessment of risk caused by the decision we made considering the fuzzy $IRR$ as the criterion.

Let $\tilde{A}$ be fuzzy value and $A$ be rectangular fuzzy value defined on the support of $\tilde{A}$ and represented by the characteristic function $\eta_A(x) = 1, x \in A; \eta_A(x) = 0, x \notin A$. Obviously, such rectangular value is not a fuzzy value at all, but it is asymptotic limit (object) we obtain when fuzziness of $\tilde{A}$ tends to zero. Hence, it seems quite natural to define a measure of fuzziness of $\tilde{A}$ as its distinction from $A$. To do this we define primarily the measure of non fuzziness as

$$MNF(\tilde{A}) = \int_0^1 f(\alpha)((A_{\alpha 2} - A_{\alpha 1})/(A_{02} - A_{01}))d\alpha$$

where $f(\alpha)$ is some function of $\alpha$, e.g., $f(\alpha) = 1$ or $f(\alpha) = \alpha$. Of course, last expression makes sense only for the fuzzy or interval values, i.e., only for non zero width of support $A_{02} - A_{01}$. It is easy to see that if $\tilde{A} \to A$ then $MNF(\tilde{A}) \to 1$. Obviously, the measure of fuzziness can be defined as $MF(\tilde{A}) = 1 - MNF(\tilde{A})$.

We can say that rectangular value $A$ defined on the support of $\tilde{A}$ is a more uncertain object than $\tilde{A}$. Really, only what we know about $A$ is that all $x \in A$ belong to $A$ with equal degrees, whereas the membership function, $0 \le \mu(x) \le 1$, characterizing the fuzzy value $\tilde{A}$ brings more information to the description and as a consequence, represents a more certain object. Therefore, we can treat the measure of non fuzziness, $MNF$, as the uncertainty measure. Hence, if some decision is made concerning the fuzzy $IRR$, the uncertainty and, consequently, the risk of such decision can be calculated as $MNF(IRR)$. Since the fuzzy $IRR$ and correspondent risk assessment can play the role of local criteria, the multiple criteria problem arises. Generally, an investment evaluation is a multiple criteria task. The methods for its solving in a fuzzy setting are discussed in (Dimova et al. 2006).

# 6 Conclusion

The aim of the present paper is to analyze the problem of calculation of $IRR$ in fuzzy setting and to introduce a new numerical method for obtaining the $IRR$ directly as fuzzy value. The only way to obtain a consistent and well-understandable result is to use a method based on reducing the initial fuzzy interval problem to a set of crisp interval tasks on its $\alpha$-cuts. It is clear that such approach implies the use of numeric methods, which is inevitable if initial form of fuzzy intervals is complex enough. A new method for solving interval equations on $\alpha$-cuts allows to obtain a restricted fuzzy-interval solution of crisp interval problem. The problem of $IRR$ estimation in a fuzzy setting is considered in a framework of more general problem of fuzzy equations solving. The general method for solving of interval equations based on a new approach, named "interval extended zero" method, is elaborated. The method is illustrated with numerical example.

The method proposed can be be employed for the solution of linear as well as non-linear systems of interval and fuzzy-interval equations.

Utilization of the method allows to obtain a result in a form of the fuzzy number (interval) with corresponding membership function.

Fuzzy-interval solution can be used directly or can be defuzzified as well, if the crisp final result is needed. It must be emphasized that for practical purposes the most useful approach is a reduction of fuzzy $IRR$ to some crisp intervals by means of defuzzification. The width of such interval can serve as a natural measure of risk connected with using of *IRR* in financial analysis in fuzzy setting.

# References

Abbasbandy, S., Asady, B.: Newton's method for solving fuzzy nonlinear equations. Applied Mathematics and Computation 159, 349–356 (2004)

Belletante, B., Arnaud, H.: Choisir ses investissements. Chotar et Assosies Editeurs, Paris (1989)

Babusiaux, D., Pierru, A.: Capital budgeting, project valuation and financing mix: Methodological proposals. Europian Journal of Operational Research 135, 326–337 (2001)

Bogle, H.F., Jehenck, G.K.: Investment analysis: us oil and gas producers score high in university survey. In: Proceedings of Hydrocarbon Economics and Evaluation Symposium, Dallas, pp. 234–241 (1985)

Brigham, E.F.: Fundamentals of financial management. Dryden Press, New York (1992)

Buckley, J.J.: The fuzzy mathematics of finance. Fuzzy Sets and Systems 21, 257–273 (1987)

Buckley, J.J., Qu, Y.: Solving linear and quadratic fuzzy equations. Fuzzy Sets and Systems 38, 43–59 (1990)

Buckley, J.J., Eslami, E.: Neural net solutions to fuzzy problems: The quadratic equation. Fuzzy Sets and Systems 86, 289–298 (1997)

Buckley, J.J., Eslami, E., Hayashi, Y.: Solving fuzzy equations using neural nets. Fuzzy Sets and Systems 86, 271–278 (1997)

Calzi, M.L.: Towards a general setting for the fuzzy mathematics of finance. Fuzzy Sets and Systems 35, 265–280 (1990)

Chansa-ngavej, C., Mount-Campbell, C.A.: Decision criteria in capital budgeting under uncertainties: implications for future research. International Journal of Production Economics 23, 25–35 (1991)

Caprani, O., Madsen, K.: Mean value forms in interval analysis. Computing 25, 147–154 (1980)

Chen, S.: An empirical examination of capital budgeting techniques: impact of investment types and firm characteristics. Engineering Economist 40, 145–170 (1995)

Chiu, C.Y., Park, C.S.: Fuzzy cash flow analysis using present worth criterion. Engineering Economist 39, 113–138 (1994)

Choobineh, F., Behrens, A.: Use of intervals and possibility distributions in economic analysis. Journal of Operational Reserach Society 43, 907–918 (1992)

Cleary, J.C.: Logical Arithmetic. Future Computing Systems 2, 125–149 (1987)

Dimova, L., Sevastianov, P., Sevastianov, D.: MCDM in a fuzzy setting: investment projects assessment application. International Journal of Production Economics 100, 10–29 (2006)

Dimova, L., Sevastianov, D., Sevastianov, P.: Application of fuzzy sets theory, methods for the evaluation of investment efficiency parameters. Fuzzy economic review 5, 34–48 (2000)

Gardnes, E., Mielgo, H., Trepat, A.: Modal intervals: Reasons and ground semantics. In: Nickel, K. (ed.) Interval mathematics 212. LNCS. Springer, Berlin (1985)

Hansen, E.: A generalized interval arithmetic. In: Nickel, K. (ed.) Interval Mathematics. LNCS, vol. 29. Springer, Heidelberg (1975)

Jaulin, L., Kieffir, M., Didrit, O., Walter, E.: Applied Interval Analysis. Springer, London (2001)

Kahraman, C., Tolga, E., Ulukan, Z.: Justification of manufacturing technologies using fuzzy benefit/cost ratio analysis. International Journal of Production Economics 66, 45–52 (2000)

Kahraman, C., Ulukan, Z.: Continous compounding in capital budgeting using fuzzy concept. In: Proceedings of 6th IEEE International Conference on Fuzzy Systems (FUZZ-IEEE 1997), Bellaterra (1997a)

Kahraman, C., Ulukan, Z.: Fuzzy cash flows under inflation. In: Proceedings of 7th IEEE International Fuzzy Systems Association World Congress (IFSA 1997), Univeristy of Economics, Prague Czech Republic Bellaterra (1997b)

Kahraman, C.: Fuzzy versus probabilistic benefit/cost ratio analysis for public work projects. Int. J. Appl. Math. Comp. Sci. 11, 705–718 (2001)

Kahraman, C., Ruan, D., Tolga, E.: Capital budgeting techniques using discounted fuzzy versus probabilistic cash flows. Information Sciences 142, 57–76 (2002)

Kaufmann, A., Gupta, M.: Introduction to fuzzy-arithmetic theory and applications. Van Nostrand Reinhold, New York (1985)

Kuchta, D.: Fuzzy capital budgeting. Fuzzy Sets and Systems 111, 367–385 (2000)

Liang, P., Song, F.: Computeraided risk evaluation system for capital investment. Omega 22, 391–400 (1994)

Markov, S.M.: A non-standard subtraction of intervals. Serdica 3, 359–370 (1977)

Moore, R.E.: Interval analysis. Prentice-Hall, Englewood Cliffs (1966)

Perrone, G.: Fuzzy multiple criteria decision model for the evaluation of AMS. Comput. Integrated Manufacturing Systems 7, 228–239 (1994)

Wadman, D., Schneider, M., Schnaider, E.: On the use of interval mathematics in fuzzy expert system. International Journal of intelligent Systems 9, 241–259 (1994)

Wang, X., Kerre, E.E.: Reasonable properties for the ordering of fuzzyquantities (I)-(II). Fuzzy Sets and Systems 112, 375–385, 387-405 (2001)

Ward, T.L.: Discounted fuzzy cash flow analysis. In: Proceedings of Fall Industrial Engineering Conference, pp. 476–481 (1985)

Sendov, B.: Segment arithmetic and segment limit. C.R. Acad. Bulgare Sci. 30, 958–995 (1977)

Sevastianov, P., Sevastianov, D.: Risk and capital budgeting parameters evaluation from the fuzzy sets theory position. Reliable software 1, 10–19 (in Russian) (1997)

Sevastianov, P., Dimova, L., Zhestkova, E.: Methodology of the multicriteria quality estimation and its software realizing. In: Proceedings of the Fourth International Conference on New Information Technologies 'NITe', Minsk, pp. 50–54 (2000)

Sevastianov, P., Rog, P.: A probabilistic approach to fuzzy and interval ordering. Artificial and Computational Intelligence 7, 147–156 (2003)

Sevastianov, P., Rog, P.: Fuzzy modeling of manufacturing and logistic systems. Mathematics and Computers in Simulation 63, 569–585 (2003)

Sevastianov, P., Rog, P.: Two-objective method for crisp and fuzzy interval comparison in optimization. Computers and Operation Research 33(1), 115–131 (2006)

Yager, R.A.: On the measure of fuzziness and negation. Part 1. Membership in the Unit Interval. Int. J. Gen. Syst. 5, 221–229 (1979)

Yager, R.R., Detyniecki, M., Bouchon -Meunier, B.: A context-dependent method for ordering fuzzy numbers using probabilities. Information Sciences 138, 237–255 (2001)

Zadeh, L.A.: Fuzzy sets. Inf. Control 8, 338–353 (1965)

Zimmermann, H.J., Zysno, P.: Latest connectives in human decision making. Fuzzy Sets and Systems 4, 37–51 (1980)

# Fuzzy Benefit/Cost Analysis and Applications

Cengiz Kahraman and İhsan Kaya

Istanbul Technical University, Department of Industrial Engineering,
34367, Macka, İstanbul, Turkey

**Abstract.** Benefit-cost analysis is a systematic evaluation of the economic advantages (benefits) and disadvantages (costs) of a set of investment alternatives. The objective of a benefit-cost analysis is to translate the effects of an investment into monetary terms and to account for the fact that benefits generally accrue over a long period of time while capital costs are incurred primarily in the initial years. In this chapter, benefit cost ratio (BCR) analysis is analyzed under fuzzy environment. Fuzzy continuous payments and fuzzy discrete payments are summarized briefly and two applications are illustrated.

## 1 Introduction

Benefit-Cost Analysis is an economic tool to aid social decision-making, and is typically used by governments to evaluate the desirability of a given intervention in markets. The aim is to gauge the efficiency of the intervention relative to the status quo. The costs and benefits of the impacts of an intervention are evaluated in terms of the public's willingness to pay for them (benefits) or willingness to pay to avoid them (costs). Inputs are typically measured in terms of opportunity costs - the value in their best alternative use. The guiding principle is to list all of the parties affected by an intervention, and place a monetary value of the effect it has on their welfare as it would be valued by them.

During benefit-cost analysis, monetary values may also be assigned to less tangible effects such as the various risks which could contribute to partial or total project failure; loss of reputation, market penetration, long-term enterprise strategy alignments, etc. This is especially true when governments use the technique, for instance to decide whether to introduce business regulation, build a new road or offer a new drug on the state healthcare. In this case, a value must be put on human life or the environment, often causing great controversy. The benefit-cost principle says, for example, that we should install a guardrail on a dangerous stretch of mountain road if the dollar cost of doing so is less than the implicit dollar value of the injuries, deaths, and property damage thus prevented. Benefit-cost calculations typically involve using time value of money formula. This is usually done by converting the future expected streams of costs and benefits to a present value amount.

The purpose of benefit cost analysis is to give management 'a reasonable picture of the costs, benefits and risks associated with a given project so that it can be compared to other investment opportunities' (Davis 1999). Benefit cost analysis has been traditionally applied to fields including policies, programs, projects, regulations, demonstrations and other government interventions (Boardman et al. 2001). Many capital

---

C. Kahraman (Ed.): Fuzzy Engineering Economics with Appl., STUDFUZZ 233, pp. 129–143, 2008.
springerlink.com © Springer-Verlag Berlin Heidelberg 2008

investment projects (Murphy and Simon 2001) such as budget planning, dams and airports construction, and safety and environmental programs planning have adopted the CBA approach to compare the costs and benefits. Similar to LCCA, the CBA technique relies on the NVP method as the basis for analysis (Boardman et al. 2001, Davis 1999, Murphy and Simon 2001). In the use of the CBA technique, attention should be paid to the following aspects. First, project costs and benefits at each stage of the life cycle are discounted into the present values. Second, some studies suggest that information such as development and operating costs, and tangible benefits by time period must be obtained before performing the CBA (Davis 1999).

The benefit-cost ratio can be defined as the ratio of the equivalent value of benefits to the equivalent value of costs. The equivalent values can be present values, annual values, or future values. The benefit-cost ratio (BCR) is formulated as

$$BCR = \frac{B}{C} \tag{1}$$

where $B$ represents the equivalent value of the benefits associated with the project and $C$ represents the project's net cost (Blank and Tarquin 1989). A $B/C$ ratio greater than or equal to 1.0 indicates that the project evaluated is economically advantageous.

In $B/C$ analyses, costs are not preceded by a minus sign. The objective to be maximized behind the $B/C$ ratio is to select the alternative with the largest net present value or with the largest net equivalent uniform annual value, because $B/C$ ratios are obtained from the equations necessary to conduct an analysis on the incremental benefits and costs. Suppose that there are two mutually exclusive alternatives. In this case, for the incremental $BCR$ analysis ignoring disbenefits the following ratios must be used:

$$\frac{\Delta B_{2-1}}{\Delta C_{2-1}} = \frac{\Delta PVB_{2-1}}{\Delta PVC_{2-1}} \tag{2}$$

or

$$\frac{\Delta B_{2-1}}{\Delta C_{2-1}} = \frac{\Delta EUAB_{2-1}}{\Delta EUAC_{2-1}} \tag{3}$$

where $\Delta B_{2-1}$ is the incremental benefit of Alternative 2 relative to Alternative 1, $\Delta C_{2-1}$ is the incremental cost of Alternative 2 relative to Alternative 1, $\Delta PVB_{2-1}$ is the incremental present value of benefits of Alternative 2 relative to Alternative1, $\Delta PVC_{2-1}$ is the incremental present value of costs of Alternative 2 relative to Alternative 1, $\Delta EUAB_{2-1}$ is the incremental equivalent uniform annual benefits of Alternative 2 relative to Alternative 1 and $\Delta EUAC_{2-1}$ is the incremental equivalent uniform annual costs of Alternative 2 relative to Alternative 2.

Thus, the concept of $B/C$ ratio includes the advantages of both $NPV$ and net $EUAV$ analyses. Because it does not require to use a common multiple of the alternative lives (then $B/C$ ratio based on equivalent uniform annual cash flow is used) and it is a more understandable technique relative to rate of return analysis for many financial managers, $B/C$ analysis can be preferred to the other techniques such as present value analysis, future value analysis, rate of return analysis.

Wilhelm and Parsaei (1991) used a fuzzy linguistic approach to justify a computer integrated manufacturing system. Kahraman et al. (1995) used a fuzzy approach based on the fuzzy present value analysis for the manufacturing flexibility. Buckley (1987), Ward (1985), Chiu and Park (1994), Wang and Liang (1995), Kahraman and Tolga (1995), Kahraman et al. (2002), Kahraman and Bozdağ (2003), Sheen (2005) are among the authors who deal with the fuzzy present worth analysis, the fuzzy benefit/cost ratio analysis, the fuzzy future value analysis, the fuzzy payback period analysis, and the fuzzy capitalized value analysis.

The rest of this chapter is organized as follows. In section 2, fuzzy benefit cost ratio analysis is summarized. For this purpose, discrete and continuous compounding are analyzed in fuzzy environment. Section 3 has two applications about fuzzy discrete and continuous compounding. The last section emphasizes that the results of fuzzy benefit cost ratio analysis.

## 2 Fuzzy Benefit/Cost Ratio Analysis

### 2.1 Discrete Compounding

In the case of fuzziness, the steps of the fuzzy $B/C$ analysis are given in the following:

- *Step 1:* Calculate the overall fuzzy measure of benefit-to-cost ratio and eliminate the alternatives that have

$$\tilde{B}/\tilde{C} = \left( \frac{\sum_{t=0}^{n} B_t^{l(y)}\left(1+r^{r(y)}\right)^{-t}}{\sum_{t=0}^{n} C_t^{l(y)}\left(1+r^{r(y)}\right)^{-t}}, \frac{\sum_{t=0}^{n} B_t^{r(y)}\left(1+r^{l(y)}\right)^{-t}}{\sum_{t=0}^{n} C_t^{l(y)}\left(1+r^{r(y)}\right)^{-t}} \right) \prec \tilde{1} \tag{4}$$

where $\tilde{r}$ is the fuzzy interest rate and $r(y)$ and $l(y)$ are the right and left side representations of the fuzzy interest rates and $\tilde{I}$ is $(1, 1, 1)$, and $n$ is the crisp life cycle.

- *Step 2:* Assign the alternative that has the lowest initial investment cost as the defender and the next lowest acceptable alternative as the challenger.
- *Step 3:* Determine the incremental benefits and the incremental costs between the challenger and the defender.
- *Step 4:* Calculate the $\Delta\tilde{B}/\Delta\tilde{C}$ ratio, assuming that the largest possible value for the cash in year t of the alternative with the lowest initial investment cost is less than the least possible value for the cash in year t of the alternative with the next-lowest initial investment cost.

The fuzzy incremental BCR is

$$\frac{\Delta\tilde{B}}{\Delta\tilde{C}} = \left( \frac{\sum_{t=0}^{n} \left(B_{2t}^{l(y)} - B_{1t}^{r(y)}\right)\left(1+r^{r(y)}\right)^{-t}}{\sum_{t=0}^{n} \left(C_{2t}^{r(y)} - C_{1t}^{l(y)}\right)\left(1+r^{l(y)}\right)^{-t}}, \frac{\sum_{t=0}^{n} \left(B_{2t}^{r(y)} - B_{1t}^{l(y)}\right)\left(1+r^{l(y)}\right)^{-t}}{\sum_{t=0}^{n} \left(C_{2t}^{l(y)} - C_{1t}^{r(y)}\right)\left(1+r^{r(y)}\right)^{-t}} \right) \tag{5}$$

If $\Delta\tilde{B}/\Delta\tilde{C}$ is equal or greater than $(1, 1, 1)$, Alternative 2 is preferred.

132     C. Kahraman and İ. Kaya

In the case of a regular annuity, the fuzzy $\tilde{B}/\tilde{C}$ ratio of a single investment alternative is

$$\tilde{B}/\tilde{C} = \left( \frac{A^{l(y)}\gamma\left(n,r^{r(y)}\right)}{C^{r(y)}}, \frac{A^{r(y)}\gamma\left(n,r^{l(y)}\right)}{C^{l(y)}} \right) \tag{6}$$

where $\tilde{C}$ is the first cost and $\tilde{A}$ is the net annual benefit, and $\gamma(n,r) = \left(\left((1+r)^n - 1\right)/(1+r)^n r\right)$.

The $\Delta\tilde{B}/\Delta\tilde{C}$ ratio in the case of a regular annuity is

$$\Delta\tilde{B}/\Delta\tilde{C} = \left( \frac{\left(A_2^{l(y)} - A_1^{r(y)}\right)\gamma\left(n,r^{r(y)}\right)}{C_2^{r(y)} - C_1^{l(y)}}, \frac{\left(A_2^{r(y)} - A_1^{l(y)}\right)\gamma\left(n,r^{l(y)}\right)}{C_2^{l(y)} - C_1^{r(y)}} \right) \tag{7}$$

- *Step 5:* Repeat steps 3 and 4 until only one alternative is left, thus the optimal alternative is obtained.

The cash-flow set $\{A_t = A : t = 1,2,...,n\}$ consisting of $n$ cash flows, each of the same amounts $A$, at times $1, 2,..., n$ with no cash flow at time zero, is called the equal-payment series. An older name for it is the uniform series, and it has been called an annuity, since one of the meanings of "annuity" is a set of fixed payments for a specified number of years. To find the fuzzy present value of a regular annuity $\{\tilde{A}_t = \tilde{A} : t = 1,2,...,n\}$ we will use Eq. 8. The membership function $\mu\left(x|\tilde{P}_n\right)$ for $\tilde{P}_n$ is determined by

$$f_{ni}\left(y|\tilde{P}_n\right) = f_i\left(y|\tilde{A}\right)\gamma\left(n, f_{3-i}\left(y|\tilde{r}\right)\right) \tag{8}$$

for i=1,2 and $\gamma(n,r) = \dfrac{\left(1 - (1+r)^{-n}\right)}{r}$. Both $\tilde{A}$ and $\tilde{r}$ are positive fuzzy numbers. $f_1(.)$ and $f_2(.)$ shows the left and right representations of the fuzzy numbers, respectively.

In the case of a regular annuity, the fuzzy $\tilde{B}/\tilde{C}$ ratio may be calculated as in the following:

The fuzzy $\tilde{B}/\tilde{C}$ ratio of a single investment alternative is

$$\tilde{B}/\tilde{C} = \left( \frac{A^{l(y)}\gamma\left(n,r^{r(y)}\right)}{FC^{r(y)}}, \frac{A^{r(y)}\gamma\left(n,r^{l(y)}\right)}{FC^{l(y)}} \right) \tag{9}$$

where $F\tilde{C}$ is the first cost and $\tilde{A}$ is the net annual benefit.

The $\Delta\tilde{B}/\Delta\tilde{C}$ ratio in the case of a regular annuity is

$$\Delta\tilde{B}/\Delta\tilde{C} = \left( \frac{\left(A_2^{l(y)} - A_1^{r(y)}\right)\gamma\left(n,r^{r(y)}\right)}{FC_2^{r(y)} - FC_1^{l(y)}}, \frac{\left(A_2^{r(y)} - A_1^{l(y)}\right)\gamma\left(n,r^{l(y)}\right)}{FC_2^{l(y)} - FC_1^{r(y)}} \right) \tag{10}$$

Up to this point, we assumed that the alternatives had equal lives. When the alternatives have life cycles different from the analysis period, a common multiple of the alternative lives (*CMALs*) is calculated for the analysis period. Many times, a CMALs for the analysis period hardly seems realistic (*CMALs (7, 13) =91 years*). Instead of an analysis based on present value method, it is appropriate to compare the annual cash flows computed for alternatives based on their own service lives. In the case of unequal lives, the following fuzzy $\tilde{B}/\tilde{C}$ and $\Delta\tilde{B}/\Delta\tilde{C}$ ratios will be used:

$$\tilde{B}/\tilde{C} = \left( \frac{PVB^{l(y)}\beta(n,r^{l(y)})}{PVC^{r(y)}\beta(n,r^{r(y)})}, \frac{PVB^{r(y)}\beta(n,r^{r(y)})}{PVC^{l(y)}\beta(n,r^{l(y)})} \right) \tag{11}$$

$$\Delta\tilde{B}/\Delta\tilde{C} = \left| \begin{array}{c} \left( \dfrac{PVB_2^{l(y)}\beta(n,r^{l(y)}) - PVB_1^{r(y)}\beta(n,r^{r(y)})}{PVC_2^{r(y)}\beta(n,r^{r(y)}) - PVC_1^{l(y)}\beta(n,r^{l(y)})} \right), \\ \left( \dfrac{PVB_2^{r(y)}\beta(n,r^{r(y)}) - PVB_1^{l(y)}\beta(n,r^{l(y)})}{PVC_2^{l(y)}\beta(n,r^{l(y)}) - PVC_1^{r(y)}\beta(n,r^{r(y)})} \right) \end{array} \right| \tag{12}$$

where PVB is the present value of benefits, PVC the present value of costs and

$$\beta(n,r) = \left( \frac{(1+n)^n i}{(1+r)^n - 1} \right)$$

## 2.2 Continuous Compounding

### 2.2.1 Fuzzy Continuous Payments

Now, the future value of a continuous and uniform flow of the fuzzy $\tilde{A}k$ over year k will be obtained. Consider Fig. 1.

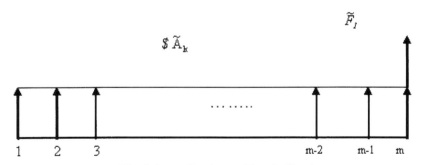

**Fig. 1.** Fuzzy Continuous Flow in Year k

$$\tilde{F}_1 = \frac{\tilde{A}_k}{M}\left(F/\tilde{A},\tilde{r}/m\%,m\right) \tag{13}$$

which reduces to

$$\tilde{F}_1 = \tilde{A}_k\left[\left(1+\frac{\tilde{r}}{m}\right)^m - 1\right] \tag{14}$$

where "$\tilde{r}$" is the fuzzy nominal interest rate. Letting "$m$" approach infinity yields,

$$F_1 = \tilde{A}_k \frac{(e^{\tilde{r}} - 1)}{\tilde{r}} \qquad (15)$$

The left and right representation of the fuzzy numbers $\tilde{A}_k$ and $\tilde{r}$ can be used in Eq. 15. Assume $\tilde{A}$ and $\tilde{r}$ are positive. Then Eq. 16 is obtained.

$$f_i^c(y|\tilde{F}_1) = f_i(y|\tilde{A}_k) \left[ \frac{(e^{f_i(y|\tilde{r})} - 1)}{f_i(y|\tilde{r})} \right] \qquad (16)$$

where $i=1, 2$. The letter $c$ above $f$ represents continuous payments.

Assume $\tilde{A}_k$ is negative and $\tilde{r}$ is positive. Then Eq. 17 is obtained.

$$f_i^c(y|\tilde{F}_1) = f_i(y|\tilde{A}_k) \left[ \frac{(e^{f_{3-i}(y|\tilde{r})} - 1)}{f_{3-i}(y|\tilde{r})} \right] \qquad (17)$$

where i=1, 2.

The formula for determining the fuzzy continuous compounding compound amount with continuous uniform payments can be given as in Eq. 18 and Eq. 19 or Eq. 20. Consider Figure 2.

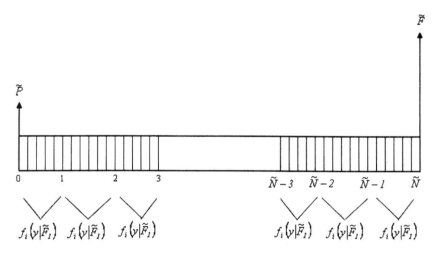

**Fig. 2.** Cash Flow Diagram for Continuous Compounding Uniform Cash Flow factors

If $\tilde{F}_1$, $\tilde{r}$, and $\tilde{N}$ are positive,

$$f_i^c(y|\tilde{F}_N) = f_i(y|\tilde{F}_1) \left[ \frac{1 - e^{f_i(y|\tilde{r})f_i(y|\tilde{N})}}{1 - e^{f_i(y|\tilde{r})}} \right] \qquad (18)$$

Fuzzy Benefit/Cost Analysis and Applications    135

or

$$f_i^c\left(y\middle|\tilde{F}_N\right)= f_i\left(y\middle|\tilde{A}_k\right)\left[\frac{e^{f_i\left(y\middle|\tilde{r}\right)f_i\left(y\middle|\tilde{N}\right)}-1}{f_i\left(y\middle|\tilde{r}\right)}\right] \tag{19}$$

where i=1, 2.

If $\tilde{A}_k$ is negative and $\tilde{r}$ and $\tilde{N}$ are positive,

$$f_i^c\left(y\middle|\tilde{F}_N\right)= f_i\left(y\middle|\tilde{A}_k\right)\left[\frac{e^{f_{3-i}\left(y\middle|\tilde{r}\right)f_{3-i}\left(y\middle|\tilde{N}\right)}-1}{f_{3-i}\left(y\middle|\tilde{r}\right)}\right] \tag{20}$$

where i=1, 2.

Similarly, the formula for determining the fuzzy continuous compounding present worth of continuous uniform payments can be given as in Eq. 21 or Eq. 23.

$$f_i^c\left(y\middle|\tilde{P}_N\right)= \frac{f_i\left(y\middle|\tilde{A}_k\right)}{f_{3-i}\left(y\middle|\tilde{r}\right)}\left[\frac{e^{f_{3-i}\left(y\middle|\tilde{r}\right)f_i\left(y\middle|\tilde{N}\right)}-1}{e^{f_{3-i}\left(y\middle|\tilde{r}\right)f_i\left(y\middle|\tilde{N}\right)}}\right] \tag{21}$$

where i=1, 2 and $\tilde{A}_k$, $\tilde{r}$ and $\tilde{N}$ are all positive. Assume that the fuzzy present value of continuous payments is obtained by Eq. (21). Then the fuzzy B/C ratio for continuous payments is calculated using Eq. (22):

$$\tilde{B}/\tilde{C} =\left(\frac{{}^c P_N{}^{l(y)}}{FC^{r(y)}},\frac{{}^c P_N^{r(y)}}{FC^{l(y)}}\right) \tag{22}$$

For negative uniform cash flows, we use Eq. (23) for B/C ratio:

$$f_i^c\left(y\middle|\tilde{P}_N\right)= \frac{f_i\left(y\middle|\tilde{A}_k\right)}{f_i\left(y\middle|\tilde{r}\right)}\left[\frac{e^{f_i\left(y\middle|\tilde{r}\right)f_{3-i}\left(y\middle|\tilde{N}\right)}-1}{e^{f_i\left(y\middle|\tilde{r}\right)f_{3-i}\left(y\middle|\tilde{N}\right)}}\right] \tag{23}$$

where i=1, 2 and $\tilde{A}_k$ is negative, $\tilde{r}$ and $\tilde{N}$ are positive.

### 2.2.2 Fuzzy Discrete Payments
The formula for determining the fuzzy compound amount that can be obtained in $N$ years under continuous compounding from a fuzzy present amount of $P$ can be given as in Eq. 24 or Eq. 25.

$$f_i\left(y\middle|\tilde{F}_N\right)= f_i\left(y\middle|\tilde{P}\right)e^{f_i\left(y\middle|\tilde{r}\right)f_i\left(y\middle|\tilde{N}\right)} \tag{24}$$

where i=1, 2 and $\tilde{P}$, $\tilde{r}$ and $\tilde{N}$ are all positive.

$$f_i\left(y\middle|\tilde{F}_N\right)= f_i\left(y\middle|\tilde{P}\right)e^{f_{3-i}\left(y\middle|\tilde{r}\right)f_{3-i}\left(y\middle|\tilde{N}\right)} \tag{25}$$

where i=1, 2 and $\tilde{P}$ is negative and $\tilde{r}$ and $\tilde{N}$ are positive.

The formula for determining the fuzzy discrete uniform series continuous compounding compound amount can be given as in Eq. 26 or Eq. 27.

$$f_i^d\left(y|\tilde{F}_N\right)= f_i\left(y|\tilde{A}\right)\left[\frac{e^{f_i\left(y|\tilde{r}\right)f_i\left(y|\tilde{N}\right)}-1}{e^{f_i\left(y|\tilde{r}\right)}-1}\right]$$

(26)

where i=1, 2 and $\tilde{A}$ is the fuzzy discrete uniform-series and $\tilde{A}$, $\tilde{r}$ and $\tilde{N}$ are all positive. The letter $d$ above $f$ represents discrete payments.

$$f_i^d\left(y|\tilde{F}_N\right)= f_i\left(y|\tilde{A}\right)\left[\frac{e^{f_{3-i}\left(y|\tilde{r}\right)f_{3-i}\left(y|\tilde{N}\right)}-1}{e^{f_{3-i}\left(y|\tilde{r}\right)}-1}\right]$$

(27)

where i=1, 2 and $\tilde{A}$ is the fuzzy discrete uniform-series and $\tilde{A}$ is negative, $\tilde{r}$ and $\tilde{N}$ are positive.

The formula for determining the present worth of fuzzy discrete uniform-series under continuous compounding can be given as in Eq. 28 or Eq. 29.

$$f_i^d\left(y|\tilde{P}_N\right)= \frac{f_i\left(y|\tilde{A}\right)}{e^{f_{3-i}\left(y|\tilde{r}\right)f_i\left(y|\tilde{N}\right)}}\left[\frac{e^{f_{3-i}\left(y|\tilde{r}\right)f_i\left(y|\tilde{N}\right)}-1}{e^{f_{3-i}\left(y|\tilde{r}\right)}-1}\right]$$

(28)

where i=1, 2 and $\tilde{A}$, $\tilde{r}$ and $\tilde{N}$ are all positive. Assume that the fuzzy present value of discrete payments is obtained by Eq. (27). Then the fuzzy B/C ratio for discrete payments is calculated using Eq. (28):

$$\tilde{B}/\tilde{C} =\left(\frac{d\,P_N^{l(y)}}{FC^{r(y)}},\frac{d\,P_N^{r(y)}}{FC^{l(y)}}\right)$$

(29)

For negative uniform cash flows, we use Eq. (29) for B/C ratio:

$$f_i^d\left(y|\tilde{P}_N\right)= \frac{f_i\left(y|\tilde{A}\right)}{e^{f_i\left(y|\tilde{r}\right)f_{3-i}\left(y|\tilde{N}\right)}}\left[\frac{e^{f_i\left(y|\tilde{r}\right)f_{3-i}\left(y|\tilde{N}\right)}-1}{e^{f_i\left(y|\tilde{r}\right)}-1}\right]$$

(30)

where i=1, 2 and $\tilde{A}$ is negative, $\tilde{r}$ and $\tilde{N}$ are all positive.

## 2.3 Ranking Methods

There are a number of methods that are devised to rank mutually exclusive projects such as Chang's method (1981), Jain's method (1976), Dubois and Prade's method (1983), Yager's method (1980), Baas and Kwakernaak's method (1977). However, certain shortcomings of some of the methods have been reported in (Bortolan 1985, Chen 1985, Kim 1990). Because the ranking methods might give different ranking results, they must be used together to obtain the true rank. Chiu and Park (1994) compare some ranking methods by using a numerical example and determine which methods give the same or very close results to one another. In Chiu and Park's (1994) paper, several dominance methods are selected and discussed. Most methods are tedious in graphic manipulation requiring complex mathematical calculation. Chiu and

Park's weighting method and Kaufmann and Gupta's (1988) three criteria method are the methods giving the same rank for the considered alternatives and these methods are easy to calculate and require no graphical representation. Therefore, these two methods will be explained in the following and will be used in the application section. Chiu and Park's (1994) weighted method for ranking TFNs with parameters (a, b, c) is formulated as

$$((a+b+c)/3)+wb \qquad (31)$$

where *w* is a value determined by the nature and the magnitude of the most promising value.

Kaufmann and Gupta (1988) suggest three criteria for ranking TFNs with parameters *(a, b, c)*. The dominance sequence is determined according to priority of:

1. Comparing the ordinary number *(a+2b+c)/4*,
2. Comparing the mode (the corresponding most promising value), b, of each TFN,
3. Comparing the range, *c-a*, of each TFN.

The preference of projects is determined by the amount of their ordinary numbers. The project with the larger ordinary number is preferred. If the ordinary numbers are equal, the project with the larger corresponding most promising value is preferred. If projects have the same ordinary number and most promising value, the project with the larger range is preferred.

## 3 Applications

In this hypothetical application, two assembly manufacturing systems will be considered, namely a transfer machine with robot workhead (RW) and a robot assembly cell (RAC). Figs. 3 and 4 show these two manufacturing systems.

The robot assembly cell uses one robot. This sophisticated robot has six degrees of freedom. Tables 1 and 2 give the data for RAC and RW, respectively. The tax rate (TR) is 40% and is not fuzzy. The crisp life cycles are 7 years for RAC and 5 years for RW. The fuzzy discount rate is (15%, 20%, 25%) per year.

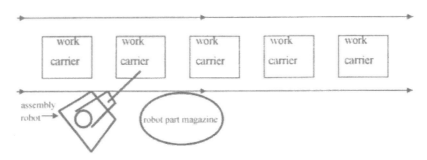

**Fig. 3.** RW transfer machine with robot workhead

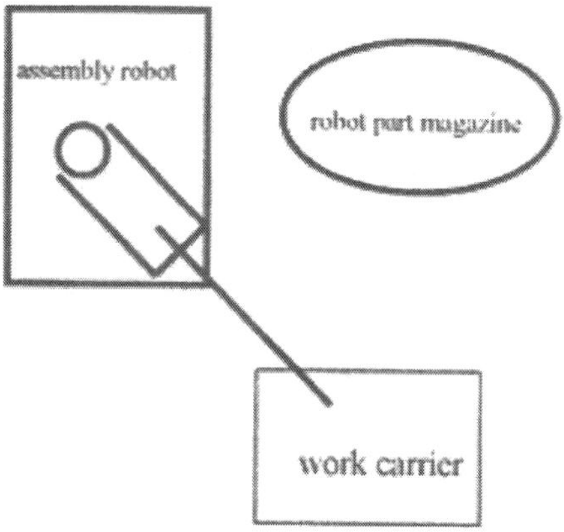


**Fig. 4.** RAC -robot assembly cell

**Table 1.** The fuzzy data for RAC

| Year, t | Revenue($) $R\tilde{E}V_t$(*10³) | Labor Cost ($) $L\tilde{A}B_t$(*10³) | Investment($) $\tilde{I}_t$ (*10³) | Depreciation($) $D\tilde{E}P_t$(*10³) | After tax cash flow ($) $ATC\tilde{F}_t$ (*10³) |
|---|---|---|---|---|---|
| 1 | (180,200,220) | (60,70,80) | (270,280,290) | (50,60,70) | (-210,-178,-146) |
| 2 | (200,210,220) | (65,75,85) | (7,8,9) | (45,50,55) | (78,93,108) |
| 3 | (220,230,240) | (70,80,90) | (7,8,9) | (35,40,45) | (83,98,113) |
| 4 | (320,330,340) | (105,115,125) | (25,30,40) | (35,40,45) | (91,115,134) |
| 5 | (350,360,370) | (115,120,125) | (10,15,20) | (30,35,40) | (127,143,161) |
| 6 | (390,400,410) | (130,135,140) | (300,310,320) | (80,85,90) | (-138,-117,-96) |
| 7 | (425,430,435) | (140,150,160) | (20,25,30) | (70,75,80) | (157,-173,189) |

**Table 2.** The Calculated $ATC\tilde{F}_t$ for RW

| Year, t | 1 | 2 | 3 | 4 | 5 |
|---|---|---|---|---|---|
| $ATC\tilde{F}_t$ (*10³ $) | (-90, -128, -94) | (109, 130, 151) | (116, 137, 158) | (127, 161, 187) | (178, 200, 225) |

To fuzzify the data for RAC and RW, it is assumed that the model parameters are triangular fuzzy numbers. The parameters like the average time for the robot to assemble a part, the part quality, the average downtime at a station due to a defective part, the number of production hours per shift, the plant efficiency, the cost of a robot, the cost of a part feeder, the cost of a robot gripper, the annual cost of a supervisor have been accepted as triangular fuzzy numbers. For example, the average time of the RAC to complete an assembly *i*, is used while calculating the number of shifts and the annual labour cost. It is formulated by

$$A\tilde{T}_i = N\tilde{P}_i \otimes \left\lfloor A\tilde{P} \oplus P\tilde{Q} \otimes A\tilde{D} \right\rfloor \tag{32}$$

Fuzzy Benefit/Cost Analysis and Applications     139

where $N\tilde{P}_i$ is the fuzzy number of parts in each assembly $i$, $A\tilde{P}$ the fuzzy average time (in seconds) for the robot to assemble one part, $P\tilde{Q}$ the fuzzy part quality (the fraction of defective parts to good parts), and $A\tilde{D}$ the fuzzy average downtime at a station due to a defective part.

The other details are not given in this chapter but the calculated results are.

In year 6, a new robot assembly cell is required. So, $ATC\tilde{F}6 < 0$. While calculating after-tax cash flow, the following formula has been used:

$$ATC\tilde{F}_t = \left(RE\tilde{V}_t - LA\tilde{B}_t\right)\left(1-TR\right) \oplus DE\tilde{P}_t \times TR - \tilde{I}_t \tag{33}$$

For example, (-210,-178,-146) is obtained as in the following way:

$$\left(RE\tilde{V}_t - LA\tilde{B}_t\right) = (180, 200, 220) - (60,70,80) = (100,130,160)$$
$$\left(1-TR\right) = 1.00 - 0.40 = 0.60$$
$$\left(RE\tilde{V}_t - LA\tilde{B}_t\right)\left(1-TR\right) = (60,78,96)$$
$$\left(DE\tilde{P}_t \times TR\right) = (50,60,70) \times 0,40 = (20,24,28)$$
$$(20,24,28) - (270,280,290) = (-270,-256,-242)$$
$$(60,78,96) \oplus (-270,-256,-242) = (-210,-178,-146)$$

Now, using Eq. 11, we will calculate $\tilde{B}/\tilde{C}$ ratios for each alternative. First, let us calculate $PV\tilde{B}_{RAC}$, $PV\tilde{B}_{RW}$, $PV\tilde{C}_{RAC}$, and $PV\tilde{C}_{RW}$.

$$PV\tilde{B}_{RAC} = \left(\left(\frac{15y+78}{(1.25-0.05y)^2} + \frac{15y+83}{(1.25-0.05y)^3} + \frac{24y+91}{(1.25-0.05y)^4} + \frac{16y+127}{(1.25-0.05y)^5} + \frac{16y+157}{(1.25-0.05y)^7}\right),\right.$$
$$\left.\left(\frac{108-15y}{(1.15+0.05y)^2} + \frac{113-15y}{(1.15+0.05y)^3} + \frac{134-19y}{(1.15+0.05y)^4} + \frac{161-18y}{(1.15+0.05y)^5} + \frac{189-16y}{(1.15+0.05y)^7}\right)\right)$$

$$PV\tilde{B}_{RW} = \left(\left(\frac{32y-210}{1.15+0.05y} + \frac{-21y-117}{(1.15+0.05y)^6}\right),\left(\frac{-32y-146}{1.25-0.05y} + \frac{-21y-96}{(1.25-0.05y)^6}\right)\right)$$

$$PV\tilde{C}_{RAC} = \left(\left(\frac{21y+109}{(1.25-0.05y)^2} + \frac{21y+116}{(1.25-0.05y)^3} + \frac{34y+127}{(1.25-0.05y)^4} + \frac{22y+178}{(1.25-0.05y)^5}\right),\right.$$
$$\left.\left(\frac{151-21y}{(1.15+0.05y)^2} + \frac{158-21y}{(1.15+0.05y)^3} + \frac{187-26y}{(1.15+0.05y)^4} + \frac{225-25y}{(1.15+0.05y)^5}\right)\right)$$

$$PV\tilde{C}_{RW} = \left(\frac{62y-190}{1.15+0.05y}, \frac{-34y-94}{1.15+0.05y}\right)$$

The approximate forms of $PV\tilde{B}_{RAC}$, $PV\tilde{B}_{RW}$, $PV\tilde{C}_{RAC}$, and $PV\tilde{C}_{RW}$ are founds as in the following taking y=0, 1 and 0, respectively:

$$PV\tilde{B}_{RAC} = \$(204.230, 282.505, 383.675),$$

$$PV\tilde{B}_{RW} = \$(239.498, 327.578, 436.848),$$

$$\left|PV\tilde{C}_{RAC}\right| = \$(153.817, 194.549, 233.191),$$

$$\left|PV\tilde{C}_{RW}\right| = \$(81.739, 106.667, 165.217).$$

$$PV\tilde{B}_{RAC} = (78.275\,y + 204.230, 383.675 - 101.17\,y),$$

$$PV\tilde{B}_{RW} = (88.08\,y + 239.498, 436.848 - 109.27\,y),$$

$$\left|PV\tilde{C}_{RAC}\right| = (40.732\,y + 153.817, 233.191 - 38.642\,y),$$

$$\left|PV\tilde{C}_{RW}\right| = (24.928\,y + 81.739, 165.217 - 58.55\,y)$$

$$V = \left[\frac{(1.15 + 0.05\,y)^5 (0.15 + 0.05\,y)}{(1.15 + 0.05\,y)^5 - 1}\right]$$

$$L = \left[\frac{(1.15 + 0.05\,y)^7 (0.15 + 0.05\,y)}{(1.15 + 0.05\,y)^7 - 1}\right]$$

$$Z = \left[\frac{(1.25 - 0.05\,y)^5 (0.25 - 0.05\,y)}{(1.25 - 0.05\,y)^5 - 1}\right]$$

$$S = \left[\frac{(1.25 - 0.05\,y)^7 (0.25 - 0.05\,y)}{(1.25 - 0.05\,y)^7 - 1}\right]$$

$$\tilde{B}/\tilde{C}_{RAC} = \left(\frac{(204.320 + 78.275\,y)L}{(233.191 - 38.642\,y)S}, \frac{(383.675 - 101.17\,y)S}{(153.817 + 40.732\,y)L}\right)$$

$$\tilde{B}/\tilde{C}_{RW} = \left(\frac{(239.498 + 88.08\,y)V}{(165.217 - 58.55\,y)Z}, \frac{(468.848 - 109.27\,y)Z}{(81.739 + 24.928\,y)V}\right)$$

$$\Delta\tilde{B}/\Delta\tilde{C} = \left(\frac{PVB_{RAC}^{l(y)}L - PVB_{RW}^{r(y)}Z}{\left|PVC_{RAC}^{l(y)}\right|S - \left|PVC_{RW}^{l(y)}\right|V}, \frac{PVB_{RAC}^{r(y)}S - PVB_{RW}^{r(y)}V}{\left|PVC_{RAC}^{l(y)}\right|L - \left|PVC_{RW}^{l(y)}\right|Z}\right)$$

$$\tilde{B}/\tilde{C}_{RAC} = (0.665, 1.452, 3.283)$$

$$\tilde{B}/\tilde{C}_{RW} = (1.163, 3.071, 6.662)$$

$$\Delta\tilde{B} = (-113.352, -31.162, 49.926)$$

$$\Delta\tilde{C} = (-24.464, -18.305, 49.384)$$

$$\Delta\tilde{B}/\Delta\tilde{C} = (-2.295, -1.702, 2.041)$$

$$\Delta\tilde{B}/\Delta\tilde{C} = (a, b, c).$$

We also need fuzzy ranking methods to compare alternatives. According to Kaufmann and Gupta's (1988) ranking method:

$$(a+2b+c)/4 = -0.915$$
$$\tilde{1} = (1,1,1)$$
$$(a+2b+c)/4 = 1$$

$\Delta \tilde{B} / \Delta \tilde{C} < \tilde{1}$ , then the preferred alternative is RW.

According to Chiu and Park's (1994) weighting method (w=0.3):

$$((a+b+c)/3)+wb = -1.16$$
$$\tilde{1} = (1,1,1)$$
$$((a+b+c)/3)+wb = 1.3$$

$\Delta \tilde{B} / \Delta \tilde{C} < \tilde{1}$ , then the preferred alternative is RW.

## 3.2 Application of Fuzzy B/C in Case of Continuous Compounding

Assume that we have the following data for an alternative:

**Table 3.** Alternative Information

| First cost | US \$ (360,000; 370,000; 380,000) |
|---|---|
| Uniform annual continuous benefit | US \$ (56,000; 60,000; 64,000) |
| Useful life, years | (10, 12, 14) |
| Annual continuous interest rate | (5%, 6%, 7%) |

$$f_1^c\left(y|\tilde{P}_N\right) = \frac{(56,000+4,000y)}{(0.05+0.01y)}\left[\frac{e^{(0.05+0.01y)(10+2y)}-1}{e^{(0.05+0.01y)(10+2y)}}\right]$$

For y=0, $f_1^c\left(y|\tilde{P}_N\right) = \frac{(56,000)}{(0.05)}\left[\frac{e^{(0.05)(10)}-1}{e^{(0.05)(10)}}\right] = $ US\$ 440,686

For y=1, $f_1^c\left(y|\tilde{P}_N\right) = \frac{(60,000)}{(0.06)}\left[\frac{e^{(0.06)(12)}-1}{e^{(0.06)(12)}}\right] = $ US\$ 513,248

and

$$f_2^c\left(y|\tilde{P}_N\right) = \frac{(64,000-4,000y)}{(0.07-0.01y)}\left[\frac{e^{(0.07-0.01y)(14-2y)}-1}{e^{(0.07-0.01y)(14-2y)}}\right]$$

For y=0, $f_2^c\left(y|\tilde{P}_N\right) = \frac{(64,000)}{(0.07)}\left[\frac{e^{(0.07)(14)}-1}{e^{(0.07)(14)}}\right] = $ US\$ 571,144

$$\tilde{B}/\tilde{C} = \left( \frac{^cP_N^{l(y)}}{FC^{r(y)}}, \frac{^cP_N^{r(y)}}{FC^{l(y)}} \right) = \left( \frac{440{,}686}{380{,}000}, \frac{513{,}000}{370{,}000}, \frac{571{,}144}{360{,}000} \right)$$

$$= (1.1597, 1.3865, 1.5865)$$

The fuzzy B/C is larger than 1.0. Therefore, this alternative is acceptable.

## 4 Conclusions

This chapter develops fuzzy *B/C* ratio analyses. The fuzzy *B/C* ratio analysis is equivalent to the fuzzy present value analysis. However, fuzzy *B/C* ratio based on equivalent uniform annual benefits and costs has the advantage of comparing alternatives having life cycles different from the analysis period, without calculating a common multiple of the alternative lives. The details of fuzzy *B/C* ratio have been presented in this chapter.

The developed fuzzy *B/C* ratio analysis only takes into account of a quantitative criterion that is the profitability, while justifying a manufacturing technology. To justify a manufacturing technology, generally, a number of quantitative and qualitative criteria have to be considered. In this case, deterministic or nondeterministic multiple criteria methods such as scoring or fuzzy linguistics should be taken into account.

## References

Baas, S.M., Kwakernaak, H.: Rating and ranking multiple aspect alternatives using fuzzy sets. Automatica 13, 47–58 (1977)

Blank, L.T., Tarquin, J.A.: Engineering economy, 3rd edn. McGraw-Hill, Inc., New York (1989)

Boardman, A.E., Greenberg, D.H., Vining, A.R., Weimer, D.L.: Cost– Benefit Analysis: Concepts and Practices, 2nd edn. Prentice Hall, New Jersey (2001)

Bortolan, G., Degani, R.: A review of some methods for ranking fuzzy subsets. Fuzzy Sets and Systems 15, 1–19 (1985)

Buckley, J.J.: The fuzzy mathematics of finance. Fuzzy Sets and Systems 21, 257–273 (1987)

Chang, W.: Ranking of fuzzy utilities with triangular membership functions. In: Proceedings of the International Conference of Policy Anal. and Inf. Systems, pp. 263–272 (1981)

Chen, S.H.: Ranking fuzzy numbers with maximizing and minimizing set. Fuzzy Sets and Systems 17, 113–129 (1985)

Chiu, C., Park, C.S.: Fuzzy cash flow analysis using present worth criterion. The Engineering Economist 39(2), 113–138 (1994)

Davis, W.S.: Cost/benefit analysis. In: Davis, W.S., Yen, D.C. (eds.) The Information System Consultant's Handbook: Systems Analysis and Design. CRC Press, Boca Raton (1999)

Dubois, D., Prade, H.: Ranking fuzzy numbers in the setting of possibility theory. Information Sciences 30, 183–224 (1983)

Jain, R.: Decision-making in the presence of fuzzy variables. IEEE Transactions on Systems, Man, and Cybernetics 6, 693–703 (1976)

Kahraman, C., Bozdağ, C.E.: fuzzy investment analyses using capital budgeting and dynamic programming techniques. In: Chen, S.H., Wang, P.P. (eds.) Computational Intelligence in Economics and Finance. Springer, Heidelberg (2003)

Kahraman, C., Ruan, D., Tolga, E.: Capital budgeting techniques using discounted fuzzy versus probabilistic cash flows. Information Sciences 142, 57–76 (2002)

Kahraman, C., Tolga, E.: The effects of fuzzy inflation rate on after-tax rate calculations. In: third Balkan Conference on Operational Research, Thessaloniki, Greece (1995)

Kahraman, C., Tolga, E., Ulukan, Z.: Fuzzy flexibility analysis in automated manufacturing systems. In: Proceedings of INRIA/IEEE Conference on Emerging Technologies and Factory Automation, Paris (1995)

Kaufmann, A., Gupta, M.M.: Fuzzy mathematical models in engineering and management science. Elsevier, Amsterdam (1988)

Kim, K., Park, K.S.: Ranking fuzzy numbers with index of optimism. Fuzzy sets and Systems 35, 143–150 (1990)

Murphy, K., Simon, S.: Using cost benefit analysis for enterprise resource planning project evaluation: a case of including intangibles. In: Van Grembergen, W. (ed.) Information Technology Evaluation Methods and Management, Idea Group, London (2001)

Sheen, J.N.: Fuzzy financial profitability analyses of demand side management alternatives from participant perspective. Information Sciences 169, 329–364 (2005)

Wang, M.J., Liang, G.S.: Benefit/cost analysis using fuzzy concept. The Engineering Economist 40(4), 359–376 (1995)

Ward, T.L.: Discounted fuzzy cash flow analysis. In: Proceedings of 1985 Fall Industrial Engineering Conference, Institute of Industrial Engineers (1985)

Wilhelm, M.R., Parsaei, H.R.: A fuzzy linguistic approach to implementing a strategy for computer integrated manufacturing. Fuzzy Sets and Systems 42, 191–204 (1991)

Yager, R.R.: On choosing between fuzzy subsets. Kybernetes 9, 151–154 (1980)

# Fuzzy Replacement Analysis

Cengiz Kahraman[1] and Murat Levent Demircan[2]

[1] Istanbul Technical University, Department of Industrial Engineering,
34367, Maçka, İstanbul, Turkey
[2] Galatasaray University, Engineering and Technology Faculty,
34357 Ortakoy Besiktas Istanbul, Turkey

**Abstract.** In an uncertain economic decision environment, our knowledge about the defender's remaining life and its cash flow information usually consist of a lot of vagueness. To describe a planning horizon which may be implicitly forecasted from past incomplete information, a linguistic description like `approximately between 8 and 10 years' is often used. Using fuzzy equivalent uniform annual cash flow analysis, a fuzzy replacement analysis for two operating systems is handled in this chapter.

## 1  Introduction

Enterprises, and even individuals, face technology replacement decisions more frequently as technology upgrades accelerate. Fleischer (1994) cited the reasons of replacement analysis in his milestone book as follows:

- The existing asset can no longer satisfy current or anticipated needs.
- Technology upgrades are available. Improved assets are less expensive, can operate with less maintenance cost and are technologically more efficient.
- The need for the purchased asset no longer exists.
- A major breakdown occurred. Existing asset can no longer operate.

Since late 50's there exist both theoretical and practical efforts about Replacement Analysis: Bellman (1955) based his replacement analysis study to conventional dynamic programming. Dreyfus (1960) referred his ancestor Bellman (1955) and discussed the mathematical formulation of replacement decisions in a more complicated and realistic way. Lake and Muhlemann (1979) developed a simulated replacement model for a particular wrapping machine in food industry in order to analyze the cost sensitivity according to changing resale figures. Their study argued that the weakness of the pure economic life models was that they ignored unpredicted situations before the end of economic life. Oakford et al. (1984) presented Dreyfus and Wagner's dynamic replacement model (1972) on several numerical examples. They stated that computational advances permitted detailed sensitivity analysis on engineering economy issues, but they also expressed that the difficulty of forecasting functional relationships between cash flows of the defender and prospective challengers would well discourage the use of new economic decision approaches.

Lohmann (1986) ameliorated the study of Oakford et al. (1984) with a stochastic approach. The study states the principal difference between deterministic and stochastic treatment of replacement problems. It argued that deterministic models could conclude the optimal sequence of future challengers to be installed after current optimal defender whereas stochastic models could not. The study determined the probability that each prospective challenger was optimal for finite and infinite horizon and it determined the corresponding cumulative distributions of economic life, net present value and equivalent finite horizon time of each alternative, as well. Al-Najjar and Alsyouf (2003) assessed the most popular maintenance approaches using a fuzzy multiple criteria decision making evaluation methodology. Their methodology led to less planned replacements and failure reduction in order to achieve higher utilization of component life.

Zadeh (1965) introduced the fuzzy set theory to deal with the uncertainty due to imprecision and vagueness. A major contribution of fuzzy set theory was its capability of representing vague data. The theory also allowed mathematical operators and programming to apply to the fuzzy domain. A fuzzy set is a class of objects with a continuum of grades of membership. Such a set is characterized by a membership function, which assigns to each object a grade of membership ranging between zero and one.

Fuzzy Set Theory has also been applied to many engineering economic areas: Buckley (1987) developed fuzzy mathematics for compound interest problems. He determined the fuzzy present value and fuzzy future value of fuzzy cash amounts, using fuzzy interest rates. Chiu and Park (1994) developed comprehensive left and right side representation of fuzzy finance. Kahraman et al. (2002, 2003) developed the fuzzy formulations of present value, equivalent uniform annual value, fuzzy future value, fuzzy benefit cost ratio, and fuzzy payback period techniques. Kahraman et al. (2000) and Kahraman (2001a, 2001b) applied fuzzy present worth and fuzzy benefit/cost ratio analyses for the justification of manufacturing technologies and for public work projects. They used triangular fuzzy numbers as illustrated in Figure 1.

Karsak (1998) developed some measures of liquidity risk supplementing fuzzy discounted cash flow analysis. Boussabaine and Elhag (1999) examined the possible application of the fuzzy set theory to the cash flow analysis in construction projects. Dimitrovski and Matos (2000) presented an approach to including non-statistical uncertainties in utility economic analysis by modeling uncertain variables with fuzzy numbers. Kuchta (2000) proposed fuzzy equivalents of all the classical capital budgeting methods.

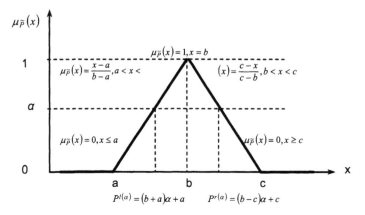

**Fig. 1.** Left and Right Representation of a TFN, $\tilde{P}$

Kai-Yuan and Chuan-Yuan (1990) expressed Campbell's statement on a street-lighting system as that $N$ street-lighting posts had to be provided continually with lampsThey fuzzified this statement such that the system failure was defined in a fuzzy way. Then a concept of fuzzy reliability was introduced in the context of the transition from the system fuzzy success state to the system fuzzy failure state, where a transition between two fuzzy states was just a fuzzy event. With the concept of fuzzy reliability they made a comparison between two replacement policies, e.g. the block replacement policy in a non-fuzzy environment and the periodic replacement policy without repair at lamp failures in a fuzzy environment, and concluded that a complex system might be as reliable as a simple one in view of the system fuzzy reliability.

Esogbue and Hearnes (1998) demonstrated a replacement analysis approach using fuzzy concepts. Baskak and Kahraman (1998) examined a fuzzy replacement analysis for a defender. The life of the defender was given by a membership function that was an intersection of the membership functions regarding the physical impairment of the defender, its obsolescence, and external economic conditions. A fuzzy economic analysis for a replacement decision was also given.

Tolga et al. (2005) developed an Operating System (OS) selection framework for decision makers (DMs). Both economic and non-economic aspects of technology selection were considered in the developed framework. The economic part of the decision process was developed by fuzzy replacement analysis. Non-economic factors and financial figures were combined using a fuzzy analytic hierarchy process (Fuzzy AHP) approach. Ping-Teng Chang (2005) presents a fuzzy methodology for replacement of equipment. Issues such as fuzzy modeling of degradation parameters and determining fuzzy strategic replacement and economic lives are extensively discussed. For the strategic purpose, addible market and cost effects from the replacements against the counterpart, i.e., the existing equipment cost and market obsolescence, are modeled fuzzily and interactively, in addition to the equipment deterioration. Both the standard fuzzy arithmetic and re-termed requisite-constraint vertex fuzzy arithmetic (or the vertex method) are applied and investigated.

The aim of this chapter is to show how a replacement analysis is made under fuzziness. To do this, fuzzy equivalent uniform annual worth analysis (EUAW) should be applied. The rest of the chapter is organized as follows. Section 2 gives a simple numerical example of classical replacement analysis. Section 3 summarizes fuzzy EUAW analysis. Section 4 presents a numerical application of operating system selection with fuzzy replacement analysis.

## 2 Classical Replacement Analysis

In the classical replacement analysis, you calculate the equivalent uniform annual cost (EUAC) for each value of the useful life (e.g., $n = 1$, $n = 2$, $n = 3$, etc.). The number of years at which the EUAC is minimized is the minimum cost life (economic useful life). If the defender is more economical, it should be retained. If the challenger is more economical, it should be installed. The following assumptions are made:

- The best challenger is available in all subsequent years and will be unchanged in economic cost.
- The period of needed service is infinitely long.

148     C. Kahraman and M.L. Demircan

If the replacement repeatability assumption holds, EUAC of the defender asset at its minimum cost life against the EUAC of the challenger at its minimum cost life is compared.

In the following, a simple numerical example is given to illustrate the steps of the analysis.

Consider the following data of a challenger:

–   $7500 initial cost (P)
–   $900 arithmetic gradient maintenance cost (G)
–   $500 uniform cost (A) and 400 arithmetic gradient operating cost (G)

Then the EUAC values are obtained as in Table 1, using equivalent uniform annual cash flow analysis. As it is seen from Table 1, the useful life of the challenger is found as 4 years.

**Table 1.** EUAC calculations for increasing values of useful life

| Year | EUAC of Capital Recovery Costs | EUAC of Maintenance and Repair Costs | EUAC of Operating Costs | EUAC Total | Interest rate |
|---|---|---|---|---|---|
| Initial year | -7500 | 0 | -500 | | |
| Arithmetic gradient | | -900 | -400 | | 8% |
| 1 | $8,100.00 | $0.00 | $500.00 | $8,600.00 | |
| 2 | $4,205.77 | $432.69 | $692.31 | $5,330.77 | |
| 3 | $2,910.25 | $853.87 | $879.50 | $4,643.62 | |
| 4 | $2,264.41 | $1,263.56 | $1,061.58 | $4,589.55 | <-----MIN |
| 5 | $1,878.42 | $1,661.82 | $1,238.59 | $4,778.84 | |
| 6 | $1,622.37 | $2,048.71 | $1,410.54 | $5,081.62 | |
| 7 | $1,440.54 | $2,424.30 | $1,577.47 | $5,442.31 | |
| 8 | $1,305.11 | $2,788.67 | $1,739.41 | $5,833.19 | |
| 9 | $1,200.60 | $3,141.93 | $1,896.41 | $6,238.94 | |
| 10 | $1,117.72 | $3,484.18 | $2,048.53 | $6,650.43 | |
| 11 | $1,050.57 | $3,815.55 | $2,195.80 | $7,061.93 | |
| 12 | $995.21 | $4,136.17 | $2,338.30 | $7,469.68 | |
| 13 | $948.91 | $4,446.19 | $2,476.08 | $7,871.18 | |
| 14 | $909.73 | $4,745.75 | $2,609.22 | $8,264.69 | |
| 15 | $876.22 | $5,035.01 | $2,737.78 | $8,649.02 | |

The graphical representation of the obtained results in Table 1 is given in Figure 2.

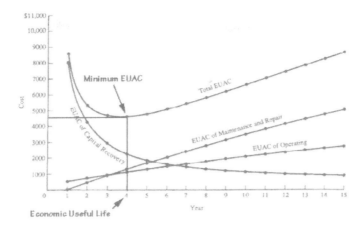

**Fig. 2.** Graph of EUAC by $n$

## 3 Fuzzy EUAW

The equivalent uniform annual worth (EUAW) means that all incomes and disbursements (irregular and uniform) must be converted into an equivalent uniform annual amount, which is the same each period. The major advantage of the method over all other methods is that it does not require making the comparison over the least common multiple of years when the alternatives have different lives. Kahraman et al. (2002) demonstrate the fuzzy EUAW as follows, where $R$ is discount rate and $T$ stands for the number of periods:

$$EUAW = A = NPV\gamma^{-1}(T,R) = NPV \times \frac{(1+R)^T R}{(1+R)^T - 1} \tag{1}$$

where $NPV$ is the net present value. In the case of fuzziness, $\widetilde{NPV}$ will be calculated and then the fuzzy $\widetilde{EUAW}(\tilde{A}_T)$ will be found. The membership function $\mu(x|\tilde{A}_T)$ for $\tilde{A}_T$ is determined by

$$f_{T_i}(\alpha|\tilde{A}_T) = f_i(\alpha|\widetilde{NPV})\gamma^{-1}(T, f_i(\alpha|\tilde{R})) \tag{2}$$

where $\tilde{R}$ represents fuzzy discount rate. TFN($\alpha$) for fuzzy EUAW is

$$\tilde{A}_T(\alpha) = \left( \frac{NPV^{l(\alpha)}}{\gamma(T, R^{l(\alpha)})}, \frac{NPV^{r(\alpha)}}{\gamma(T, R^{r(\alpha)})} \right) \tag{3}$$

Not slightly different from Kahraman et al. (2002), a fuzzy replacement framework with another $EUAW$ computation using TFNs is developed. First, replacement analyis parameters are defined: Let $\tilde{R}$ be a fuzzy discount rate and $T$ represents period. The fuzzy capital recovery factor is developed as follows;

$$\left( \tilde{A}/\tilde{P}, \tilde{R}, T \right) = \frac{\tilde{R} \otimes (1 \oplus \tilde{R})^T}{(1 \oplus \tilde{R})^T \ominus 1} = \left[ \frac{R^{l(\alpha)} \times (1 + R^{l(\alpha)})^T}{(1 + R^{r(\alpha)})^T - 1}, \frac{R^{r(\alpha)} \times (1 + R^{r(\alpha)})^T}{(1 + R^{l(\alpha)})^T - 1} \right] \tag{4}$$

Let $\tilde{I}$ be fuzzy initial investment, $\tilde{S}_T$ be fuzzy salvage value at period $T$. $\tilde{C}_T$ represents fuzzy Capital Recovery. Please note that, $\tilde{I}$ will be considered as crisp in the numerical application, it is the initial investment and does not include fuzziness. Beside, $\tilde{S}_T$ will be considered as none for each period since there exists no salvage value.

$$\tilde{C}_T = \left(\tilde{I} \ominus \tilde{S}_T\right) \otimes \left(\tilde{A}\Big/_{\tilde{P}}, \tilde{R}, T\right) \oplus \left(\tilde{R} \otimes \tilde{S}_T\right) = \left[C_T^{\,l(\alpha)}, C_T^{\,r(\alpha)}\right] \tag{5}$$

$$\tilde{C}_T = \left[ \begin{array}{c} \dfrac{\left(I^{l(\alpha)} - S_T^{\,l(\alpha)}\right) \times R^{l(\alpha)} \times \left(1 + R^{l(\alpha)}\right)^T}{\left(1 + R^{r(\alpha)}\right)^T - 1} + \left(R^{l(\alpha)} \times S_T^{\,l(\alpha)}\right), \\[3mm] \dfrac{\left(I^{r(\alpha)} - S_T^{\,r(\alpha)}\right) \times R^{r(\alpha)} \times \left(1 + R^{r(\alpha)}\right)^T}{\left(1 + R^{l(\alpha)}\right)^T - 1} + \left(R^{r(\alpha)} \times S_T^{\,r(\alpha)}\right) \end{array} \right] \tag{6}$$

Let $\tilde{F}_T$ be fuzzy operating and maintenance cost for period $T$. Whilst $\sum \tilde{P}_T$ represents cumulative fuzzy present value of $\tilde{F}_T$'s, from $t=1$ to $t=T$:

$$\sum \tilde{P}_T = \left[ P_T^{\,l(\alpha)}, P_T^{\,r(\alpha)} \right]$$

$$= \sum_{t=1}^{T} \frac{\tilde{F}_t}{\left(1 \oplus \tilde{R}\right)^t} = \left[ \sum_{t=1}^{T} \frac{F_t^{\,l(\alpha)}}{\left(1 + R^{r(\alpha)}\right)^t}, \sum_{t=1}^{T} \frac{F_t^{\,r(\alpha)}}{\left(1 + R^{l(\alpha)}\right)^t}, \right] \tag{7}$$

Let $\tilde{A}\left(\sum \tilde{P}_T\right)$ be $EUAW$ of cumulative fuzzy present value of $\tilde{F}_T$'s over $T$ period(s).

$$\tilde{A}\left(\sum \tilde{P}_T\right) = \left(\sum \tilde{P}_T\right) \otimes \left(\tilde{A}\Big/_{\tilde{P}}, \tilde{R}, T\right)$$

$$= \left[ \frac{PT^{l(\alpha)} \times R^{l(\alpha)} \times \left(1 + R^{l(\alpha)}\right)^T}{\left(1 + R^{r(\alpha)}\right)^T - 1}, \frac{P_T^{\,r(\alpha)} \times R^{r(\alpha)} \times \left(1 + R^{r(\alpha)}\right)^T}{\left(1 + R^{l(\alpha)}\right)^T - 1} \right] \tag{8}$$

Let $\tilde{W}_T$ be total $EUAW$.

$$\tilde{W}_T = \tilde{C}_T \oplus \tilde{A}\left(\sum \tilde{P}_T\right) = \left[W_T^{\,l(\alpha)}, W_T^{\,r(\alpha)}\right] = \left((W_T)_1, (W_T)_2, (W_T)_3\right) \tag{9}$$

Figure 3 illustrates the membership function of $\tilde{W}_T$

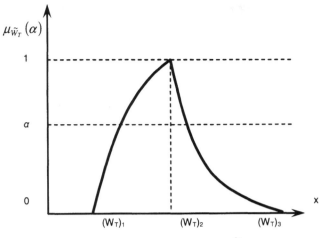

**Fig. 3.** Total Fuzzy EUAW, $\tilde{W}_T$

## 4 A Numerical Application: Operating System Selection with Fuzzy Replacement Analysis

To better understand the replacement analysis, let us represent a numerical application for the framework developed. In IT related replacement decisions, technologic and financial reasons become the leading factors. In most technology selection studies, the cost comparison parameter has been cited as one of other parameters without regarding its crucial role in Decision Makers' choice.

In this application of fuzzy replacement analysis, two operating systems, AX and BS will be evaluated by using fuzzy EUAW analysis. In the following, first operating system features and then financial assessment of Operating System will be given.

Migration issues in enterprise computing are becoming more compelling for companies. Firms want to experience benefits from server consolidation as well as IT staff reduction. In order to assess the financial aspect of enterprise computing migration, total cost of ownership (TCO) of both defender and challenger(s) should be determined. TCO is the cost figure over time taking into account the costs of acquiring and supporting the hardware and software required for each system workloads. System workloads are the group of related tasks and services of an operating system. These workloads are Networking, File Server, Print Server, Web Server and Security Applications.

Network workload provides the mere infrastructure services. It includes Dynamic Host Configuration Protocol (DHCP), Domain Name System (DNS), Windows Internet Naming Service, Internet Connection Sharing, Remote Access Connection Manager (RACM), directory and caching services, and traditional services such as routers, hubs or switches. File Server workload includes File Transfer Protocol (FTP), Network File System (NFS), and Common Internet File System (CIFS). Print Server workload covers all print stream protocols as well as Internet Printing Protocol and Line Printer Daemon services. Web Server workload includes all internet, intranet and extranet services which delivers both static and dynamic Web pages. Security workload covers Virtual Private Networking (VPN), intrusion detection services, antivirus management services, authentication, access, and authorization services.

Each workload has hardware and software related costs, which comprise the TCO. Costs include Hardware, Storage, Networking, Facilities, Software (OS), Software (Applications), Downtime, IT Staffing, IT Staff Training, and Outsourcing costs.

Hardware, Storage and Facility costs include the purchase of specified physical equipment. Software costs covers acquisition and deployment of specified software. Since challenger(s) may benefit major CPU based savings by server consolidation, Software costs need to be subcategorized CPU Licensed and User Licensed. Downtime costs are the costs needed to repair, fix, or reconfigure the malfunctioning computing services. Platform selected and the knowledge and experience level of IT

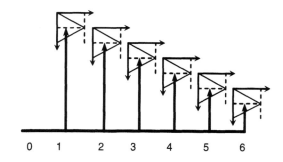

**Fig. 4.** Descending Fuzzy Cash Flows of BS

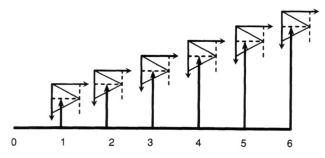

**Fig. 5.** Ascending Fuzzy Cash Flows of AX

**Table 2.** Initial investment and Fuzzy Operating and Maintenance Costs of Alternatives AX and BS

| Period | Fuzzy CFs of AX | Fuzzy CFs of BS |
|---|---|---|
| 0 | -($10,000; $10,000; $10,000 ) | -($ 30,000; $ 30,000; $ 30,000) |
| 1 | ($24,000; $26,000; $28,000) | ($8,000; $10,000; $12,000) |
| 2 | ($24,500; $26,500; $28,500) | ($7,000; $9,000; $11,000) |
| 3 | ($25,500; $27,500; $29,500) | ($6,000; $8,000; $10,000) |
| 4 | ($27,500; $29,500; $31,500) | ($5,000; $7,000; $9,000) |
| 5 | ($30,500; $32,500; $34,500) | ($4,000; $6,000; $8,000) |
| 6 | ($34,500; $36,500; $38,500) | ($3,000; $5,000; $7,000) |

staff play an important role over downtime costs. IT Staff Training costs are both training expenditures and productivity loss for time spent in training. Support and Maintenance are outsourced services, which generates a considerable cost.

For ease of computation each workload and their software and hardware based costs are combined in periodic disbursements (operating costs). We defined OS alternative AX as open source and license free. Beside, BS is a licensed OS platform, which requires higher initial investment than AX. Table 2 shows the fuzzy initial

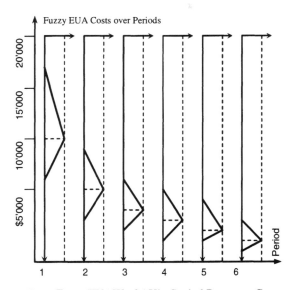

**Fig. 6.** Fuzzy EUAW of AX's Capital Recovery Costs

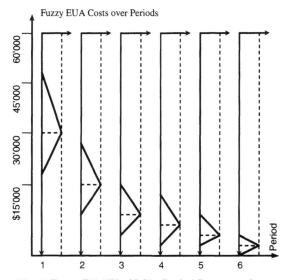

**Fig. 7.** Fuzzy EUAW of BS's Capital Recovery Costs

investment and operating costs. BS's operating costs will decrease over time since staff knowledge threshold will increase due to BS's GUI capabilities and Support Availability (Figure 4). While AX represents lower initial investment, it requires higher expertise to operate. Therefore, IT Staffing, IT Staff Training and Outsourcing costs will increase over time (Figure 5).

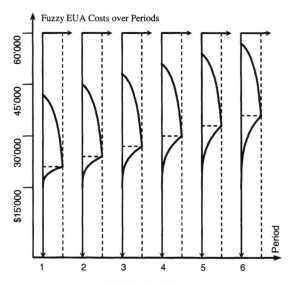

**Fig. 8.** Fuzzy EUAW of AX's Operating Costs

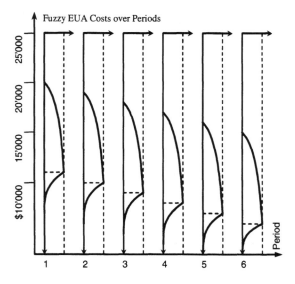

**Fig. 9.** Fuzzy EUAW of BS's Operating Costs

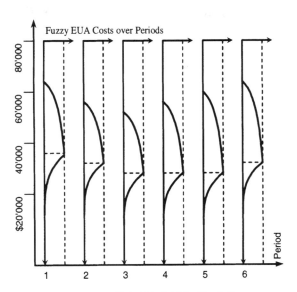

**Fig. 10.** Fuzzy EUAW of AX's Total Costs

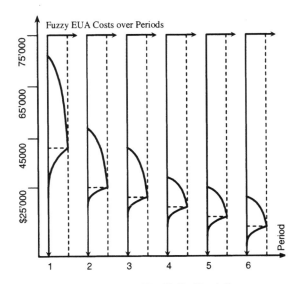

**Fig. 11.** Fuzzy EUAW of BS's Total Costs

To apply a fuzzy replacement analysis to financial figures above, fuzzy discount rate $\tilde{R}$ is set as $\tilde{R} = (3\%, 4\%, 5\%)$. Figures 6 to 9 illustrate the fuzzy EUAW of capital recovery and operating and maintenance costs. Figure 10 and Figure 11 illustrate the total fuzzy EUAWs. When compared the fuzzy EUAWs of AX's and BS's total costs, a decision maker may give moderate importance to the favor of BS. That is why AX may be the preferred Operating System.

# 5 Conclusion

Replacement is never a question of "if we replace" but rather a question of "when we replace. The key question is: "Shall we replace the defender now, or keep it for one or more years before replacing it?" The uncertainty of fuzzy cash flows and interest rates should be taken into account while giving replacement decisions. The fuzzy replacement analysis presents us the possibility of any total EUAC and thus the possibility of replacement time.

# References

Al-Najjar, B., Alsyouf, I.: Selecting the most efficient maintenance approach using fuzzy multiple criteria decision making. International Journal of Production Economics, 84–100 (2003)

Baskak, M., Kahraman, C.: Fuzzy replacement analysis. In: Fuzzy Information Processing Society – Conference of the North American (1998)

Bellman, R.: Equipment replacement policy. Journal of the Society for Industrial and Applied Mathematics 3(3), 133–136 (1955)

Boussabaine, A.H., Elhag, T.: Applying fuzzy techniques to cash flow analysis. Construction Management and Economics 17(6), 745–755 (1999)

Buckley, J.J.: The fuzzy mathematics of finance. Fuzzy Sets and Systems 21, 257–273 (1987)

Chang, P.T.: Fuzzy strategic replacement analysis. European Journal of Operational Research 160, 532–559 (2005)

Chiu, C.Y., Park, C.S.: Fuzzy cash flow analysis using present worth criterion. The Engineering Economist 39(2), 113–138 (1994)

Dimitrovski, A.D., Matos, M.A.: Fuzzy engineering economic analysis. IEEE Transactions on Power Systems 15(1), 283–289 (2000)

Dreyfus, S.E.: A generalized equipment replacement study. Journal of the Society for Industrial and Applied Mathematics 8(3), 425–435 (1960)

Dreyfus, S.E., Wagner, R.A.: The Steiner problem in graphs. Networks 1, 195–207 (1972)

Esogbue, A.O., Hearnes, W.E.: On replacement models via a fuzzy set theoretic framework. IEEE Transactions on Systems, Man, and Cybernetics, Part C 28(4), 549–560 (1998)

Fleischer, G.A.: Introduction to engineering economy. PWS Publishing Company (1994)

Kahraman, C.: Capital budgeting techniques using discounted fuzzy cash flows. In: Ruan, D., Kacprzyk, J., Fedrizzi, M. (eds.) Soft Computing for Risk Evaluation and Management: Applications in Technology, Environment and Finance. Physica Verlag, Heidelberg (2001a)

Kahraman, C.: Fuzzy versus probabilistic benefit/cost ratio analysis for public works projects. International Journal of Applied Mathematics and Computer Science 11(3), 101–114 (2001b)

Kahraman, C., Ruan, D., Bozdağ, C.E.: Optimization of multilevel investments using dynamic programming based on fuzzy cash flows. Fuzzy Optimization and Decision Making Journal 2(2), 101–122 (2003)

Kahraman, C., Ruan, D., Tolga, E.: Capital budgeting techniques using discounted fuzzy versus probabilistic cash flows. Information Sciences 42(1-4), 57–76 (2002)

Kahraman, C., Tolga, E., Ulukan, Z.: Justification of manufacturing technologies using fuzzy benefit / cost ratio analysis. International Journal of Production Economics 66(1), 45–52 (2000)

Kai-Yuan, C., Chuan-Yuan, W.: Street-lighting lamps replacement: A fuzzy viewpoint. Fuzzy Sets and Systems 37(2), 161–172 (1990)

Karsak, E.E.: Measures of liquidity risk supplementing fuzzy discounted cash flow analysis. Engineering Economist 43(4), 331–344 (1998)

Kuchta, D.: Fuzzy capital budgeting. Fuzzy Sets and Systems 111, 367–385 (2000)

Lake, D.H., Muhlemann, A.P.: An equipment replacement problem. Journal of Operational Research Society 30(5), 405–411 (1979)

Lohmann, J.R.: A stochastic replacement economic decision model. IEEE Transactions 182-194 (1986)

Oakford, R.V., Lohmann, J.R., Salazar, A.: A dynamic replacement economy decision model. IEE Transactions, 65–72 (1984)

Tolga, E., Demircan, M.L., Kahraman, C.: Operating system selection using fuzzy replacement analysis and analytic hierarchy process. International Journal of Production Economics 97(1), 89–117 (2005)

Zadeh, L.: Fuzzy sets. Information Control 8, 338–353 (1965)

# Depreciation and Income Tax Considerations under Fuzziness

Cengiz Kahraman and İhsan Kaya

Istanbul Technical University, Department of Industrial Engineering, 34367, Macka, İstanbul, Turkey

**Abstract.** Depreciation is an income tax deduction that allows a taxpayer to recover the cost of property or assets placed in service. Straight line depreciation is the most frequently used method of depreciating new equipment for financial statements. Accelerated cost recovery systems include the methods like declining balance depreciation and sum-of-years digits depreciation. Incomplete information about the future values of the parameters in an after-tax rate of return analysis causes us to use the fuzzy set theory. This chapter includes the fuzzy after-tax cash flow analyses in case of fuzzy cash flows, fuzzy depreciation, fuzzy tax rate, and fuzzy minimum attractive rate of return with numerical examples.

## 1 Introduction

Almost every business must invest in some major equipment, vehicles, machinery, or furniture in order to operate. Some businesses will require assets such as land, a building, patents, or franchise rights. Major assets that will be used in a business for more than a year are known as "capital assets" and are subject to special treatment under the tax laws. Most importantly, you generally can't deduct the entire cost of acquiring such an asset in the year you acquire it.

When a company spends money for a service or anything else that is short-lived, this expenditure is usually immediately tax deductible in some countries, and the company enjoys an immediate tax benefit. To be eligible for depreciation, an asset must have two features: (1) it has a useful life beyond the taxable year (essentially why it was capitalized in the first place), and (2) it wears out, decays, declines in value due to natural causes, or is subject to exhaustion or obsolescence. Therefore, when a company buys an asset that will last longer than one year, like a computer, car, or building, the company cannot immediately deduct the cost and enjoy an immediate tax benefit. Instead, the company must *depreciate* the cost over the useful life of the asset, taking a tax deduction for a part of the cost each year.

Depreciation is an allowance or provision made in the financial records of a business or association for "wear and tear" and "technical obsolescence "on plant and equipment. The idea of depreciation is to spread the cost of that capital asset over the period of its "useful life to the entity" that currently owns it. Theoretically, the cost of an asset should be deducted over the number of years that the asset will be used, according to the actual drop in value that the asset will suffer each year. At the end of each year, you could subtract all depreciation claimed to date from the cost of the asset, to arrive at the asset's "book value," which would be equal to its market value. At

---

C. Kahraman (Ed.): Fuzzy Engineering Economics with Appl., STUDFUZZ 233, pp. 159–171, 2008.
springerlink.com                                    © Springer-Verlag Berlin Heidelberg 2008

the end of the asset's useful life for the business, any undepreciated portion would represent the salvage value for which the asset could be sold or scrapped.

For tax purposes, depreciation is used as a deduction to reflect the diminishing value of assets. Depreciation decreases the overall worth of a business and is reflected as an operating expense or cost of doing business. There are many different systems used for depreciation and they can change with each tax year. For many items, their value is reduced to zero or near zero after several years. even if the item retains some resale value.

Depreciation expense is calculated utilizing either a straight line depreciation method or an accelerated depreciation method. The straight line method calculates depreciation by spreading the cost evenly over the life of the fixed asset. Accelerated depreciation methods such as declining balance and sum of years digits calculate depreciation by expensing a large part of the cost at the beginning of the life of the fixed asset.

In the literature, there are several works on depreciation and tax issues and few on fuzzy cases. Robinson (1987) describes and illustrates after tax analysis for both capitalization and cash flow techniques. Furthermore, slices of equated yield attributable to the main components of return from real property are demonstrated. Wang and Liang (1995) handle a fuzzy benefit/cost analysis including depreciation and tax parameters.

Contrary to previous studies, Keating and Zimmerman (2000) find managers change depreciation policies in predictable ways. They identify three dimensions of depreciation-policy changes: whether it is a method change or an estimate revision; whether it is income-increasing or decreasing; and whether it applies to new assets only or both new and existing assets. This disaggregation leads to three findings: First, a 1981 tax law altered the frequency of estimate revisions and method changes. Second, firms adopting income-increasing method changes for all assets experience worse performance than those adopting such changes only for new assets. Finally, non-income-increasing policy changes are associated with changes in investment opportunities.

Berg et al. (2001) focus on the choice between the two most commonly used methods in practice, i.e. the straight line depreciation method and accelerated depreciation method, such as the double declining balance method and the sum of years-digits method. They show how the optimal choice depends on the discount factor, the degree of uncertainty in future cash flows, and the structure of the tax system. Jacco et al. (2002) focus on the effect of a progressive tax system on optimal tax depreciation. By using dynamic optimization they show that an optimal strategy exists, and they provide an analytical expression for the optimal depreciation charges. Depreciation charges initially decrease over time, and after a number of periods the firm enters a steady state where depreciation is constant and equal to replacement investments. This way, the optimal solution trades off the benefits of accelerated depreciation (because of discounting) and of constant depreciation (because of the progressive tax system). They show that the steady state will be reached sooner when the initial tax base is lower or when the discounting effect is stronger.

Hartman et al. (2007) consider the impact of US tax depreciation rules, which differ depending upon whether a US corporation locates its assets at domestic or foreign branches. Their analysis and illustrative examples demonstrate that US depreciation law can indeed have a non-trivial impact on location and sourcing decisions, with direct ownership of foreign assets appearing relatively less attractive once depreciation law is taken into account. More broadly, their results demonstrate that comprehensive asset location and ownership decisions require a detailed understanding of international tax law, rather than just a simple recognition of differences in tax rates among countries.

Under incomplete information, the fuzzy set theory helps to convert human's linguistic evaluations into numerical outcomes. While a tax rate may be linguistically expressed as in "I expect the tax rate per year to be around 30% within the next 5 years", it can be represented by a triangular fuzzy number as (28%, 30%, 32%). A linguistic expression like "an annual receipt between $20,000 and $24,000" can be incorporated into calculations using a trapezoidal fuzzy number. The objective of this chapter is to show how this is done using fuzzy numbers.

The rest of this chapter is organized as follows: Section 2 summarizes the depreciation methods and gives some numerical applications. Section 3 includes the depreciation and tax issues together under fuzziness. Finally, Section 4 gives the conclusions.

## 2 Depreciation Methods

In the following, straight line depreciation and sum-of-years digit depreciation are explained (Newnan and Lavelle 1998).

### 2.1 Straight Line (SL) Depreciation

This method assumes a constant depreciation value per year.

Assuming the price of a depreciating asset is $P$ and its salvage value after $N$ years is $S$.

$$\text{Annual Depreciation} = \frac{P - S}{N} \tag{1}$$

Suppose an asset's price is $20,000 and it has a salvage value of $5,000 in five years. The annual depreciation would be

$$[(\$20,000\text{-}\$5,000)/5] = \$3,000$$

### 2.2 Sum-of-Years Digits Depreciation

An asset often loses more of its value early in its lifetime. A method that exhibits this dynamic is desirable.

Assume an asset depreciates from price $P$ to salvage value $S$ in $N$ years. First compute the value: sum-of-years = $1+2+ \dots +N$. The depreciation for the years after the asset's purchase is:

For the $i$th year of the asset's use this equation generalizes to

$$\text{Annual Depreciation} = \frac{N+1-i}{\text{sum} - \text{of} - \text{years}} \; (P\text{-}S) \tag{2}$$

For our example $N=5$ and the sum of years is $1+2+3+4+5=15$. The depreciation during the first year is

$$(\$20,000\text{-}\$5,000)(5/15) = \$5,000$$

162    C. Kahraman and İ. Kaya

**Table 1.** Sum-of-years General Example

| Year number | Annual depreciation |
|---|---|
| first | $\dfrac{N}{\text{sum} - \text{of} - \text{years}}$ (P-S) |
| second | $\dfrac{N-1}{\text{sum} - \text{of} - \text{years}}$ (P-S) |
| third | $\dfrac{N-2}{\text{sum} - \text{of} - \text{years}}$ (P-S) |
| $\vdots$ | $\vdots$ |
| final | $\dfrac{1}{\text{sum} - \text{of} - \text{years}}$ (P-S) |

Table 2 describes how declining balance would depreciate the asset.

**Table 2.** Sum-of-years Example

| Year | Depreciation | Year-end Value |
|---|---|---|
| 1 | ($20,000-$5,000)(5/15) = $5,000 | $15,000 |
| 2 | ($20,000-$5,000)(4/15) = $4,000 | $11,000 |
| 3 | ($20,000-$5,000)(3/15) = $3,000 | $8,000 |
| 4 | ($20,000-$5,000)(2/15) = $2,000 | $6,000 |
| 5 | ($20,000-$5,000)(1/15) = $1,000 | $5,000 |

### 2.3  Declining Balance (DB)

Recall that the straight line method assumes a constant depreciation value. Conversely, the declining balance method assumes a constant depreciation rate per year. And like the Sum-of-years method, more depreciation tends to occur earlier in the asset's life.

Assume the price of a depreciating asset is $P$ and its salvage value after $N$ years is $S$. You could assume the asset depreciates by a factor of $\dfrac{1}{N}$ (or a rate of $\dfrac{100}{N}$ %). This method is known as Single Declining Balance. In an equation this looks like:

$$\text{Annual Depreciation} = \frac{1}{N} \times \text{Previous year's value} \qquad (3)$$

So for our example, the depreciation during the first year is $20,000/5 = $4,000. Table 3 describes how Declining Balance would depreciate the asset.

Depreciation and Income Tax Considerations under Fuzziness 163

**Table 3.** Declining Balance Example

| Year | Depreciation | Year-end Value |
|------|-------------|----------------|
| 1 | [$20,000 /5] = $4,000 | $16,000 |
| 2 | [$16,000/5] = $3,200 | $12,800 |
| 3 | [$12,800/5] = $2,560 | $10,240 |
| 4 | [$10,240/5] = $2,048 | $8,192 |
| 5 | [$8,192/5] = $1,638.40 | $6,553.60 |

### 2.3.1 DB Factor

We could also accelerate the depreciation by increasing the factor (and hence the rate) at which depreciation occurs. Other commonly accepted depreciation rates are $\frac{200}{N}$ % (called double declining balance as the depreciation factor becomes $\frac{2}{N}$) and $\frac{150}{N}$ %. Investment analysis enables you to choose between these three types for declining balance: 2 (with $\frac{200}{N}$ % depreciation), 1.5 (with $\frac{150}{N}$ %), and 1 (with $\frac{100}{N}$ %).

### 2.3.2 Declining Balance and the Salvage Value

The Declining Balance method assumes that depreciation is faster earlier in an asset's life; this is what you wanted. But notice the final value is greater than the salvage value. Even if the salvage value were greater than $6,553.60 the final year-end value would not change. The salvage value never enters the calculation. so there is no way for the salvage value to force the depreciation to assume its value. Newnan and Lavelle (1998) describe two ways to adapt the declining balance method to assume the salvage value at the final time. One way is as follows:

Suppose you call the depreciated value after $i$ years $V(i)$. This sets $V(0)=P$ and $V(N)=S$.

- If $V(N)>S$ according to the usual calculation for $V(N)$. redefine $V(N)$ to equal $S$.
- If $V(i)<S$ according to the usual calculation for $V(i)$ for some $i$ (and hence for all subsequent $V(i)$ values), you can redefine all such $V(i)$ to equal $S$.

This alteration to declining balance forces the depreciated value of the asset after $N$ years to be $S$ and keeps $V(i)$ no less than $S$.

### 2.3.3 Conversion to SL

The second (and preferred) way to force declining balance to assume the salvage value is by Conversion to Straight Line. If $V(N)>S$. the first way redefines $V(N)$ to equal $S$; you can think of this as converting to the straight line method for the last time step.

If the $V(N)$ value supplied by DB is appreciably larger than $S$. then the depreciation in the final year would be unrealistically large. An alternate way is to compute the DB and SL step at each timestep and take whichever step gives a larger depreciation (unless DB drops below the salvage value).

Once SL assumes a larger depreciation, it continues to be larger over the life of the asset. This forces the value at the final time to equal the salvage value as SL forces this.

164     C. Kahraman and İ. Kaya

## 2.4  Comparison of Depreciation Methods

Tables 4-7 display the depreciation for four depreciation methods (Newnan and Lavelle 1998).

- Under depreciation table, realize a 5-year class MACRS Depreciation actually lasts 6 years.
- The declining balance method is double declining balance with conversion to straight line.

**Table 4.** Straight Line Depreciation

| Year | sbvalue | Depreciation | ebvalue |
|------|---------|--------------|---------|
| 1999 | 20,000 | 3,000 | 17,000 |
| 2000 | 17,000 | 3,000 | 14,000 |
| 2001 | 14,000 | 3,000 | 11,000 |
| 2002 | 11,000 | 3,000 | 8,000 |
| 2003 | 8,000 | 3,000 | 5,000 |

**Table 5.** Sum of Years Digits Depreciation

| Year | sbvalue | Depreciation | ebvalue |
|------|---------|--------------|---------|
| 1999 | 20,000 | 5,000 | 15,000 |
| 2000 | 15,000 | 4,000 | 11,000 |
| 2001 | 11,000 | 3,000 | 8,000 |
| 2002 | 8,000 | 2,000 | 6,000 |
| 2003 | 6,000 | 1,000 | 5,000 |

**Table 6.** Depreciation Table

| Year | sbvalue | Depreciation | ebvalue |
|------|---------|--------------|---------|
| 1999 | 20,000 | 4,000 | 16,000 |
| 2000 | 16,000 | 6,400 | 9,600 |
| 2001 | 9,600 | 3,840 | 5,760 |
| 2002 | 5,760 | 2,304 | 3,456 |
| 2003 | 3,000 | 2,304 | 1,152 |
| 2004 | 1,152 | 1,152 | 0 |

**Table 7.** Declining Balance

| Year | sbvalue | Depreciation | ebvalue |
|------|---------|--------------|---------|
| 1999 | 20,000 | 8,000 | 12,000 |
| 2000 | 12,000 | 4,800 | 7,200 |
| 2001 | 7,200 | 2,200 | 5,000 |
| 2002 | 5,000 | 0 | 5,000 |
| 2003 | 5,000 | 0 | 5,000 |

Depreciation and Income Tax Considerations under Fuzziness     165

# 3 Consideration of Depreciation and Tax under Fuzzy Case

## 3.1 Sum of Years Digits Depreciation

Let the previous example to be reconsidered for this case. The first cost is $20,000 the salvage value is $5,000. The useful life is 5 years and the sum of years digits depreciation is used. The crisp case is first given by using a 30% tax rate. As it is seen from

**Table 8.** Consideration of depreciation and tax – crisp case

| Year | BTCF | SYD Depreciation | Taxable cash flow | Tax | ATCF |
|------|------|------------------|-------------------|-----|------|
| 0 | -20,000 | | | | -20,000 |
| 1 | 7,000 | 5,000 | 2,000 | 600 | 6,400 |
| 2 | 9,000 | 4,000 | 5,000 | 1,500 | 7,500 |
| 3 | 11,000 | 3,000 | 8,000 | 2,400 | 8,600 |
| 4 | 13,000 | 2,000 | 11,000 | 3,300 | 9,700 |
| 5 | 15,000 | 1,000 | 14,000 | 4,200 | 10,800 |
| 5 | SV=5,000 | | | | 5,000 |
| BTCF-ROR 41.74% | | | | ATCF-ROR | 32.07% |

**Table 9.** Consideration of depreciation and tax – fuzzy case

| | Fuzzy BTFC | | | SYD Depreciation | | |
|------|------|------|------|------|------|------|
| 0 | -18,000 | -20,000 | -22,000 | | | |
| 1 | 6,300 | 7,000 | 7,700 | 4,500 | 5,000 | 5,500 |
| 2 | 8,100 | 9,000 | 9,900 | 3,600 | 4,000 | 4,400 |
| 3 | 9,900 | 11,000 | 12,100 | 2,700 | 3,000 | 4,400 |
| 4 | 11,700 | 13000 | 14300 | 1,800 | 2,000 | 2,200 |
| 5 | 13,500 | 15,000 | 16,500 | 900 | 1,000 | 1,100 |
| 5 | 4,500 | 5,000 | 5,500 | | | |
| BTCF-ROR | 41.74% | 41.74% | 41.74% | | | |

**Table 10.** Consideration of depreciation and tax – fuzzy case (continues)

| Taxable cash flow | | | Tax (0.28; 0.30; 0.32) | | | ATCF | | |
|------|------|------|------|------|------|------|------|------|
| | | | | | | -18,000 | -20,000 | -22,000 |
| 800 | 2,000 | 3,200 | 224 | 600 | 1,024 | 5,276 | 6,400 | 7,476 |
| 3,700 | 5,000 | 6,300 | 1,036 | 1,500 | 2,016 | 6,084 | 7,500 | 8,864 |
| 5,500 | 8,000 | 9,400 | 1,540 | 2,400 | 3,008 | 6,892 | 8,600 | 10,560 |
| 9,500 | 11,000 | 12,500 | 2,660 | 3,300 | 4,000 | 7,700 | 9,700 | 11,640 |
| 12,400 | 14,000 | 15,600 | 3,472 | 4,200 | 4,992 | 8,508 | 10,800 | 13,028 |
| | | | | | | 4,500 | 5,000 | 5,500 |
| | | | | | ATCF-ROR | 27.45% | 32.07% | 35.72% |

Table 8, the rate of returns (ROR) of the before-tax cash flow (BTCF) and the after-tax cash flow (ATCF) are calculated as 41.74% and 32.07%, respectively. The fuzzy minimum attractive rate of return (MARR) is (15%, 17%, 19%).

Now the fuzzy case will be examined using a 10% fuzzification rate and a fuzzy tax rate (0.28, 0.30, 0.32). Table 9 shows this calculation.

Table 11 shows the calculated ROR values for some fuzzification ratios.

From Table 11 it can be seen that after a fuzzification ratio of 0.25, the considered investment has a possibility of negative net present worth.

**Table 11.** ROR values for various fuzzification ratios

| Fuzzification Rates | CRISP BTCF-ROR | ATCF-ROR | BTCF-ROR | FUZZY ATCF-ROR | | |
|---|---|---|---|---|---|---|
| 0.01 | 0.4174 | 0.3207 | 0.4174 | 31.02% | 32.07% | 33.33% |
| 0.02 | 0.4174 | 0.3207 | 0.4174 | 30.66% | 32.07% | 33.62% |
| 0.03 | 0.4174 | 0.3207 | 0.4174 | 30.29% | 32.07% | 33.90% |
| 0.04 | 0.4174 | 0.3207 | 0.4174 | 29.92% | 32.07% | 34.18% |
| 0.05 | 0.4174 | 0.3207 | 0.4174 | 29.53% | 32.07% | 34.45% |
| 0.10 | 0.4174 | 0.3207 | 0.4174 | 27.45% | 32.07% | 35.72% |
| 0.15 | 0.4174 | 0.3207 | 0.4174 | 29.92% | 32.07% | 34.18% |
| 0.20 | 0.4174 | 0.3207 | 0.4174 | 22.35% | 32.07% | 37.91% |
| 0.25 | 0.4174 | 0.3207 | 0.4174 | 19.16% | 32.07% | 38.87% |
| 0.30 | 0.4174 | 0.3207 | 0.4174 | 15.38% | 32.07% | 39.74% |
| 0.35 | 0.4174 | 0.3207 | 0.4174 | 10.80% | 32.07% | 40.55% |
| 0.40 | 0.4174 | 0.3207 | 0.4174 | 5.09% | 32.07% | 41.29% |
| 0.45 | 0.4174 | 0.3207 | 0.4174 | -2.31% | 32.07% | 41.98% |
| 0.50 | 0.4174 | 0.3207 | 0.4174 | -12.61% | 32.07% | 42.62% |

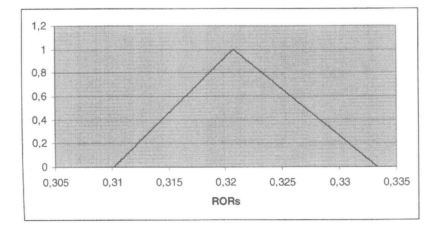

**Fig. 1.** Possible RORs and Their Membership Degrees with SYD; FR=0.01

### 3.1.1 Graphical Illustrations

Fig.1 illustrates the fuzzy ROR value for a fuzzification rate of 0.01. The least possible value is larger than the largest possible value of MARR. This means that at this fuzzification level, there is no possibility of a negative net present worth.

Fig. 2 illustrates the fuzzy ROR value for a fuzzification rate of 0.02.

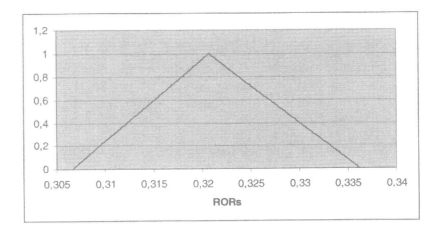

**Fig. 2.** Possible RORs and Their Membership Degrees with SYD; FR=0.02

Fig. 3 illustrates the fuzzy ROR value for a fuzzification rate of 0.10.

Fig. 4 illustrates the fuzzy ROR value for a fuzzification rate of 0.30. The least possible value is not anymore larger than the largest possible value of MARR. This means that at this fuzzification level, there is a possibility of a negative net present worth.

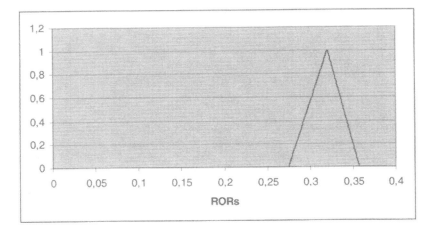

**Fig. 3.** Possible RORs and Their Membership Degrees with SYD; FR=0.10

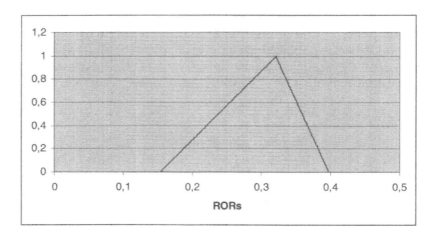

**Fig. 4.** Possible RORs and Their Membership Degrees with SYD; FR=0.30

### 3.2 Straight Line (SL) Depreciation

When SL depreciation is preferred, the following table is obtained in the crisp case:

**Table 12.** SL Depreciation – Crisp Case

| Year | BTCF | Straight Line Depreciation | Taxable cash flow | Tax | ATCF |
|---|---|---|---|---|---|
| 0 | -20,000 | | | | -20,000 |
| 1 | 7,000 | 3,000 | 4,000 | 1,200 | 5,800 |
| 2 | 9,000 | 3,000 | 6,000 | 1,800 | 7,200 |
| 3 | 11,000 | 3,000 | 8,000 | 2,400 | 8,600 |
| 4 | 13,000 | 3,000 | 10,000 | 3,000 | 10,000 |
| 5 | 15,000 | 3,000 | 12,000 | 3,600 | 11,400 |
| 5 | SV=5,000 | | | | 5,000 |
| BTCF-ROR | 41.74% | | | ATCF-ROR | 31.21% |

Under fuzziness, the ATCF-ROR value can be calculated as follows.

As it can be seen from Table 14, the investment project is acceptable since the least possible value, 26.63%, is larger than the largest possible value of the fuzzy MARR.

Figure 5 shows the membership function of the fuzzy ATCF-ROR value for FR=10%.

Now, with a 30% fuzzification ratio, the fuzzy ATCF-ROR is calculated as (16.45%, 31.21%, 37.90%) as in Table 15, which means that there is a slight possibility of a negative net present worth.

Depreciation and Income Tax Considerations under Fuzziness    169

**Table 13.** SL Depreciation – Fuzzy Case with a 10% Fuzzification Ratio

|   | Fuzzy BTFC | | | Straight Line Depreciation (10% fuzzification) | | |
|---|---|---|---|---|---|---|
| 0 | -18,000 | -20,000 | -22,000 | | | |
| 1 | 6,300 | 7,000 | 7,700 | 2,700 | 3,000 | 3,300 |
| 2 | 8,100 | 9,000 | 9,900 | 2,700 | 3,000 | 3,300 |
| 3 | 9,900 | 11,000 | 12,100 | 2,700 | 3,000 | 3,300 |
| 4 | 11,700 | 13,000 | 14,300 | 2,700 | 3,000 | 3,300 |
| 5 | 13,500 | 15,000 | 16,500 | 2,700 | 3,000 | 3,300 |
| 5 | 4,500 | 5,000 | 5,500 | | | |
| BTCF-ROR | 41.74% | 41.74% | 41.74% | | | |

**Table 14.** SL Depreciation – Fuzzy Case (continues)

| Taxable cash flow | | | Tax (0,28; 0,30; 0,32) | | | ATCF | | |
|---|---|---|---|---|---|---|---|---|
| | | | | | | -18,000 | -20,000 | -22,000 |
| 3,000 | 4,000 | 5,000 | 840 | 1,200 | 1,600 | 4,700 | 5,800 | 6,860 |
| 4,800 | 6,000 | 7,200 | 1,344 | 1,800 | 2,304 | 5,796 | 7,200 | 8,556 |
| 6,600 | 8,000 | 9,400 | 1,848 | 2,400 | 3,008 | 6,892 | 8,600 | 10,252 |
| 8,400 | 10,000 | 11,600 | 2,352 | 3,000 | 3,712 | 7,988 | 10,000 | 11,948 |
| 10,200 | 12,000 | 13,800 | 2,856 | 3,600 | 4,416 | 13,584 | 16,400 | 19,144 |
| | | | | | | 4,500 | 5,000 | 5,500 |
| | | | | | ATCF-ROR | 26.63% | 31.21% | 34.59% |

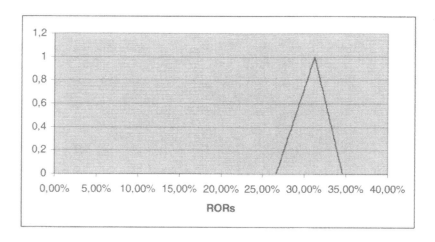

**Fig. 5.** Possible RORs and Their Membership Degrees with SL; FR=0.10

**Table 15.** SL Depreciation – Fuzzy Case with a 30% Fuzzification Ratio

|   | Fuzzy BTFC | | | Straight Line Depreciation (10% fuzzification) | | |
|---|---|---|---|---|---|---|
| 0 | -14,000 | -20,000 | -26,000 | | | |
| 1 | 4,900 | 7,000 | 9,100 | 2,700 | 3,000 | 3,300 |
| 2 | 6,300 | 9,000 | 11,700 | 2,700 | 3,000 | 3,300 |
| 3 | 7,700 | 11,000 | 14,300 | 2,700 | 3,000 | 3,300 |
| 4 | 9,100 | 13,000 | 16,900 | 2,700 | 3,000 | 3,300 |
| 5 | 10,500 | 15,000 | 19,500 | 2,700 | 3,000 | 3,300 |
| 5 | 3,500 | 5,000 | 6,500 | | | |
| BTCF-ROR | 41.74% | 41.74% | 41.74% | | | |

**Table 16.** SL Depreciation – Fuzzy Case with a 30% Fuzzification Ratio (continues)

| Taxable cash flow | | | Tax (0,28; 0,30; 0,32) | | | ATCF | | |
|---|---|---|---|---|---|---|---|---|
| | | | | | | -14,000 | -20,000 | -26,000 |
| 1,600 | 4,000 | 6,400 | 448 | 1,200 | 2,048 | 2,852 | 5,800 | 8,652 |
| 3,000 | 6,000 | 9,000 | 840 | 1,800 | 2,880 | 3,420 | 7,200 | 10,860 |
| 4,400 | 8,000 | 11,600 | 1,232 | 2,400 | 3,712 | 3,988 | 8,600 | 13,068 |
| 5,800 | 10,000 | 14,200 | 1,624 | 3,000 | 4,544 | 4,556 | 10,000 | 15,276 |
| 7,200 | 12,000 | 16,800 | 2,016 | 3,600 | 5,376 | 8,624 | 16,400 | 23,984 |
| | | | | | | 3,500 | 5,000 | 6,500 |
| | | | | | ATCF-ROR | 16.45% | 31.21% | 37.90% |

Figure 6 shows the membership function of the fuzzy ATCF-ROR value for FR=30%.

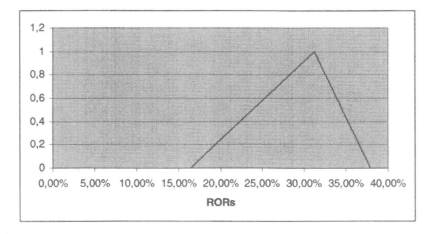

**Fig. 6.** Possible RORs and Their Membership Degrees with SL; FR=0.30

# 4 Conclusions

Depreciation allows for the wear and tear on a fixed asset and must be deducted from your income. You must claim depreciation on fixed assets used in your business that have a useful lifespan of more than 12 months. Not all fixed assets can be depreciated. Land is a common example of a fixed asset that cannot be depreciated. This chapter considered the after-tax cash flow analysis including fuzzy cash flows and fuzzy depreciation payments. A fuzzy ROR is obtained from the fuzzy ATCF. Two different depreciation methods are involved in the analysis: Straight line depreciation and accelerated cost recovery system. The fuzzification ratio was changed from 1% to 50%. In the analyses, capital gains and capital loses have not been taken into account. A capital gain is the difference between what you paid for an investment and what you received when you sold that investment. If you made a profit on the investment, then you have a capital gain. If you lost money on the investment, then you have a capital loss. The further research may handle capital loses and capital gains.

# References

Berg, M., Waegenaere, A.D., Wielhouwer, J.L.: Optimal tax depreciation with uncertain future cash flows. European Journal of Operational Research 132, 197–209 (2001)

Hartman, J.C., Liedtka, S.L., Snyder, L.V.: The impact of US tax depreciation law on asset location and ownership decisions. Computers & Operations Research 34(12), 3560–3568 (2007)

Jacco, L., Wielhouwer, A.D., Waegenaere, P., Kort, M.: Optimal tax depreciation under a progressive tax system. Journal of Economic Dynamics & Control 27, 243–269 (2002)

Joseph, C.H., Stephen, L.L., Lawrence, V.S.: The impact of US tax depreciation law on asset location and ownership decisions. Computers & Operations Research 34, 3560–3568 (2007)

Keating, A.S., Zimmerman, J.L.: Depreciation-policy changes: tax, earnings management, and investment opportunity incentives. Journal of Accounting and Economics 28(3), 359–389 (1999)

Wang, M.J., Liang, G.S.: Benefit/Cost analysis using fuzzy concept. Engineering Economist 40(4), 359–377 (1995)

Newnan, D.G., Lavelle, J.P.: Essentials of engineering economic analysis. Oxford University Press, Oxford (1998)

Robinson, J.: After tax cash flow analysis. Journal of Property Valuation and Investment 5(1), 18–29 (1987)

Scott, K., Jerold, L.Z.: Depreciation-policy changes: tax, earnings management, and investment opportunity incentives. Journal of Accounting and Economics 28, 359–389 (2000)

# Effects of Inflation under Fuzziness and Some Applications

Cengiz Kahraman[1], Tufan Demirel[2], and Nihan Demirel[2]

[1] Istanbul Technical University, Department of Industrial Engineering, 34367,
Macka, İstanbul, Turkey
[2] Yıldız Technical University, Department of Industrial Engineering, İstanbul, Turkey

**Abstract.** This chapter presents the ways of incorporating the parameter *fuzzy inflation* to the engineering economy analyses. Inflation is a financial parameter difficult to estimate. The fuzzy set theory gives us the possibility of converting linguistic expressions about inflation estimates to numerical values. In the chapter, discounted cash flow techniques including these fuzzy expressions and some numerical examples are given. The obtained results show the interval of the worst and the best possible outcomes when fuzzy inflation rates are taken into account.

## 1 Introduction

Inflation is an increase in the amount of money necessary to obtain the same amount of product or service before the inflated price was present. With the inflation in any time prices rise and the purchasing power decreases, it takes more dollars for the same amount of goods or services. Deflation is the opposite of inflation. It has the opposite effects, with deflation prices decrease and the purchasing power increases. With the deflation, it takes fewer dollars in the future to buy the same amount of goods or services as it does today. The governments can be face to face with the inflation much more commonly than deflation at national economy (Blank and Tarquin 2002, Sharp-Bette and Park 1990, Degarmo et al. 1990, Gönen 1990, Young 1993).

Most people are undoubtedly aware that inflation has to do with price increases. What is perhaps less well-known is that the meaning of the word inflation has changed somewhat over time. Originally the word inflation was used to describe a characteristic of money – that its value was eroded. This happens when all prices in an economy rise at the same rate over time. When all prices rise at the same rate, households' incomes (for example wages) increase as much as their expenses. This means that households have to pay more for the same quantity of goods. However, neither household consumption nor its actual value (utility) is affected when all prices rise at the same rate.

Over time, however, the meaning of the word *inflation* has changed somewhat. Today it is often used synonymously with the words *price increase* and can thereby describe any kind of price rises, not just increases in all prices. For example, one often hears of wage inflation, domestic inflation or imported inflation. None of these terms mean an increase in all prices. Rather, they refer to rises in the prices of certain specific goods or services.

The most common and most well-known measure of inflation is the change in the consumer price index - the CPI. The CPI is a so-called cost-of-living index or

---

C. Kahraman (Ed.): Fuzzy Engineering Economics with Appl., STUDFUZZ 233, pp. 173–182, 2008.
springerlink.com        © Springer-Verlag Berlin Heidelberg 2008

compensation index. This means that the CPI measures how consumers' cost of living changes over time. If consumers' incomes increase at the same rate as their cost of living, their utility will be unchanged over time. The CPI is often used for exactly this purpose - as a basis for adjusting pensions or determining how compensation clauses in different agreements should be interpreted.

To adjust for the effects of inflation in project evaluation, most authors prefer to use a general index, such as the Consumers Price Index. The reason is as follows: Since it is the investors' real income or purchasing power that we seek to enhance, there is a slight advantage in choosing an index of Consumer Goods' prices such as the CPI.

The changeable value of currency is the reason of inflation. With the inflation the currency value goes down. The inflation and deflation can be occurred as higher prices for food, cars, and other purchased commodities and services for the people. On the other hand for the business and government, inflation has eroded the purchasing power of savings and earnings, if interest rates and salary raises have not kept pace with general price trends.

Inflation types are shown below:

- Cost push inflation: Increases in producers' costs that are passed along to customers, sometimes with disproportionate escalations that push prices up.
- Demand-pull inflation: Excessive spending power of consumers, sometimes obtained at the expense of savings that pulls prices up.

When the literature is searched, we can see that there are few works on fuzzy inflation. Kahraman and Tolga (1995) examine the effects of fuzzy inflation rate on after-tax rate calculations. De and Goswami (2006) present an EOQ model with fuzzy inflation rate and fuzzy deterioration rate when a delay in payment is permissible.

## 2 Relation between Inflation and Interest

Inflation affects everyone with some degree. The degree of inflation affects the consequences when inflation is mild, the economy prospers. When inflation is moderate, increased demand pulls prices still higher. When inflation is severe, prices rise much faster than wages do. When inflation is hyperinflation (this is the most dangerous level of inflation), this uncontrolled inflation destroys a nation's economy (Park 2006). Hyperinflation is a problem in countries where political instability, overspending by the government, weak international trade balances, etc., is present (Sharp-Bette and Park 1990). When the government has inflation in its economy, first of all it pays attention to credit restrictions, wage controls, contraction of the money supply, reduction in demand by raising taxes, increased demand by reducing taxes, enlarged supply of goods through greater productivity stimulated by investment incentives and wage-price guidelines backed by political persuasion to manage inflation and its effects on the economy (Park 2006). On the other hand the government can redefine the currency in terms of the currency of another country, control banks and corporations, and control of the flow of capital into and out of the country in order to decrease inflation (Sharp-Bette and Park 1990).

Most of the engineering economists are more interested in the effects of inflation than its causes and corrections. Due to inflation, a dollar assumes different values at different times. The dollars do not have the same value for yesterday, today and tomorrow. The engineering economists have to predict the dollars' values and organize their works by concerning with inflation and also they have to research what would happen if inflation became high level such as moderate or severe.

## 2.1 Inflation Measurement

The measurement of inflation is difficult because the prices of different goods and services do not increase or decrease by the same amount, they change at different times by the different amounts. The calculation of a general inflation rate can be changed by geographical differences in prices and different buying habits of consumers.

The whole sale price index, producer price index, and consumer price index are used for the measurement of inflation rates. The consumer price index is the most commonly used technique for the inflation rate measurement.

- The whole sale price index (WPI): This index measures inflation at the wholesale level for both consumer and industrial goods.
- Producer price index (PPI): This index measures average changes in prices received by producers of all commodities. On the other hand PPI is a composite that measures changes in prices paid for selected goods and services used by producers (Blank and Tarquin 2002).
- Consumer price index (CPI): For calculation of CPI, prices for goods are obtained monthly and are averaged according to demographic distributions. Then the prices are weighted according to the expenditure proportions of the typical family (Park 2006). CPI is a composite that measures changes in prices paid for selected goods and services used by ultimate consumers (Blank and Tarquin 2002). In CPI seven different things are measured as fixed market basket of goods, foods and beverages, housing, apparel and upkeep, transportation, medical care, entertainment, and other (education, personal care, etc.). This index shows the effect of retail price changes on a selected standard of living.

## 2.2 Impact of Inflation

The analysts have to be interested in the impact of inflation on economic evaluations. There are two basic methods for researchers in the literature for consideration of inflation in their calculations.

- Constant (real) dollars: All cash flows are converted to money units that have constant purchasing power for eliminating inflation effects. It is generally easier to estimate future costs in constant dollars because the estimator is familiar with today's values. On the other hand there are two weaknesses limiting the usefulness of the constant dollar: tax effects are ignored and no provision is made for differences in escalation rates among price and cost components (Biermann and Smidt 1990). It is denoted "R$".
- Future dollars: In the amount of money units that are called future dollars actually exchanged at the time of each transaction for estimating cash flows. Future dollars

are sometimes called then-current dollars, nominal dollars, or actual dollars. It is denoted as "A$"

It is a simple matter to convert estimates in real dollar flow to actual dollar flow when inflation is assumed to be a constant rate. To make comparisons between monetary amounts which occur in different time periods, the different-valued dollars must first be converted to constant value dollars in order to represent the same purchasing power over time (Sharp-Bette and Park 1990).

As an economic evaluation, when the rate of inflation increases, there is a corresponding increase in the market interest rate. Inflation is differential rather than uniform. Goods' and services' prices do not always change proportionately. Including the effect of inflation is a second-order refinement for economic evaluations; the first-order refinement was the inclusion of the effect of taxes on basic cash flow.

For eliminating weaknesses of the constant dollar approach, the analysts can use after-tax actual cash flow comparisons. In the literature, there are three different comparisons for actual cash flows:

- No responsive charges in after-tax analysis
- Multiple inflation rates in an after-tax analysis
- After-tax modified cash flow comparison.

In the literature there are three different rates about inflation:

- Real or inflation- free interest rate ($i$): This is the real interest rate which interest earned when the effects of changes in the value of currency (inflation) have been removed. The real interest rate presents an actual gain in put chasing power. Increase in real purchasing power expressed as a percent per period, or the interest rate at which R$ outflow is equivalent to R$ inflow (Blank and Tarquin 2002).

$$i = \frac{i_f - f}{1+f} \tag{1}$$

- Inflation-adjusted interest rate ($i_f$): This is the market interest rate which is the inter rate that has been adjusted to take inflation into account. This rate is a combination of the real interest rate and the inflation rate and, therefore it changes as the inflation rate changes. Increase in dollar amount to cover real interest and inflation expressed as a percent per period; it is the interest rate at which A$ outflow is equivalent to A$ inflow (Blank and Tarquin 2002).

$$i_f = i + f + if \tag{2}$$

- Inflation rate ($f$): This is a measure of the rate of changing in the value of the currency. On the other hand, general inflation rate is the overall rate for an individual or organization.

### 2.3 Tax Consideration

If a tax is paid for the interest received from an investment, Equation (2) becomes

$$i_f = (i + f + i \times f)/(1 - v) \tag{3}$$

where $v$ is the tax rate.

## 2.4 Future Worth Calculations Adjusted for Inflation

In future worth calculations, a future amount $F$ can have any one of four different interpretations (Sharp-Bette and Park 1990):

- Case 1. Actual amount accumulated: $F$, the actual amount of money accumulated, is obtained using the inflation adjusted (market) index rate.

$$P = F(1+i_f)^{-n} = F(P/F, i_f, n) \tag{4}$$

$$F = P(1+i_f)^n = P(F/P, i_f, n) \tag{5}$$

- Case 2. Constant-value with purchasing power: The purchasing power future dollars is determined by using the market rate if to calculate $F$ then deflating the future amount through division by $(1+f)^n$.

$$F = \left[P(1+i_f)^n\right]/(1+f)^n = \left[P(F/P, i_f, n)\right]/(1+f)^n \tag{6}$$

- Case 3. Future amount required no interest: This case recognizes that prices increase when inflation is present. Simply put, future dollars are worth less, so more are needed. No interest rate is considered at all in this case.

$$F = P(1+f)^n = P(F/P, f, n) \tag{7}$$

- Case 4. Inflation and real interest: Maintaining purchasing power and earning interest must account for both increasing prices and the time value of money. If the growth of capital is to keep up, funds must grow at a rate equal to or above the real interest rate $i$ plus a rate equal to the inflation rate $f$.

## 2.5 Capital Recovery Calculations Adjusted for Inflation

Since future dollars have less buying power than today's dollars, it is obvious that more dollars will be required to recover the present investment. This suggests the use of the inflated interest rate in the A/P formula (Sharp-Bette and Park 1990):

$$A = F(A/F, i_f, n) \tag{8}$$

$$A = P(A/P, i_f, n) \tag{9}$$

# 3 Fuzzy Inflation-Adjusted Interest Rate

The fuzzy inflation-adjusted interest is calculated by

$$\tilde{i}_f \cong \tilde{i} + \tilde{f} + \tilde{i} \times \tilde{f} \tag{10}$$

When $\tilde{i}$ and $\tilde{f}$ are represented by triangular fuzzy numbers (TFNs) $(i_1, i_2, i_3)$ and $(f_1, f_2, f_3)$, respectively, $\tilde{i}_f$ becomes

$$\tilde{i}_f \cong (i_1 + f_1 + i_1 f_1, \quad i_2 + f_2 + i_2 f_2, \quad i_3 + f_3 + i_3 f_3) \tag{11}$$

or using $\alpha -$ cuts, $\tilde{i}_f$ can be defined as

$$\tilde{i}_f \cong \begin{bmatrix} (i_2 - i_1 + f_2 - f_1 + i_2 f_2 - i_1 f_1)\alpha + i_1 + f_1 + i_1 f_1; \\ (i_2 - i_3 + f_2 - f_3 + i_2 f_2 - i_3 f_3)\alpha + i_3 + f_3 + i_3 f_3 \end{bmatrix} = \left( i_f^{l(y)}, i_f^{r(y)} \right) \tag{12}$$

where $i_f^{l(y)}$ : the left representation of the fuzzy inflation-adjusted interest rate at time t

and $i_f^{r(y)}$ the right representation of the fuzzy inflation-adjusted interest rate at time t.

## 4 Fuzzy Inflation Rates and Present Worth Calculations

Fuzzy present worth can be calculated by the following equation:

$$\tilde{PW} \cong \sum_{t=1}^{n} \tilde{C}_t \left( P/F, \tilde{i}_f, t \right) - \tilde{FC} \tag{13}$$

where $\tilde{C}_t$ is the fuzzy net cash flow in period $t$; $\tilde{i}_f$ is the fuzzy inflation-adjusted interest rate, and $FC$ is the fuzzy first cost.

Chiu and Park (1994) propose a present value formulation of a fuzzy cash flow. The result of the present value is also a fuzzy number with nonlinear membership function. The present value can be approximated by a TFN. Chiu and Park (1994)'s formulation is

$$\tilde{PW} = \begin{bmatrix} \sum_{t=0}^{n} \left( \dfrac{\max(C_t^{l(y)}, 0)}{\prod\limits_{t'=0}^{t} (1 + i_{ft'}^{r(y)})} + \dfrac{\min(C_t^{l(y)}, 0)}{\prod\limits_{t'=0}^{t} (1 + i_{ft'}^{l(y)})} \right), \\ \sum_{t=0}^{n} \left( \dfrac{\max(C_t^{r(y)}, 0)}{\prod\limits_{t'=0}^{t} (1 + i_{ft'}^{l(y)})} + \dfrac{\min(C_t^{r(y)}, 0)}{\prod\limits_{t'=0}^{t} (1 + i_{ft'}^{r(y)})} \right) \end{bmatrix} \tag{14}$$

where $C_t^{l(y)}$ : the left representation of the net cash at time $t$, $C_t^{r(y)}$ : the right representation of the net cash at time $t$, $i_{f_t}^{l(y)}$ : the left representation of the fuzzy inflation-adjusted interest rate at time $t$, $i_{f_t}^{r(y)}$ : the right representation of the fuzzy inflation-adjusted interest rate at time $t$.

Buckley's (1987) membership function for $\widetilde{PW}_n$,

$$\mu(x|\widetilde{PW}_n) = (pw_{n1}, f_{n1}(y|\widetilde{PW}_n)/pw_{n2}, pw_{n2}/f_{n2}(y|\widetilde{PW}_n), pw_{n3}) \tag{15}$$

is determined by

$$f_{ni}(y|\widetilde{PW}_n) = \sum_{t=1}^{n} f_i(y|\tilde{C}_t)(1 + f_k(y|\tilde{i}_f))^{-n} \tag{16}$$

for $i = 1,2$ where $k = i$ for negative $\tilde{C}$ and $k=3-i$ for positive $\tilde{C}$.

## 5 Fuzzy Inflation Rates and Future Worth Calculations

Chiu and Park's (1994) formulation for the fuzzy future value has the same logic of fuzzy present value formulation:

$$\begin{bmatrix} \sum_{t=0}^{n-1}[\max(C_t^{l(y)},0) \prod_{t'=t+1}^{n}(1+i_{ft'}^{l(y)}) + \min(C_t^{l(y)},0) \prod_{t'=t+1}^{n}(1+i_{ft'}^{r(y)})] + C_n^{l(y)}, \\ \sum_{t=0}^{n-1}[\max(C_t^{r(y)},0) \prod_{t'=t+1}^{n}(1+i_{ft'}^{r(y)}) + \min(C_t^{r(y)},0) \prod_{t'=t+1}^{n}(1+i_{ft'}^{l(y)})] + C_n^{r(y)} \end{bmatrix} \tag{17}$$

Buckley's (1987) membership function $\mu(x|\widetilde{FW})$ is determined by

$$f_i(y|\widetilde{FW}_n) = \sum f_i(y|\tilde{C}_t)(1 + f_i(y|\tilde{i}_f))^n \tag{18}$$

For the uniform cash flow series, $\mu(x|\widetilde{FW})$ is determined by

$$f_{ni}(y|\widetilde{FW}) = f_i(y|\tilde{A})\beta(n, f_i(y|\tilde{i}_f)) \tag{19}$$

where $i=1, 2$; $\tilde{A}$ is the fuzzy equivalent uniform annual cash flow and $\beta(n,i_f) = (((1+i_f)^n - 1)/i_f)$ and $\tilde{A} \succ 0$ and $\tilde{i}_f \succ 0$.

## 6 Fuzzy Inflation Rates and Capital Recovery Calculations

The EUAV means that all incomes and disbursements (irregular and uniform) must be converted into an equivalent uniform annual amount, which is the same each period. The major advantage of this method over all the other methods is that it does not require making the comparison over the least common multiple of years when the alternatives have different lives (Kahraman et al. 2002). The general equation for this method is

$$EUAV = A = PW\gamma^{-1}(n,i_f) = PW[\frac{(1+i_f)^n i_f}{(1+i_f)^n - 1}] \tag{20}$$

180     C. Kahraman, T. Demirel, and N. Demirel

where $PW$ is the net present value. In the case of fuzziness, $\tilde{PW}$ will be calculated and then the fuzzy $EU\tilde{A}V$ ($\tilde{A}_n$) will be found. The membership function $\mu(x|\tilde{A}_n)$ for $\tilde{A}_n$ is determined by

$$f_{ni}(y|\tilde{A}_n) = f_i(y|\tilde{PW})\gamma^{-1}(n, f_i(y|\tilde{i}_f))$$

(21)

and TFN(y) for fuzzy $EUAV$ is

$$\tilde{A}_n(y) = (\frac{\tilde{PW}^{l(y)}}{\gamma(n, i_f{}^{l(y)})}, \frac{\tilde{PW}^{r(y)}}{\gamma(n, i_f{}^{r(y)})})$$

(22)

## 7   Some Applications

In the following, three applications will be given:

### Application 1

Compute the fuzzy real interest rate if the fuzzy inflation-adjusted interest rate and the fuzzy inflation rate are (15%, %16, %17) and (1%, 2%, 3%), respectively.

$$(0.15, 0.16, 0.17) = (0.01, 0.02, 0.03) + \tilde{i} + (0.01, 0.02, 0.03)\tilde{i}$$

$$(1.01, 1.02, 1.03)\tilde{i} = (0.12, 0.14, 0.16) \text{ and } \tilde{i} \cong (0.1165, 0.1373, 0.1584)$$

### Application 2

A \$ (-14,000, -12,000, -10,000) investment will return annual benefits \$(2,650, 2,775, 2,900) for six years with no salvage value at the end of six years. Compute the fuzzy net present worth ($NPW$) of the cash flow using a fuzzy inflation-adjusted interest rate of (7.12%, 10.25%, 13.42%) per year.

$$f_{6,1}(y|\tilde{NPW}) = \sum_{j=0}^{6} f_{j,1}(y|\tilde{F}_j)[1 + f_2(y|\tilde{i}_f)]^{-j} - \tilde{FC}_1$$

(23)

$$f_{6,2}(y|\tilde{NPW}) = \sum_{j=0}^{6} f_{j,2}(y|\tilde{F}_j)[1 + f_1(y|\tilde{i}_f)]^{-j} - \tilde{FC}_2$$

(24)

For y = 0, $f_{6,1}(y|\tilde{PW}) = \$ - 3{,}525.57$

For y = 1, $f_{6,1}(y|\tilde{PW}) = f_{6,2}(y|\tilde{P}) = \$ - 24.47$

For y = 0, $f_{6,2}(y|\tilde{PW}) = \$ + 3{,}786.34$

Effects of Inflation under Fuzziness and Some Applications     181

**Table 1.** Fuzzy cash flow

| Year | Before-tax cash flow | Depreciation | Cash flows to be taxed | Tax amounts | After-tax cash flow (ATCF) | ATCF in to-day's dollars |
|------|----------------------|--------------|------------------------|-------------|----------------------------|--------------------------|
| 0 | (-18, -16, -14) | | | | (-18, -16, -14) | (-18, -16, -14) |
| 1 | (5, 6, 7) | (0.5, 0.6, 0.7) | (4.3, 5.4, 6.5) | (1.29, 1.62, 1.95) | (3.05, 4.38, 5.71) | (2.933, 4.252, 5.598) |
| 2 | (6, 7, 8) | (0.6, 0.7, 0.8) | (5.2, 6.3, 7.4) | (1.56, 1.89, 2.22) | (3.78, 5.11, 6.44) | (3.635, 4.961, 6.314) |
| 3 | (7, 8, 9) | (0.7, 0.8, 0.9) | (6.1, 7.2, 8.3) | (1.83, 2.16, 2.49) | (4.51, 5.84, 7.17) | (4.337, 5.670, 7.029) |
| 4 | (8, 9, 10) | (0.8, 0.9, 1.0) | (7.0, 8.1, 9.2) | (2.10, 2.43, 2.76) | (5.24, 6.57, 7.90) | (5.038, 6.379, 7.745) |

The possibility of $\widetilde{PW} = 0$ for this triangular fuzzy number can be calculated using a linear interpolation:

$$x = -3,810.81y + 3,786.34$$

For $x = 0$, Poss($\widetilde{PW} = 0$) = 0.9936.

**Application 3**

Consider the following before-tax fuzzy inflated cash flows and fuzzy depreciation payments ($\times\$1000$). Using a tax rate of 30%, a fuzzy real interest rate of (6%, 7%, 8%) and a fuzzy annual inflation rate of (2%, 3%, 4%), compute the fuzzy net present worth of the after-tax cash flows.

$$f_{4,1}\left(y\middle|\widetilde{NPW}\right) = \sum_{j=0}^{4} f_{j,1}(y\middle|\widetilde{F}_j)[1 + f_2(y\middle|\widetilde{i}_f)]^{-j} - \widetilde{FC}_1$$

$$= (2.933 + 1.319y)(1.08 - 0.01y)^{-1} + (3.635 + 1.326y)(1.08 - 0.01y)^{-2} +$$

$$(4.337 + 1.333y)(1.08 - 0.01y)^{-3} + (5.038 + 1.341y)(1.08 - 0.01y)^{-4} - (18 - 2y)$$

$$f_{4,2}\left(y\middle|\widetilde{NPW}\right) = \sum_{j=0}^{4} f_{j,2}(y\middle|\widetilde{F}_j)[1 + f_1(y\middle|\widetilde{i}_f)]^{-j} - \widetilde{FC}_2$$

$$= (5.598 - 1.346y)(1.06 + 0.01y)^{-1} + (6.314 - 1.353y)(1.06 + 0.01y)^{-2} +$$

$$(7.029 - 1.359y)(1.06 + 0.01y)^{-3} + (7.745 - 1.366y)(1.06 + 0.01y)^{-4} - (14 + 2y)$$

For $y = 0$, $f_{4,1}\left(y\middle|\widetilde{NPW}\right) = -\$5,022$

For $y = 1$, $f_{4,1}\left(y\middle|\widetilde{NPW}\right) = f_{4,2}\left(y\middle|\widetilde{NPW}\right) = \$1,802$

For $y = 0$, $f_{4,2}\left(y\middle|\widetilde{NPW}\right) = \$8,937$

## 8 Conclusions

Inflation is a process of continuous (persistent) increase in the price level. Inflation results in a decrease of the value of money. Inflation is an increase in the prices of all goods and services not only of a particular good or service. An increase in the price of one good is not inflation. Inflation is an ongoing process, not a one-time jump in the price level. Inflation is one of the hardest parameters in financial analysis to estimate. Fuzzy set theory presents an excellent tool to handle this uncertainty. The expressions like "an inflation rate around 5% per annum" or "an inflation rate between 4% and 6% per annum" can be incorporated into the discounted cash flow analysis. This chapter illustrates this with numerical examples including depreciation and tax parameters.

## References

Bierman, H., Smidt, S.: The capital budgeting decision, economic analysis of investment projects. Mc Millan Inc. (1990)

Blank, L., Tarquin, A.: Engineering economy. McGraw-Hill, New York (2002)

Buckley, J.J.: The fuzzy mathematics of finance. Fuzzy Sets and Systems 21, 257–273 (1987)

Chiu, C.Y., Park, C.S.: Fuzzy cash flow analysis using present worth criterion. Engineering Economist 39(2), 113–138 (1994)

De, K.S., Goswami, A.: An EOQ model with fuzzy inflation rate and fuzzy deterioration rate when a delay in payment is permissible. International Journal of Systems Scienc 37(5), 323–335 (2006)

Degarmo, E.P., Sullivan, W.G., Bontadelli, J.A.: Engineering economy. Mac Millan (1990)

Gönen, T.: Engineering economics for engineering managers. John Wiley and Sons, Chichester (1990)

Kahraman, C., Ruan, D., Tolga, E.: Capital budgeting techniques using discounted fuzzy versus probabilistic cash flows. Information Sciences 142(1), 57–76 (2002)

Kahraman, C., Tolga, E.: The effects of fuzzy inflation rate on after-tax rate calculations. In: Third Balkan Conference on Operational Research Thessaloniki, Greece (1995)

Kahraman, C., Tolga, E., Ulukan, Z.: Justification of manufacturing technologies using fuzzy benefit/cost ratio analysis. International Journal of Production Economics 66, 45–52 (2000)

Park, C.S.: Contemporary engineering economics. Prentice Hall, Englewood Cliffs (2006)

Sharp-Bette, G.P., Park, C.S.: Advanced engineering economics. Wiley, New York (1990)

Young, D.: Modern engineering economy. John Wiley and Sons, Chichester (1993)

# Fuzzy Sensitivity Analysis and Its Application

Toshihiro Kaino[1], Kaoru Hirota[2], and Witold Pedrycz[3]

[1] School of Business, Aoyama Gakuin University
[2] Interdisciplinary Graduate School of Science and Engineering, Tokyo Institute of Technology
[3] Department of Electrical and Computer Engineering, University of Alberta

**Abstract.** In this study, we are concerned with the concept, properties and algorithms of differentiation of the Choquet integral. The differentiation of the Choquet integral of a nonnegative measurable function is studied in the setting of sensitivity analysis. It is shown that the differentiation of Choquet integral reflects a way in which the aggregation is sensitive to the function being aggregated. The aggregation function using fuzzy integral could be viewed central to data-mining, the business process mining, web-marketing, e-commerce and the other pattern-matching problems in economics. Next, the differentiation of the Choquet integral is extended to the differentiation of the generalized t-conorm integral. Four types of the differentiation of the generalized t-conorm integrals (namely, Choquet integral, Sugeno integral, and the other generalized t-conorm integrals) are discussed and compared with regard to their sensitivity properties. Lastly, the Choquet integral is applied to the credit risk analysis (long-term debt rating ) to make clear the significance of them, especially, this differentiation is shown to be an effective sensitivity analysis tool which gives us how much evaluation of each index influence to the total evaluation on the corporations.

## 1 Introduction

The concept of a fuzzy measure and fuzzy integral have been introduced by Sugeno (1972) and the functional (the Choquet integral) defined by Choquet (1953) has been revalued in terms of the fuzzy measure. These fuzzy integrals are applied to multi-criteria evaluation and prediction problems (Liou and Tzeng 2008, Chen et al. 2008, Hu and Tseng 2007). While fuzzy measures and fuzzy integrals have been studied quite intensively, the concept of differentiation, with a few exceptions (Kaino and Hirota 1999, Wu and Li 2001), still deserves careful attention. Kaino and Hirota (1999, 2000, 2004, 2005) started investigations on differentiation of fuzzy relations and the Choquet Integral and applied the findings to problems of decision making and credit risk analysis. In this paper, it is shown that the differentiation of the Choquet integral reflects a way in which the aggregation becomes sensitive to the function being aggregated. The aggregation function using fuzzy integral could be viewed central to data-mining, the business process mining, web-marketing, e-commerce and the other pattern-matching problems in economics. And, the differentiation of the Choquet integral is applied to the identification of the fuzzy measure. Furthermore, the concept of differentiation of the Choquet integral is extended to the constructs involving generalized t-conorms. With this regard, we also present some comparative analysis. Throughout the study, we present a number of numeric illustrative examples. Lastly, the Choquet integral is applied to the credit risk analysis (long-term debt rating) to

C. Kahraman (Ed.): Fuzzy Engineering Economics with Appl., STUDFUZZ 233, pp. 183–216, 2008.
springerlink.com © Springer-Verlag Berlin Heidelberg 2008

184     T. Kaino, K. Hirota, and W. Pedrycz

make clear the significance of them, especially, this differentiation is shown to be an effective tool which gives us how much evaluation of each index influence to the total evaluation on the corporations.

## 2 Sensitivity Analysis of Differentiation of the Choquet Integral as an Aggregation Function in Data-Mining

### 2.1 Interval Limited Choquet Integral

In what follows, we briefly review some basic concepts of fuzzy measures and integrals, cf. (Choquet 1953).

**Definition 1(Choquet 1953)**
Let $(X, \Im, g)$ be a fuzzy measure space. Let $f$ be a nonnegative measurable function. The Choquet integral of the function $f$ with respect to the fuzzy measure $g$ is defined as

$$(C)\int_X fdg := \int_0^{+\infty} g(\{x \mid f(x) \geq r\})dr. \tag{1}$$

Accordingly, the Choquet integral of a nonnegative simple function is expressed as presented by the following proposition.

**Proposition 1(Choquet 1953)**
Let $(X, \Im, g)$ be a fuzzy measure space. Let $f$ be a nonnegative simple function

$$f(x) = \sum_{i=1}^n r_i \chi_{D_i}(x). \tag{2}$$

where

$$\left( 0 < r_1 < r_2 < \cdots < r_n, \quad \chi_{D_i}(x) = \begin{cases} 0 \, [x \notin D_i] \\ 1 \, [x \in D_i] \end{cases}, \quad \text{if } i \neq j \text{ then } D_i \cap D_j = \phi \right)$$

The Choquet integral of the nonnegative simple function $f$ computed with respect to the fuzzy measure $g$ is given in the form

$$(C)\int_X fdg := \sum_{i=1}^n (r_i - r_{i-1})g(A_i). \tag{3}$$

$$(A_i := \cup_{j=1}^n D_j, r_0 := 0)$$

Kaino and Hirota (1999) defined the Choquet integral limited by an interval as follows

**Definition 2 (Kaino and Hirota 1999)**
Let $(X, \Im, g)$ be a fuzzy measure space. Consider $f$ to be a nonnegative measurable function. The $[0, r_*]$ limited Choquet integral of a nonnegative measurable function $f$ with respect to a fuzzy measure $g$ is defined as

$$F(r_*) := (C)\int_X f \wedge r_* dg = \int_0^{+\infty} g(\{x \mid f(x) \wedge r_* \geq r\})dr =: \int_0^{r_*} g(\{x \mid f(x) \geq r\})dr. \tag{4}$$

Let $0 < {}^\forall a \le {}^\forall c \le {}^\forall b(a, b, c \in \Re)$. Then the $[a, b]$ limited Choquet integral of a non-negative measurable function $f$ with respect to the fuzzy measure $g$ is defined as

$$\int_a^b g(\{x \mid f(x) \ge r\}) dr := F(b) - F(a) \tag{5}$$

The following property holds true,

$$\int_a^b g(\{x \mid f(x) \ge r\}) dr = \int_a^c g(\{x \mid f(x) \ge r\}) dr + \int_c^b g(\{x \mid f(x) \ge r\}) dr. \tag{6}$$

## 2.2 Differentiation of the Choquet Integral

Now let us define a process of differentiation of the Choquet integral (Kaino and Hirota 1999) based on a nonnegative measurable function.

**Definition 3 (Kaino and Hirota 1999)**
Let $(X, \Im, g)$ be a fuzzy measure space. For any nonnegative measurable function $f$, if

$$\lim_{\Delta r \to +0} \frac{F(r + \Delta r) - F(r)}{\Delta r} =: D^+ F(r) = F_+^{'}(r) \tag{7}$$

exists, then this limit is called an upper differential coefficient of the $[0, r]$ limited Choquet integral $F$ of $f$ with respect to $g$ at $r$. Similarly, if the limit

$$\lim_{\Delta r \to -0} \frac{F(r + \Delta r) - F(r)}{\Delta r} =: D^- F(r) = F_-^{'}(r) \tag{8}$$

exists, then it called a lower differential coefficient of the $[0, r]$ limited Choquet integral $F$ of $f$ with respect to $g$ at $r$. If and only if both the upper differential co-efficient and the lower differential coefficient of $F$ at $r$ exist and are equal to each other, we use the notation

$$DF(r) = F'(r) = \frac{dF(r)}{dr}, \tag{9}$$

Then $DF(r)$ is called a differential coefficient of the $[0, r]$ limited Choquet integral $F$ of $f$ with respect to $g$ at $r$. If $\dfrac{dF(r)}{dr}$ exists on a certain interval of $r$, then we say that $F(r)$ is differentiable on this interval. If $F'(r)$ is also differentiable, then the derived function of it, denoted by $\dfrac{d^2 F(r)}{dr^2}$, or $F''(r)$, is called a second derived function of $F$. Similarly, the $n$-th derived function of $f$ at $r$ is denoted by $\dfrac{d^n F(r)}{dr^n}$ or $F^{(n)}(r)$. If $F^{(n)}(r)$ exists, then $F(r)$ is called $n$-th differentiable. Moreover, if $\dfrac{dF(r)}{dr}$ exists, then $\dfrac{dF(r)}{dr} = g\{x \mid f(x) \ge r\}$.

An illustrative example helps clarify the main concept of the differential coefficient.

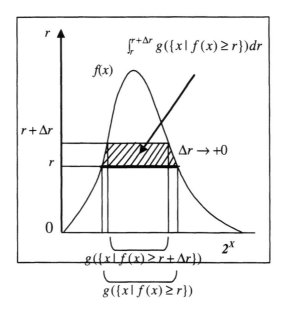

**Fig. 1.** Upper differential coefficient of the [0, r] limited Choquet integral

**Example 1(Kaino and Hirota 1999)**
Consider the following Choquet integral:

$$(C)\int_X f d\mu := \sum_{i=1}^{n}(r_i - r_{i-1})g(A_i) \quad (10)$$

$$(A_i := \cup_{j=1}^{n} D_j, r_0 := 0)$$

The differentiation of the [0, r] limited Choquet integral is realized as follows:

$$r_i < {}^\forall r < r_{i+1}, \frac{dF(r)}{dr} = g(A_{i+1}) \quad (11)$$

If $r = r_i$, then we define
$$\begin{cases} D^+F(r) = g(A_{i+1}) \\ D^-F(r) = g(A_i) \end{cases} \quad (12)$$

The following theorem presents the main features of the integral.

**Theorem 1(Kaino and Hirota 1999)**
If the $[0, r]$ limited Choquet integrals $F(r)$ and $G(r)$ of $f$ and $g$, respectively, are differentiable then $cF(r)$ (where $c$ is a constant), $F(r)+G(r)$, $F(r) \cdot G(r)$, $\frac{F(r)}{G(r)}(G(r) \neq 0)$ are also differentiable and the following relationships are satisfied.

(1) $(F(r)+c)' = F'(r)$ (where $c$ is a constant) \quad (13)

(2) $(cF(r))' = cF'(r)$ (where $c$ is a constant) \quad (14)

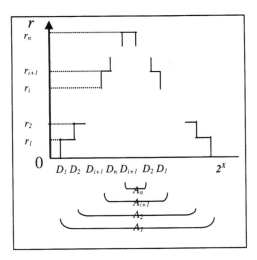

**Fig. 2.** Nonnegative simple function

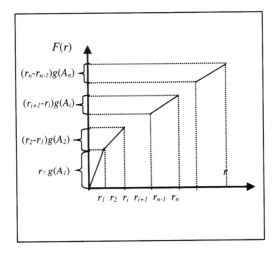

**Fig. 3.** Relation between $r$ and $F(r)$

$$(3)\ (F(r) \pm G(r))' = F'(r) \pm G'(r) \tag{15}$$

$$(4)\ (F(r) \cdot G(r))' = F'(r) \cdot G(r) + F(r) \cdot G'(r) \tag{16}$$

$$(5)\ \left(\frac{F(r)}{G(r)}\right)' = \frac{F'(r) \cdot G(r) - F(r) \cdot G'(r)}{G(r)^2} \tag{17}$$

Proof: Omit for paper length limitation.

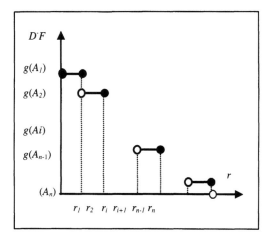

**Fig. 4.** Relation between $r$ and $D^-F(r)$

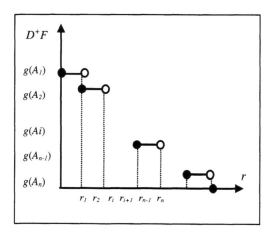

**Fig. 5.** Relation between $r$ and $D^+F(r)$

### Remark 1 (Kaino and Hirota 1999)

Let $f$ and $g$ be nonnegative measurable functions. Consider $h(x) = f_A(x) + f_B(x)$.

If $H'(r)$, $F_A'(r)$, $F_B'(r)$ are derived functions of the $[0, r]$ limited Choquet integrals of $f_A$, $f_B$, $h$, respectively, with respect to $g$ at $r$, and exist, the equation

$$H'(x) = F_A'(x) + F_B'(x) \tag{18}$$

is not always satisfied.

### Proof

We show a counter example to demonstrate that this property does not always hold.

If $A \cap B = \phi$, $f_A = t_A \cdot \chi_A$, $f_B = t_B \cdot \chi_B$ (where $t_A > t_B$, $t_A, t_B \in \Re$, $\chi_A$, $\chi_B$ are characteristic functions of sets A, B, $H'(r)$, $F_A'(r)$, $F_B'(r)$ exist, and $r < t_B$, then the following relationship can be obtained:

$$H'(r) = g(A \cup B), \quad F_A'(r) = g(A), \quad F_B'(r) = g(B). \tag{19}$$

Now, if $A \cap B = \phi$ then $g(A \cup B) = g(A) + g(B)$ is not always concluded. So, the equation (19) is not always satisfied.

## 2.3 Sensitivity Analysis as an Aggregation Function in Data-Mining

It is shown that the differentiation of the Choquet integral can reflect a way in which the aggregation is sensitive to the function $f$ being aggregated (refer to Example 2).

In advance, two interesting properties can be observed.

**Remark 2**
Let $(X, \Im, g)$ be a fuzzy measure space. Let $f$ be a nonnegative measurable function. If $F'(r)$ is a derived function of the $[0, r]$ limited Choquet integral of $f$ with respect to $g$

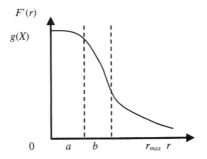

(1) High sensitivity for low values of "$r$"

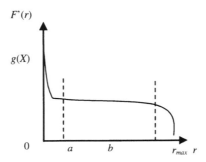

(2) Balanced sensitivity which is more or less equal over the entire range of values of "$r$" in the unit interval

**Fig. 6.** Some examples of the derived function $F'(r)$

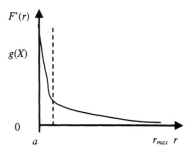

(3) Low sensitivity at almost all values of "$r$"

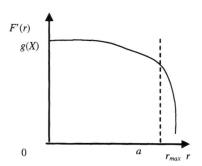

(4) Highly variable sensitivity for the entire range of values of "$r$"

**Fig. 6.** (*continued*)

at $r$, $F'(r)$ should be a non-increasing function of $r$. Here, $F'(0)= g(X)$ and $F'(r_{max})=0$ (where $r_{max}$ is the largest value of $f$).

**Proof**

Let $r_1 < r_2$ ($r_1, r_2 \in \Re$), then the inclusion $\{x \mid f(x) \wedge r_2 \geq r\} \subset \{x \mid f(x) \wedge r_1 \geq r\}$ holds given that $f$ is a non-negative measurable function.

Now, if $\{x \mid f(x) \wedge r_2 \geq r\} \subset \{x \mid f(x) \wedge r_1 \geq r\}$ , then $g(\{x \mid f(x) \wedge r_2 \geq r\}) < g(\{x \mid f(x) \wedge r_1 \geq r\})$ because that the fuzzy measure $g$ must be monotonic. Then the derived function $F'(r) = \dfrac{dF(r)}{dr} = g\{x \mid f(x) \geq r\}$ should be a non-increasing function of $r$. Here, $F'(0) = g\{x \mid f(x) \geq 0\} = g(X)$ , $F'(r_{max}) = g\{x \mid f(x) \geq r_{max}\} = g(\phi) = 0$ .(*end*)

In (1) of Figure 6, the derived function $F'(r)$ quickly decreases between the range of $a$ and $b$ of $r$. It exhibits high sensitivity for low values of $r$ and low sensitivity for high value of $r$.

In the second case shown in Figure 6, the derived function $F'(r)$ is almost flat when the values of $r$ range between $a$ and $b$ of its changes occur at the boundaries of the unit interim that is $a(\approx 0)$ and $b(\approx g(X))$.

## Fuzzy Sensitivity Analysis and Its Application

In the third case, the derived function $F'(r)$ rapidly decreases at 0 of $r$ and shows a moderate decrease for the values of $r$ between $a$ and $r_{max}$.

Finally, the fourth case illustrated in this figure shows that the derived function $F'(r)$ is highly sensitivity between the values of $r$ positioned in-between 0 and $a$.

### Remark 3

Let $(X, \Im, g)$ be a fuzzy measure space. Let $f$ be a nonnegative measurable function and $g(\cdot)$ be a $\lambda$-fuzzy measure as $g(X) = 1$. Let $D^- F(r)$ and $D^+ F(r)$ be a lower differentiation and an upper differentiation of the $[0, r]$ limited (discrete) Choquet integral of $f$ with respect to $g$ at $r$, respectively. Here, $r_i = f(x_i)$ and $r_0 = f(x_0) = 0$, the subscript $(i)$ indicates that the indices have been permuted so that $0 \leq r_1 \leq r_2 \leq \cdots \leq r_n \leq 1$, $A_i = \{x_i, \cdots, x_n\}$, and $n = |X|$. If $g(x_n)$ assumes high values, then $D^- F(r)$ and $D^+ F(r)$ $\left(0 \leq^{\forall} r < r_n\right)$ will also exhibit high values.

### Proof

From (11) and (12), we have $D^- F(r) = D^+ F(r) = g(A_{i+1})$ ( $r_i <^{\forall} r < r_{i+1}$ ), $D^- F(r) = g(A_i)$ and $D^+ F(r) = g(A_{i+1})$ $(r = r_i)$.

Here, $r_{n-1} \leq r < r_n$, $D^- F(r) = g(A_n) = g(x_n)$ and $g(x_n)$ assumes some high value close to 1. Now, in virtue of monotonicity of the fuzzy measure $g(\cdot)$ we have $g(A_i) \geq g(A_{i+1}) \geq \cdots \geq g(A_n) = g(x_n)$. Hence $D^- F(r)$ and $D^+ F(r)$ assume higher values than those assumed by $g(x_n)$.

### Example 2

The Choquet integral can be regarded as a useful aggregation vehicle (Marichal 2000, Chen and Xiao 2005, Kwak and Pedrycz 2005) when combining several classifiers in data mining. Consider a two-class problem. The fuzzy measure $g(\cdot)$ describes the quality of each classifier (its quantification can be easily captured by computing the corresponding classification rate). The obtained degrees of membership to class $\omega_i$ are given as

$$\begin{cases} f(x_1) = 0.2 \\ f(x_2) = 0.3 \\ f(x_3) = 0.4 \\ f(x_4) = 0.7 \\ f(x_5) = 0.9 \end{cases} \tag{20}$$

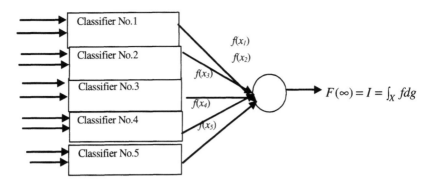

**Fig. 7.** Fusion of classifiers with the use of fuzzy integration

Then, fuzzy measure $g(\cdot)$ is treated as $\lambda$-fuzzy measure with $g(X) = 1$. The values of the fuzzy measures for each classifier $g(x_1), g(x_2), \cdots, g(x_5)$ are given as follows:

$$\begin{cases} g(x_1) = 0.57 \\ g(x_2) = 0.42 \\ g(x_3) = 0.14 \\ g(x_4) = 0.28 \\ g(x_5) = 0.28 \end{cases} \quad (21)$$

The value of the Choquet integral represents a degree of class membership forming the result of the fusion of the results generated by the individual classifier. In this way, we obtain

$$\begin{aligned} F(\infty) &= I = \int_X f dg \\ &= 0.2 \cdot g(X) + 0.3 \cdot g(\{x_2, x_3, x_4, x_5\}) + 0.4 \cdot g(\{x_3, x_4, x_5\}) \\ &\quad + 0.7 \cdot g(\{x_4, x_5\}) + 0.9 \cdot g(\{x_5\}) \\ &= 0.2 \cdot 1 + 0.3 \cdot 0.8066 + 0.4 \cdot 0.5845 + 0.7 \cdot 0.5008 + 0.9 \cdot 0.28 \\ &= 1.27843 \end{aligned} \quad (22)$$

In the sequel, we have

$$D^-F(r) = \begin{cases} g(X) = 1 & (0 \le r \le 0.2) \\ g(\{x_2, x_3, x_4, x_5\}) = 0.8066 & (0.2 < r \le 0.3) \\ g(\{x_3, x_4, x_5\}) = 0.5845 & (0.3 < r \le 0.4) \\ g(\{x_4, x_5\}) = 0.5008 & (0.4 < r \le 0.7) \\ g(\{x_5\}) = 0.28 & (0.7 < r \le 0.9) \\ g(\phi) = 0 & (0.9 < r) \end{cases} \quad (23)$$

$$D^+F(r) = \begin{cases} g(X) = 1 & (0 \le r < 0.2) \\ g(\{x_2, x_3, x_4, x_5\}) = 0.8066 & (0.2 \le r < 0.3) \\ g(\{x_3, x_4, x_5\}) = 0.5845 & (0.3 \le r < 0.4) \\ g(\{x_4, x_5\}) = 0.5008 & (0.4 \le r < 0.7) \\ g(\{x_5\}) = 0.28 & (0.7 \le r < 0.9) \\ g(\phi) = 0 & (0.9 \le r) \end{cases} \quad (24)$$

Plots of the sensitivity measure are contained in Figure 8 and 9 respectively.

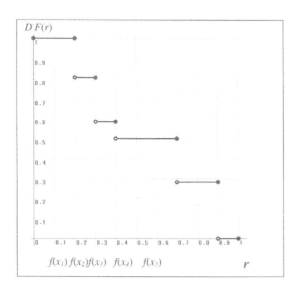

**Fig. 8.** Lower differentiation

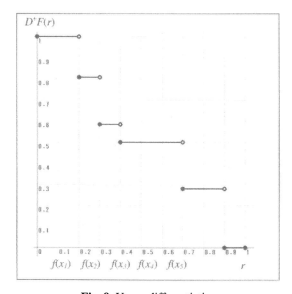

**Fig. 9.** Upper differentiation

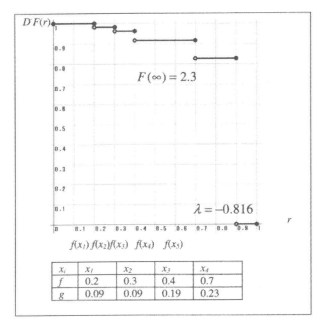

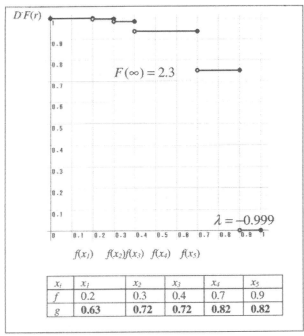

**Fig. 10.** Highly level of sensitivity of the fuzzy integral occurring for all values of "$r$"

Fuzzy Sensitivity Analysis and Its Application   195

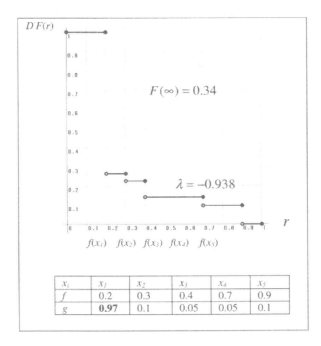

**Fig. 11.** Low sensitivity at almost all values

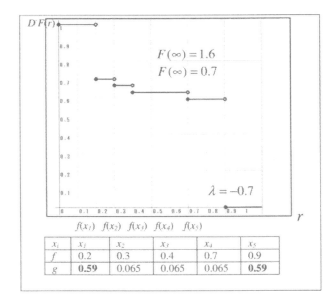

**Fig. 12.** Balanced sensitivity which is more or less equal over the entire range of values of "$r$" in the unit interval

196 T. Kaino, K. Hirota, and W. Pedrycz

If the quality of classifier $g(\cdot)$ attains high values at the significant degrees of class membership of the classified pattern $f(\cdot)$, then the sensitivity of the aggregation function will be high values and all classified results $f(\cdot)$ will be strongly reflected by the values of the aggregation function.

The sensitivity values obtained when using the lower differentiation of the Choquet integral for some cases of fuzzy measure $g(\cdot)$ are shown in Figure 11, and 12.

Figure 11 shows that all classification results except for the lowest class assignment will not be reflected into the aggregation function. Figure 12 shows the balanced sensitivity whose values are more or less equal over the entire range of values of "$r$" in the unit interval.

Therefore, the differentiation of the Choquet integral give us a useful information as to the assessment of the quality of classifier of aggregation function.

## 3 Differentiation of t-Conorm Integral

In this section, we discuss differentiation of some interesting generalizations of the fuzzy integral coming in the form of so-called t-conorm integrals.

### 3.1 Interval Limited t-Conorm Integral

Murofushi and Sugeno (1991) defined the following t-system.

**Definition 4 (Murofushi and Sugeno 1991, Narukawa and Torra 2006)**
A t-conorm system for integration (or t-system for short) is a quadruplet $(F, M, I, \nabla)$ consisting of the following components

$$F = ([0,1], \perp_1),$$

$$M = ([0,1], \perp_2),$$

$$I = ([0,1], \perp_3).$$

where $\perp_i, i = 1,2,3$ $\perp_1$, are continuous t-conorms which are $\vee$ or Archimedean,

$\nabla : F \times M \to I$ is a non-decreasing operator that satisfies the following properties:

$$(M1) \quad \nabla \text{ is left continuous on } (0,1] \tag{25}$$

$$(M2) \quad a\nabla x = 0 \text{ if and only if } a = 0 \text{ or } x = 0 \tag{26}$$

$$(M3) \text{ if } x \perp_2 y < 1 \text{ then } a\nabla(x \perp_2 y) = (a\nabla x) \perp_3 (a\nabla y) \tag{27}$$

$$(M4) \quad \text{if } a \perp_1 b < 1 \text{ then } (a \perp_1 b)\nabla x = (a\nabla x) \perp_3 (b\nabla x) \tag{28}$$

The t-system is expressed by $(\perp_1, \perp_2, \perp_3, \nabla)$ instead of $(F, M, I, \nabla)$. For example, a t-system is expressed by $(+_\lambda, \vee, \hat{+}, \cdot)$ in the case where $\perp_1 = +_\lambda, \perp_2 = \vee, \perp_3 = \hat{+}, \nabla = \cdot$ (the product operation), and by $(\vee, \vee, \vee, \wedge)$ in the case where $\perp_1 = \perp_2 = \perp_3 = \vee$ and $\nabla = \wedge$. For a given t-conorm $\perp$, we define an operation $-_\perp$ by

$$a -_\perp b := \inf\{c \mid b \perp c \geq a\} \tag{29}$$

for all $(a,b) \in [0,1]^2$. Note that if $\perp = \vee$ (viz. the maximum operator), we have

$$a -_\perp b = \begin{cases} a & \text{if } a \geq b, \\ 0 & \text{if } a < b. \end{cases} \tag{30}$$

Using the operator $-_\perp$, a function $f : X \to [0,1]$ can be expressed as

$$f = \perp_{1i=1}^{n} (r^{(i)} -_\perp r^{(i-1)}) 1_{A_i} \tag{31}$$

where $A_i := \{(i),\ldots,(n)\}$ $for$ $i = 1,\ldots,n$ and $f^{(0)} := 0$.

Murofushi and Sugeno (1991) also defined the generalized t-conorm integral.

### Definition 5 (Murofushi and Sugeno 1991, Narukawa and Torra 2006)

Let $(X, \mathfrak{I}, g)$ be a fuzzy measure space $(\perp_1, \perp_2, \perp_3, \nabla)$ denotes a certain t-system. For the function $f : X \to [0,1]$ $(f = \perp_{1i=1}^{n} (r^{(i)} -_\perp r^{(i-1)}) 1_{A_i})$, the generalized t-conorm integral is defined as follows:

$$(GT) \int f \nabla dg := \perp_{3i=1}^{n} (r^{(i)} -_{\perp_1} r^{(i-1)}) \nabla g(A_i). \tag{32}$$

If $g$ is a normal $\perp_2$ decomposable fuzzy measure (Dubois and Prade 1982), the generalized t-conorm integral coincides with the t-conorm integral.

### Definition 6 (Murofushi and Sugeno 1991, Narukawa and Torra 2006)

Let $(X, \mathfrak{I}, g)$ be a fuzzy measure space and $(\perp_1, \perp_2, \perp_3, \nabla)$ be a t-system. Let $f : X \to [0,1]$ $(f = \perp_{1i=1}^{n} (r^{(i)} -_\perp r^{(i-1)}) 1_{A_i})$ be a nonnegative measurable function. If $\forall r_* \in [0,1]$, then $f \wedge r_*$ is also measurable. The $[0, r_*]$ limited generalized t-conorm integral of a nonnegative measurable function $f$ with respect to a fuzzy measure $g$ is defined as

$$F_{GT}(r_*) := (GT) \int f \nabla r_* \nabla dg := \perp_{3i=1}^{n} (r^{(i)} -_{\perp_1} r^{(i-1)}) \nabla g(\{x \mid f(x) \nabla r_* \geq r\}). \tag{33}$$

Let $0 < {}^\forall a \leq {}^\forall c \leq {}^\forall b \leq 1$ $(a,b,c \in \mathfrak{R})$. Then the $[a,b]$ limited generalized t-conorm integral of a nonnegative measurable function $f$ with respect to the fuzzy measure $g$ is defined as

$$\perp_{3i=a}^{b} (r^{(i)} -_{\perp_1} r^{(i-1)}) \nabla g(\{x \mid f(x) \nabla r_* \geq r\}) := F_{GT}(b) - F_{GT}(a) \tag{34}$$

The following property holds true,

$$\perp_{3i=a}^{b} (r^{(i)} -_{\perp_1} r^{(i-1)}) \nabla g(\{x \mid f(x) \nabla r* \geq r\})$$

$$= \perp_{3i=a}^{c} (r^{(i)} -_{\perp_1} r^{(i-1)} \nabla g(\{x \mid f(x) \nabla r* \geq r\}) \perp_3$$

$$\perp_{3i=c}^{b} (r^{(i)} -_{\perp_1} r^{(i-1)}) \nabla g(\{x \mid f(x) \nabla r* \geq r\}). \tag{35}$$

### 3.2 Differentiation of t-Conorm Integral

**Definition 7**

Let $(X, \mathfrak{I}, g)$ be a fuzzy measure space. For any nonnegative measurable function $f$, if the limit (if it exists)

$$\lim_{\Delta r \to +0} \frac{F_{GT}(r + \Delta r) - F_{GT}(r)}{\Delta r} =: D^+ F_{GT}(r) = F_{GT+}'(r). \tag{36}$$

is called an upper differential coefficient of the $[0, r]$ limited generalized t-conorm integral $F_{GT}$ of $f$ with respect to $g$ at $r$. Similarly, the limit (if it exists)

$$\lim_{\Delta r \to -0} \frac{F_{GT}(r + \Delta r) - F_{GT}(r)}{\Delta r} =: D^- F_{GT}(r) = F_{GT-}'(r) \tag{37}$$

is called a lower differential coefficient of the $[0, r]$ limited generalized t-conorm integral $F_{GT}$ of $f$ with respect to $g$ at $r$. If and only if both the upper differential coefficient and the lower differential coefficient of $F_{GT}$ at $r$ exist and are equal, they are denoted by

$$DF_{GT}(r) = F_{GT}'(r) = \frac{dF_{GT}(r)}{dr} \tag{38}$$

The above expression is called a differential coefficient of the $[0, r]$ limited generalized t-conorm integral $F_{GT}$ of $f$ with respect to $g$ at $r$. If $\dfrac{dF_{GT}(r)}{dr}$ exists on a certain interval of $r$, then $F_{GT}(r)$ is called to be differentiable on this interval and the derived function $\dfrac{dF_{GT}(r)}{dr}$ is defined.

### 3.3 Comparison with Differentiation of t-Conorm Integrals in Sensitivity Analysis of Aggregation Function in Data Mining

In the case of $\perp_1 = \perp_2 = \perp_3 = +$ and $\nabla = \cdot$, the generalized t-conorm integral coincides with the Choquet integral. Furthermore, in the case of $\perp_1 = \perp_2 = \perp_3 = \vee$ and $\nabla = \wedge$, the generalized t-conorm integral becomes the Sugeno integral. Sensitivity analysis of differentiation of the Choquet integral as an aggregation function has been introduced in Section 2.3. Here, we are concerned with the sensitivity analysis of differentiation

Fuzzy Sensitivity Analysis and Its Application 199

of Sugeno integral and some other generalized t-conorm integral being regarded as aggregation functions. The other examples of the generalized t-conorm integral being considered deal with the following cases:

$$\begin{cases} \perp_1 = \perp_2 = \perp_3 = \perp_{Luk} \ (\perp_{Luk} \ (a,b) = (a+b) \wedge 1) \\ \nabla = \nabla_{Luk} \ (\nabla_{Luk} \ (a,b) = 0 \vee (a+b-1)) \end{cases} \tag{39}$$

$$\begin{cases} \perp_1 = \perp_2 = \perp_3 = \perp_{prob} \ (\perp_{prob} \ (a,b) = a+b-a \cdot b) \\ \nabla = \cdot \end{cases} \tag{40}$$

Let us start with the definition of the Sugeno integral (Sugeno 1972).

**Definition 8 (Sugeno 1972)**
Let $(X, \mathfrak{I}, g)$ be a fuzzy measure space. Let $f$ be a nonnegative measurable function. The Sugeno integral of the function $f$ with respect to the fuzzy measure $g$ is expressed as

$$(S)\int_X fdg := \bigvee_{r \in [0,1]} (r \wedge g(\{x \mid f(x) \geq r\})) \tag{41}$$

Accordingly, the Sugeno integral of a nonnegative simple function is given in the form

$$f(x) = \sum_{i=1}^{n} r_i \chi_{D_i}(x) \tag{42}$$

$$\left( 0 < r_1 < r_2 < \cdots < r_n, \quad \chi_{D_i}(x) = \begin{cases} 0 \ [x \notin D_i] \\ 1 \ [x \in D_i] \end{cases}, \quad \text{if } i \neq j \text{ then } D_i \cap D_j = \phi \right)$$

$$(S)\int_X fdg := \bigvee_{i=1}^{n} (r_i \wedge g(A_i)) \tag{43}$$

$$(A_i := \bigcup_{j=i}^{n} D_i, \quad r_0 := 0)$$

The Sugeno integral limited by an interval is defined accordingly.

**Definition 9**
Let $(X, \mathfrak{I}, g)$ be a fuzzy measure space. Let $f$ be a nonnegative measurable function. If $\forall r_* \in [0,1]$, then $f \wedge r_*$ is also measurable. The $[0, r_*]$ limited Sugeno integral of a nonnegative measurable function $f$ with respect to a fuzzy measure $g$ is defined as

$$F_S(r_*) := (S)\int_X f \wedge r_* dg = \bigvee_{r \in [0,1]} (r \wedge g(\{x \mid f(x) \wedge r_* \geq r\}))$$

$$=: \bigvee_{r \in [0, r_*]} (r \wedge g(\{x \mid f(x) \geq r\})). \tag{44}$$

Let $0 <^\forall a \leq^\forall c \leq^\forall b \leq 1$ $(a,b,c \in \mathfrak{R})$. Then the $[a,b]$ limited Sugeno integral of a nonnegative measurable function $f$ with respect to the fuzzy measure $g$ is

$$\bigvee_{r \in [a,b]} (r \wedge g(\{x \mid f(x) \geq r\})) dr := F_S(b) - F_S(a) \tag{45}$$

The following property is satisfied,

$$\bigvee_{r \in [a,b]} (r \wedge g(\{x \mid f(x) \geq r\}))$$

$$= \bigvee_{r \in [a,c]} (r \wedge g(\{x \mid f(x) \geq r\})) \vee \bigvee_{r \in [c,b]} (r \wedge g(\{x \mid f(x) \geq r\})). \tag{46}$$

Let us now introduce the concept of differentiation of the Sugeno integral.

**Definition 10**

Let $(X, \mathfrak{I}, g)$ be a fuzzy measure space. For any nonnegative measurable function $f$, the limit (assuming that it exists)

$$\lim_{\Delta r \to +0} \frac{F_S(r + \Delta r) - F_S(r)}{\Delta r} =: D^+ F_S(r) = F_{S+}'(r) \tag{47}$$

is called an upper differential coefficient of the $[0, r]$ limited Sugeno integral $F_S$ of $f$ with respect to $g$ at $r$. Similarly, if

$$\lim_{\Delta r \to -0} \frac{F_S(r + \Delta r) - F_S(r)}{\Delta r} =: D^- F_S(r) = F_{S-}'(r) \tag{48}$$

exists, this limit is called a lower differential coefficient of the $[0, r]$ limited Sugeno integral $F_S$ of $f$ with respect to $g$ at $r$. If and only if both the upper differential coefficient and the lower differential coefficient of $F_S$ at $r$ exist and are equal, they are denoted by

$$DF_S(r) = F_S'(r) = \frac{dF_S(r)}{dr} \tag{49}$$

and is called a differential coefficient of the $[0, r]$ limited Sugeno integral $F_S$ of $f$ with respect to $g$ at $r$. If $\dfrac{dF_S(r)}{dr}$ exists on a certain interval of $r$, then $F_S(r)$ is called to be differentiable on this interval and the derived function $\dfrac{dF_S(r)}{dr}$ is defined.

Now, if $\dfrac{dF_S(r)}{dr}$ exists, then

$$\frac{dF_S(r)}{dr} = \begin{cases} 1 [r \leq g(\{x \mid f(x) \geq r\})] \\ 0 [r > g(\{x \mid f(x) \geq r\})] \end{cases}. \tag{50}$$

# Example 3
Let the function $f$ be a nonnegative simple function as in Figure 2. Consider the Sugeno integral in the following form

$$(S)\int_X f d\mu := \bigvee_{i=1}^{n} (r_i \wedge g(A_i)) \tag{51}$$

$$(A_i := \bigcup_{j=i}^{n} D_j, \quad r_0 := 0)$$

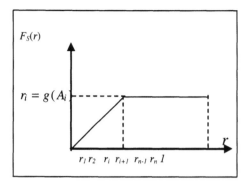

**Fig. 13.** Relation between $r$ and $F_S(r)$

The $[0, r]$ limited Sugeno integral is obtained as illustrated in Figure 13 that is

$$F_S(r) = \begin{cases} r & (g(A_i) \geq r_i) \\ r_i & (g(A_{i+1}) < r_{i+1}) \end{cases} \tag{52}$$

The differentiation of the $[0, r]$ limited Sugeno integral is illustrated in Figure 14.

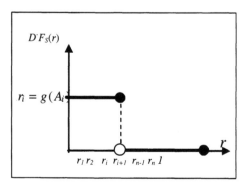

**Fig. 14.** Relation between $r$ and $D\,F_S(r)$

Suppose now that $g(A_i) \geq r_i, g(A_{i+1}) < r_{i+1}$,

$$0 < {}^\forall r < r_i, \quad \frac{dF_S(r)}{dr} = 1, \tag{53}$$

$$r = r_i, \text{ then } \begin{cases} D^+ F_S(r) = 0 \\ D^- F_S(r) = 1 \end{cases}, \tag{54}$$

$$r_i < {}^\forall r \leq 1, \quad \frac{dF_S(r)}{dr} = 0. \tag{55}$$

Here the derived function of the $[0, r]$ limited Sugeno integral of the triangular function $f$ is presented in the following example.

**Example 4**

Let $(X, \mathfrak{S}, g)$ be a fuzzy measure space.

$$g([a,b]) = \begin{cases} 16 - 6a - b^2 & (1 \leq a \leq 2, \ 2 \leq b \leq 3) \\ 0 & \text{(Others)} \end{cases} \tag{56}$$

$$f(x) = \begin{cases} x - 1 & (1 \leq x \leq 2) \\ 3 - x & (2 \leq x \leq 3) \\ 0 & \text{(Others)} \end{cases} \tag{57}$$

$$\{x \mid f(x) \geq r\} = \begin{cases} X & (r = 0) \\ [r+1, 3-r] & (0 < r \leq 1) \\ \phi & (1 < r) \end{cases} \tag{58}$$

Then

$$\mu(\{x \mid f(x) \geq r\}) = \begin{cases} 1 - r^2 & (0 \leq r \leq 1) \\ 0 & (1 < r) \end{cases} \tag{59}$$

$$F_S(r) = \bigvee_{r \in [0, r]} (r \wedge \mu(\{x \mid f(x) \geq r\}))$$

$$= \bigvee_{r \in [0, r]} (r \wedge (1 - r^2)) \tag{60}$$

$$= \begin{cases} r & (r \leq \dfrac{-1 + \sqrt{5}}{2}) \\ \dfrac{-1 + \sqrt{5}}{2} & (r > \dfrac{-1 + \sqrt{5}}{2}) \end{cases}$$

Thus, the derived function of the Sugeno integral of $f$ with respect to $g$ is expressed as follows:

$$F_S'(r) = \begin{cases} 1 & (r \leq \dfrac{-1 + \sqrt{5}}{2}) \\ 0 & (r > \dfrac{-1 + \sqrt{5}}{2}) \end{cases} \tag{61}$$

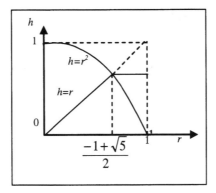

**Fig. 15.** $r$ and $g(\{x|f(x) \geq r\})$

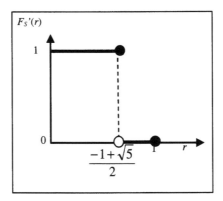

**Fig. 16.** Derived function $F_S'(r)$

Four types of the differentiation of the generalized t-conorm integrals (Choquet integral, Sugeno integral, and the other generalized t-conorm integrals) are introduced on the sensitivity analysis as an aggregation function using Example 2.

**Example 5**
The Choquet integral, the Sugeno integral, and some other generated integral are applied to the aggregation function as encountered in the case of classifier fusion, refer to Figure 7. Then, the obtained degrees of membership to class $\omega_i$ are given as

$$\begin{cases} f(x_1) = 0.2 \\ f(x_2) = 0.3 \\ f(x_3) = 0.4 \\ f(x_4) = 0.7 \\ f(x_5) = 0.9 \end{cases} \quad (62)$$

The confidence levels of the individual classifies, $g(x_1), g(x_2), \cdots, g(x_5)$ are quantified in the following manner

$$\begin{cases} g(x_1) = 0.57 \\ g(x_2) = 0.42 \\ g(x_3) = 0.14 \\ g(x_4) = 0.28 \\ g(x_5) = 0.28 \end{cases} \tag{63}$$

We obtain

$$\begin{aligned} F_C(\infty) &= I_C = (C)\int_X f dg \\ &= 0.2 \cdot g(X) + 0.3 \cdot g(\{x_2, x_3, x_4, x_5\}) + 0.4 \cdot g(\{x_3, x_4, x_5\}) \\ &\quad + 0.7 \cdot g(\{x_4, x_5\}) + 0.9 \cdot g(\{x_5\}) \\ &= 0.2 \cdot 1 + 0.3 \cdot 0.8066 + 0.4 \cdot 0.5845 + 0.7 \cdot 0.5008 + 0.9 \cdot 0.28 \\ &= 1.27843. \end{aligned} \tag{64}$$

The fusion of the classifiers realized by means of the Sugeno integral leads to the following result

$$\begin{aligned} F_S(\infty) &= I_S = (S)\int_X f dg \\ &= (0.2 \wedge g(X)) \vee (0.3 \wedge g(\{x_2, x_3, x_4, x_5\})) \vee (0.4 \wedge g(\{x_3, x_4, x_5\})) \\ &\quad \vee (0.7 \wedge g(\{x_4, x_5\})) \vee (0.9 \wedge g(\{x_5\})) \\ &= (0.2 \wedge 1) \vee (0.3 \wedge 0.8066) \vee (0.4 \wedge 0.5845) \vee (0.7 \wedge 0.5008) \vee (0.9 \wedge 0.28) \\ &= 0.5008. \end{aligned} \tag{65}$$

When considering various t-conorms. (here, $\perp_1 = \perp_2 = \perp_3 = \perp_{Luk}$ [$\perp_{Luk}(a,b) = (a+b) \wedge 1$], the corresponding integrals come in the form,

Lukasiwitz co-norm:

$$\nabla = \nabla_{Luk} [\nabla_{Luk}(a,b) = 0 \vee (a+b-1)]$$

$$\begin{aligned} F_{Luk}(\infty) &= I_{Luk} = (Luk)\int_X f dg \\ &= (0.2 \nabla_{Luk} g(X)) \perp_{Luk} (0.3 \nabla_{Luk} g(\{x_2, x_3, x_4, x_5\})) \perp_{Luk} (0.4 \nabla_{Luk} g(\{x_3, x_4, x_5\})) \perp_{Luk} (0.7 \nabla_{Luk} g(\{x_4, x_5\})) \\ &\quad \perp_{Luk} (0.9 \nabla_{Luk} g(\{x_5\})) \\ &= (0 \vee (0.2+1-1)) \perp_{Luk} (0 \vee (0.3+0.8066-1)) \perp_{Luk} (0 \vee (0.4+0.5845-1)) \perp_{Luk} (0 \vee (0.7+0.5008-1)) \\ &\quad \perp_{Luk} (0 \vee (0.9+0.28-1)) \\ &= ((((((0.2+0.1066) \wedge 1) + 0) \wedge 1) + 0.2008) \wedge 1) + 0.18) \wedge 1 \\ &= 0.6874. \end{aligned} \tag{66}$$

Probabilistic sum:

$$(\perp_1 = \perp_2 = \perp_3 = \perp_{prob} [\perp_{prob}(a,b) = a+b-a\cdot b] \text{ and } \nabla = \cdot)$$

$$F_{prob}(\infty) = I_{prob} = (prob)\int_X fdg$$
$$= (0.2 \cdot g(X)) \perp_{prob} (0.3 \cdot g(\{x_2, x_3, x_4, x_5\}))$$
$$\perp_{prob} (0.4 \cdot g(\{x_3, x_4, x_5\}))$$
$$\perp_{prob} (0.7 \cdot g(\{x_4, x_5\})) \perp_{prob} (0.9 \cdot g(\{x_5\}))$$
$$= (0.2 \cdot 1) \perp_{prob} (0.3 \cdot 0.8066) \perp_{prob} (0.4 \cdot 0.5845) \tag{67}$$
$$\perp_{prob} (0.7 \cdot 0.5008) \perp_{prob} (0.9 \cdot 0.28)$$
$$= (((0.2 \perp_{prob} 0.24198) \perp_{prob} 0.2338) \perp_{prob} 0.35056)$$
$$\perp_{prob} 0.252$$
$$= 0.774289.$$

The differentiation of these fuzzy integrals gives rise to the expressions

$$D^- F_C(r) = \begin{cases} g(X) = 1 & (0 \leq r \leq 0.2) \\ g(\{x_2, x_3, x_4, x_5\}) = 0.8066 & (0.2 < r \leq 0.3) \\ g(\{x_3, x_4, x_5\}) = 0.5845 & (0.3 < r \leq 0.4) \\ g(\{x_4, x_5\}) = 0.5008 & (0.4 < r \leq 0.7) \\ g(\{x_5\}) = 0.28 & (0.7 < r \leq 0.9) \\ g(\phi) = 0 & (0.9 < r \leq 1) \end{cases} \tag{68}$$

$$D^- F_S(r) = \begin{cases} 1 & (0 \leq r \leq 0.5008) \\ 0 & (0.5008 < r \leq 1) \end{cases} \tag{69}$$

$$D^- F_{Luk}(r) = \begin{cases} 1 & (0 \leq r < 0.3) \\ 0 & (0.3 < r \leq 0.4992) \\ 1 & (0.4992 < r \leq 0.7) \\ 0 & (0.7 < r \leq 0.72) \\ 1 & (0.72 < r \leq 0.9) \\ 0 & (0.9 < r < 1) \end{cases} \tag{70}$$

$$D^- F_{prob}(r) = \begin{cases} 1 & (0 \leq r \leq 0.2) \\ 0.64528 & (0.2 < r \leq 0.3) \\ 0.35445 & (0.3 < r \leq 0.4) \\ 0.23269 & (0.4 < r \leq 0.7) \\ 0.084491 & (0.7 < r \leq 0.9) \\ 0 & (0.9 < r \leq 1) \end{cases} \tag{71}$$

The pertinent plots of these results are covered in Figures 17-20.

Figure 17 and 20 show that the aggregation function using the Choquet integral, generalized t-conorm integral with probabilistic sum are reflected by the all classified results. Figure 18 shows that the aggregation function using Sugeno integral is reflected by some classified results which is larger than the value of Sugeno integral. Figure 19 shows that the aggregation function using the generalized t-conorm integral with Lukasiwitz co-norm is not reflected by the classified results with $f(\cdot) + g(\cdot) < 1$. Therefore, the differentiation of generalized t-conorm integral give us a rich information to choose the appropriate generalized t-conorm integral as an aggregation function.

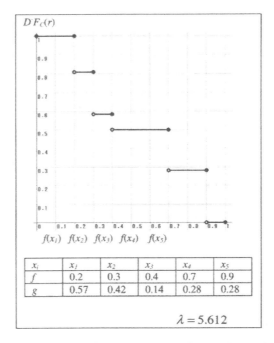

**Fig. 17.** Lower differentiation of the Choquet integral

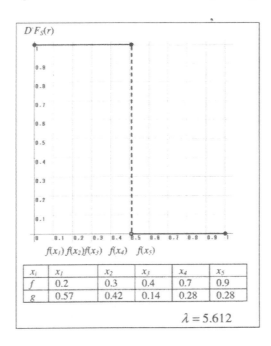

**Fig. 18.** Lower differentiation of the Sugeno integral

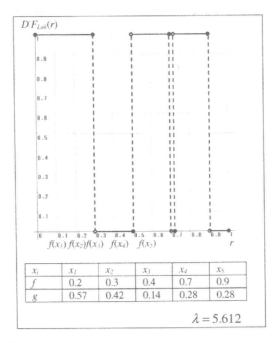

**Fig. 19.** Lower differentiation of generalized t-conorm integral $F_{Luk}$

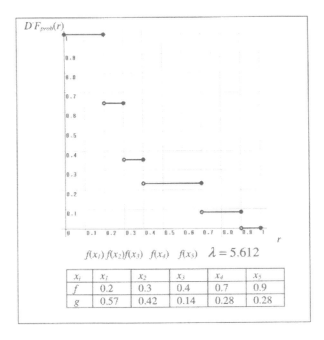

**Fig. 20.** Lower differentiation of generalized t-conorm integral $F_{prob}$

# 4 Apply to Credit Risk Analysis

Although a credit risk is one of the historical risks in a financial market, compared with a market risk, the scientific management technology is not developed enough. In a corporate bond market, a credit spread goes up from a bankruptcy of Yamaichi Securities in 1997, and the Hokkaido Takushoku Bank. And the credit risk is increasing by the MMF extensive cancellation, the MYCAL bankruptcy and the Enron bankruptcy in 2001. Such a background demands the establishment of a scientific technique of credit-risk management. As an econometric model of a credit risk, Z score model (Altman 1969) by the multi-variable-analysis technique and a ZETA credit risk model (Altman et al. 1977) are proposed by Altman and others. These have been used till today. Although such an econometrics technique has been studied for a long time (Hand and Henley 1997), it is criticized as the experiential model lacking theoretical backgrounds, and has been compared with theoretical approach (Scott 1981). Moreover, new approaches based on neural networks (Coats and Fant 1993, Vellido et al. 1999) have been proposed. However, there is no capability for neural network model to explain the factor used as the basis of credit judgment. There are the EDF model (KMV Corporation 1995) of KMV based on Merton (1974) model, the Longstaff, & Schwartz model (1995), and the Jarrow, Lando & Turnbull model (1997) as other credit risk models. Although the research on credit risk measurement progresses, the credit rating provider's role has not changed. And, the credit risk information of the credit rating provider is esteemed by the market participant (Moody's Investors Service 1999). For instance, Moody's publishes market-leading credit opinions, deal research and commentary that reach more than 3,000 institutions and 22,000 subscribers around the world. It is researched to make clear the rating determination process by the credit rating providers, and the influence factor of rating (Kobayashi 2000). Moreover, Kaino and Hirota (2000) proposed the model that can deal well with the qualitative information corresponding to a credit analyst's sensitivity and quantitative information as financial information.

## 4.1 Long Term Debt Rating Model

Firstly, the long-term debt ratings model is identified by the real interval limited Choquet integral. Here, the input of this model is the quantitative indicator and qualitative indicator of the each corporations., and the output of this model is the Moody's long-term debt ratings of the each corporations. The importance of each indicator, fuzzy measures $g$, is determined by the neural network method, where the open financial statements (Toyo Keizai Inc. 1998) and the analysis data (Daiamond Inc. 1997) are used for input data, and the Moody's debt ratings (Moody's Japan KK 1999) are used for output data. Generally, the long-term debt ratings of each rating institutions are determined by the analyst's experience and know-how. Then, it is very difficult for each rated corporations to find how to raise their rating results, clearly. So, after the identification of the long-term debt ratings model using the real interval limited Choquet integral, the advisory system to raise the rating using differentiation of the Choquet integral is proposed.

According to the efficient market hypothesis, financial data is reflecting only the information on the past of the company. Therefore, in order to raise the generality of a model, it is necessary to use the indicator explaining the present market condition. Moreover, in order to raise the precision of rating, the characteristic for every type of industry for an indicator must be taken into account. In the proposed model, the indicators of Table 1 was selected in consideration of the indicators, which are raising the result with the credit risk measurement model (Standard & Poor's Corporation 1986) by Moody's Investors Service as a financial indicators. Since there are ROE and ROA that are the indicators of profitability, business profit and cash flow are not selected and the logarithm equity capital that expresses a company scale is added for the new model's indicators.

**Table 1.** Quantitative & qualitative indicators

| |
|---|
| 9 quantitative indicators |
| leverage, interest coverage, ROE, ROA, logarithm equity capital, current assets ratio, fixed assets ratio, inventory turnover, turnover of receivables |
| 3 qualitative indicators |
| market condition, corporate image, type-of-industry trend |

Since the characteristic is different for every type of industry about current assets ratio, fixed assets ratio, inventory turnover and turnover of receivables, it is assumed that the value of each indicator resembles the normal distribution. Moreover, about other indices, it is assumed that the normal distribution is resembled in all companies. And each financial indicator is changed into the standard normal distribution of N (0, 1), and it changes into the value of [0, 1] using an accumulation distribution function. As the market condition indicator, the five-year stock price average growth rate is used. Moreover, a capital stock growth rate is used as the type-of-industry trend indicator. And, Nikkei NEEDS CASMA (Toyo Keizai Inc. 1998, Nihon Keizai Shimbun Inc. 2000) is used as the corporate image indicator.

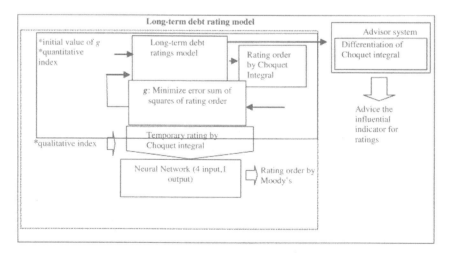

**Fig. 21.** Long-term debt rating system

In case credit capability is judged, the credit rating provider is using as the important composition element in rating determination not only the quantitative factor of analysis of a financial indicators but qualitative factors, such as information acquired from an interview with the management of a company etc. If there is a rating request, that company's financial information will be analyzed. Secondly, that company is interviewed and determined the credit rating by the credit rating provider. Then, in consideration of the above actual credit rating process, we proposed the 2-layer type long-term debt rating model.

This long-term debt rating model is identified by the actual data of the term ended March, 2000 of 21 companies. Then, we simulated by the data of the term ended March, 2001 of 46 companies (Moody's Investors Service 1999, Toyo Keizai Inc. 1998, Nihon Keizai Shimbun Inc. 2000, Nihon Keizai Shimbun Inc. 2000, Nihon Keizai Shimbun Inc. 2001). The importance of each indicator in the first layer is determined by the Steepest Descent method. Let be each financial index, and let be the value of each financial indicators changed to [0, 1] using an accumulation distribution function. The temporary rating of each corporation is given as the Choquet integral.

$$F(\infty) = (C)\int_X f dg . \tag{72}$$

And $F$ is arranged in descending order and let $y_i$ be the ranking. The importance $g(x_i)$ of each index is determined by making the following $L(g(x_i))$ into the minimum using steepest descent method.

$$L(g(x_i)) = \sum_{j=1}^{21}(Y_j - y_j)^2 \tag{73}$$

Here, the following local optimal solutions of the indicator's importance are given as the result of 5,000 calculations (where the initial value is given as $g(x_i) = 0.4$)

**Table 2.** Local optimal solution of importance $g$

| | |
|---|---|
| g(Interest coverage)=0.41 | g(leverage)=0.2 |
| g(logarithm equity capital)=0.78 | g(ROE)=0.4 |
| g(current assets ratio)=0.51 | g(ROA)=0.05 |
| g(inventory turnover)=0.5 | g(fixed assets ratio)=0.1 |
| g(turnover of receivables)=0.36 | |

**Table 3.** Quantification of rating

| Aa | A | Baa |
|---|---|---|
| 0.8~ | 0.6~0.8 | 0.4~0.6 |
| Ba | B | Caa |
| 0.2~0.4 | 0.1~0.2 | ~0.1 |

Fuzzy Sensitivity Analysis and Its Application 211

In the second layer, rating is outputted using the provisional rating and the qualitative index by Choquet integral using neural network technique. Based on the value of [0, 1], rating of an output value was quantified, as shown in Table 3.

Moreover, since there were few samples of experiment data here, the model of a neural network was expressed with the form of alignment combination of one neuron.

$$F(x_i) = \sum_{j=1}^{4} g_j f(x_{ij}) \tag{74}$$

## 4.2 Simulation Result

In this simulation, comparison of the results form two-layer type model and the one-layer type model using the proposed indicators is carried out. As a valuation basis, the rate which it is in each notch (error) of rating precision, the rate by accumulation, the total number of notches, and the precision about each rating were adopted. The precision of rating improved 22.7% compared with the one-layer type model (see table 4). The error of the one-layer type model and the two-layer type model became a maximum of 1 notch, and accumulation precision also became 100%.

**Table 4.** Simulation result

|        | 1 layer model | | 2 layer model | |
|--------|------|------|------|------|
| Notch  | Rate | Acc. | Rate | Acc. |
| $\pm 0$ | 48% | 48% | 59% | 59% |
| $\pm 1$ | 52% | 100% | 41% | 100% |
| $\pm 2$ | 0% | 100% | 0% | 100% |
| $\pm 3$ | 0% | 100% | 0% | 100% |

Regarding the precision about each rating class, the 2-layer type model had the low precision in Aa class compared with the one-layer type model. However, in the precision about Baa (adequate security) class and Ba(questionable security) class, the 2-layer type model improved 57.1% compared with the one-layer type model (Table 5).

**Table 5.** Precision about each rating class

| Rating class | No of corporate | 1 layer type Model | 2 layer type Model |
|--------|------|------|------|
| Aa | 9 | 56% | 33% |
| A | 11 | 64% | 72% |
| Baa | 16 | 38% | 56% |
| Ba | 4 | 25% | 50% |
| B | 4 | 50% | 75% |
| Caa | 1 | 100% | 100% |

## Table 6. Temporary rating & qualitative indicators (the 2$^{nd}$ layer)

| temporary rating | corporate image | market condition | type-of-Industry trend |
|---|---|---|---|
| 0.81 | 0.37 | $-0.02$ | 0.22 |

Now, the validity of using a type-of-industry trend indicator and a market condition indicator and stratifying two times and performing analysis of a fixed quantity indicator and a qualitative indicator was shown. Since there is few data used in this experiment, standard deviation of each indicators becomes large. Therefore, in case the value of each index is changed into the value of [0, 1], the error has come out. It is necessary to conduct the experiment using larger-scale data. Well, let check the weight of each indicators of the second layer. "Market condition" is not to influence the rating because of its negative value. It is a possible to raise the accuracy of rating by changing into the another indicators (distance default, etc.).

### 4.3 Sensitivity Analysis Using Differentiation of the Choquet Integral

Differentiation of the Choquet integral will be applied to the sensitivity analysis (as an advisory system) to raise the ratings of 5 corporations (Toyota, Canon, Mitsui Fudosan, Kanematsu, Fujita) within 46 corporations. Firstly, in the case of Toyota, the temporary rating $F(\infty)$ of Toyota is calculated by the definition of the Choquet integral as

$$
\begin{aligned}
F(\infty) &= (C)\int_X fdg = \sum_{i=1}^{n}(r_i - r_{i-1})g(A_i) \\
&= r_1 g(\{A_1\}) + (r_2 - r_1)g(\{A_2\}) + (r_3 - r_2)g(\{A_3\}) + \cdots + (r_9 - r_8)g(\{A_9\}) \\
&= r_1 g(X) + (r_2 - r_1)g(\{x_2, x_3, \cdots, x_9\}) + (r_3 - r_2)g(\{x_3, \cdots, x_9\}) \\
&\quad + \cdots + (r_9 - r_8)g(\{x_9\}) = 1.38
\end{aligned}
\tag{75}
$$

Now, the upper differential coefficient and the lower differential coefficient at each $r_i$ are derived using the $\lambda$-fuzzy measure ($\lambda = -0.5$) as Table 7.

Equation (75) can be replaced by equation (76).

$$
\begin{aligned}
F(\infty) &= \{D^- F(r_1) - D^+ F(r_1)\}r_1 + \{D^- F(r_2) - D^+ F(r_2)\}r_2 \\
&\quad \cdots \{D^- F(r_8) - D^+ F(r_8)\}r_8 + \{D^- F(r_9) - D^+ F(r_9)\}r_9
\end{aligned}
\tag{76}
$$

Now, on equation (77), it is noticed that the coefficient of each $r_i$ is "the lower differential coefficient $(D^- F(r_i))$ – the upper differential coefficient $(D^+ F(r_i))$ )". So, this coefficient is considered to be the change of the total evaluation of the each corporation as the evaluation of each index change slightly.

$$
V(x_i) := \frac{\partial F(\infty)}{\partial F(r_i)} =: D^- F(r_i) - D^+ F(r_i)
\tag{77}
$$

## Fuzzy Sensitivity Analysis and Its Application 213

**Table 7.** Differential coefficient of the Choquet integral

| | | | |
|---|---|---|---|
| $D^+F(r_1)$ | 1.67 | $D^-F(r_1)$ | 1.70 |
| $D^+F(r_2)$ | 1.65 | $D^-F(r_2)$ | 1.67 |
| $D^+F(r_3)$ | 1.57 | $D^-F(r_3)$ | 1.65 |
| $D^+F(r_4)$ | 1.46 | $D^-F(r_4)$ | 1.57 |
| $D^+F(r_5)$ | 1.27 | $D^-F(r_5)$ | 1.46 |
| $D^+F(r_6)$ | 1.25 | $D^-F(r_6)$ | 1.27 |
| $D^+F(r_7)$ | 1.09 | $D^-F(r_7)$ | 1.25 |
| $D^+F(r_8)$ | 0.78 | $D^-F(r_8)$ | 1.09 |
| $D^+F(r_9)$ | 0 | $D^-F(r_9)$ | 0.78 |

Hence, let $V(x_i)$ be the change of the total evaluation of the corporations as the evaluation of index change slightly. So, each $V(x_i)$ will be calculated as:

$$V(x_1) = D^-F(r_1) - D^+F(r_1) = 1.7 - 1.67 = 0.03$$

$$V(x_2) = D^-F(r_2) - D^+F(r_2) = 1.67 - 1.65 = 0.02$$

$$V(x_3) = D^-F(r_3) - D^+F(r_3) = 1.65 - 1.57 = 0.08$$

$$V(x_4) = D^-F(r_4) - D^+F(r_4) = 1.57 - 1.46 = 0.11$$

$$V(x_5) = D^-F(r_5) - D^+F(r_5) = 1.46 - 1.27 = 0.19 \qquad (78)$$

$$V(x_6) = D^-F(r_6) - D^+F(r_6) = 1.27 - 1.25 = 0.02$$

$$V(x_7) = D^-F(r_7) - D^+F(r_7) = 1.25 - 1.09 = 0.16$$

$$V(x_8) = D^-F(r_8) - D^+F(r_8) = 1.09 - 1.78 = 0.31$$

$$V(x_9) = D^-F(r_9) - D^+F(r_9) = 0.78 - 0 = 0.78$$

For instance, $V(x_3) = 0.08$ means that the temporary rating of the corporation will increase 0.08 points as the evaluation of ROE increase from 0.46 points to 0.56 points, where the evaluations of the other specifications are fixed. In the case of Toyota, if the importance of each index is applied to decide which index's value should be improved in order to raise the total evaluation (rating), then the following results will be given.

## (a) Toyota

$V(x_9$:turnover of receivables$) > V(x_8$:inventory turnover$) > V(x_5$:logarithm equity capital$) > V(x_7$:fixed assets ratio$) > V(x_4$:ROA$) > V(x_3$:ROE$) > V(x_1$:leverage$) > V(x_2$:interest coverage$) > V(x_6$:current assets ratio$)$

It will be the best way to improve the rating of " turnover of receivables" so that the temporary rating of the corporation is maximized. Similarly, the results of other 4 corporations are derived from $V$ function as follows:

## (b) Canon

$V(x_5$:logarithm equity capital$) > V(x_6$:current assets ratio$) > V(x_9$:turnover of receivables$) > V(x_3$:ROE$) > V(x_8$:inventory turnover$) > V(x_2$:interest coverage$) > V(x_4$:ROA$) > V(x_1$:leverage$) > V(x_7$:fixed assets ratio$)$

## (c) Mitsui Fudosan

$V(x_5$:logarithm equity capital$) > V(x_6$:current assets ratio$) > V(x_9$:turnover of receivables$) > V(x_8$:inventory turnover$) > V(x_3$:ROE$) > V(x_2$:interest coverage$) > V(x_1$:leverage$) > V(x_7$:fixed assets ratio$) > V(x_4$:ROA$)$

## (d) Kanematsu

$V(x_3$:ROE$) > V(x_2$:interest coverage$) > V(x_6$:current assets ratio$) > V(x_5$:logarithm equity capital$) > V(x_8$:inventory turnover$) > V(x_9$:turnover of receivables$) > V(x_1$:leverage$) > V(x_7$:fixed assets ratio$) > V(x_4$:ROA$)$

## (e) Fujita

$V(x_9$:turnover of receivables$) > V(x_2$:interest coverage$) > V(x_8$:inventory turnover$) > V(x_6$:current assets ratio$) > V(x_5$:logarithm equity capital$) > V(x_3$:ROE$) > V(x_7$:fixed assets ratio$) > V(x_1$:leverage$) > V(x_4$:ROA$)$

# 5 Conclusions

In this study, we are concerned with the concept, properties and algorithms of differentiation of the Choquet integral. The differentiation of the Choquet integral of a nonnegative measurable function is studied in the setting of sensitivity analysis. It is shown that the differentiation of the Choquet integral reflects a way in which the aggregation becomes sensitive to the function being aggregated. The aggregation function using fuzzy integral could be viewed central to data-mining, the business process mining, web-marketing, e-commerce and the other pattern-matching problems. In the presented example showing a fusion of classifiers being realized with the use of fuzzy integration we showed some interesting relationships between the sensitivity of the results and the quality of the classifiers. In this sense, the differentiation of the Choquet integral being an interesting perspective on the design the quality of classifier of aggregation function.

Next, the differentiation of the Choquet integral is extended to the differentiation of the generalized t-conorm integral. The four types of the differentiation of the generalized t-conorm integrals (Choquet integral, Sugeno integral, and the other generalized t-conorm integrals) are discussed and compared with regard to their sensitivity properties. The differentiation of generalized t-conorm integral give us a useful information to choose the appropriate generalized t-conorm integral as an aggregation function in data-mining.

Lastly, the Choquet integral is applied to the credit risk analysis (long-term debt rating) to make clear the significance of them, especially, this differentiation is shown to be an effective tool which gives us how much evaluation of each index influence to the total evaluation on the corporations. When the importance $g$ and the evaluations $f$ of each indicator's values of the corporations are given, the Choquet integral means the total evaluation of the corporation. Here, the fuzzy measure, which is given as the importance of an each quantitative and qualitative data, is derived from a neural network method.

Hence, the partial derivative of $F(\infty)$ with respect to $r_i$ ("the lower differential coefficient $(D^- F(r_i)) -$ the upper differential coefficient $(D^+ F(r_i))$") is considered to be the change of the total evaluation of the corporation as the indicator's values change slightly, and this differentiation of the Choquet integral is also considered to be an effective sensitivity analysis tool of the various decision making problems in social science which includes a lot of ambiguity.

# References

Altman, E.I.: Financial ratios, discriminant analysis and the prediction of corporate bankruptcy. Journal of Finance 23, 189–209 (1969)

Altman, E.I., Haldeman, R.G., Narayama, P.: ZETA analysis: a new model to identify bankruptcy risk of corporations. Journal of Banking and Finance 1, 29–54 (1977)

Chen, S.H., Yang, C.C., Lin, W.T., Yeh, T.M.: Performance evaluation for introducing statistical process control to the liquid crystal display industry. International Journal of Production Economics 111(1), 80–92 (2008)

Chen, X., Xiao, Z.J.G.: Nonlinear fusion for face recognition using fuzzy integral. Communications in Nonlinear Science and Numerical Simulation 12, 823–831 (2005)

Choquet, G.: Theory of capacities. Ann. Inst. Fourier 5, 131–295 (1953)

Coats, P., Fant, K.: Recognizing financial distress patterns using a neural network tool. Financial Management 22(3), 142–155 (1993)

Daiamond Inc: Ranking of Japanese corporations. Daiamond Inc 18, 26–74 (1997)

Dubois, D., Prade, H.: A class of fuzzy measures based on triangular norms. Internat. J. Gen. Systems 8, 43–61 (1982)

Hand, D.J., Henley, W.E.: Statistical classification methods in consumer credit scoring: a review. J.R. Statist. Soc. A 3, 523–541 (1997)

Hu, Y.C., Tseng, F.M.: Functional-link net with fuzzy integral for bankruptcy prediction. Neurocomputing 70(16-18), 2959–2968 (2007)

Jarrow, A.R., Lando, D., Turnbull, S.M.: A Markov model for the term structure of credit risk spreads. Review of Financial Studies 10(2), 481–523 (1997)

Kaino, T., Hirota, K.: Differentiation of the Choquet integral of nonnegative measurable function.In: Proc. of FUZZ IEEE 1999, Seoul (1999)

Kaino, T., Hirota, K.: Y-CG derivative of fuzzy relations and its application to sensitivity analysis. The International Journal of Fuzzy Systems 1(2), 129–132 (1999)

Kaino, T., Hirota, K.: Differentiation of the Choquet integral and its application to long-term debt ratings. Journal of Advanced Computational Intelligence 4(1), 66–75 (2000)

Kaino, T., Hirota, K.: Composite fuzzy measure and its application to investment decision making problem. Journal of Advanced Computational Intelligence 8(3), 252–259 (2004)

Kaino, T., Urata, K., Yoshida, S., Hirota, K.: Improved debt rating model using Choquet integral. Journal of Advanced Computational Intelligence 9(6), 615–621 (2005)

KMV Corporation. Introduction credit monitor, version 4, San Francisco (1995)

Kobayashi: Testing the ordered profit model and its application to corporate bond rating data. institute for Monetary and Economic Studies Discussion, (2000) paper No.2000-J-17

Kwak, K., Pedrycz, W.: Face recognition: A study in information fusion using fuzzy integral. Pattern Recognition Letters 26, 719–733 (2005)

Liou, J.J.H., Tzeng, G.H.: A non-additive model for evaluating airline service quality. Journal of Air Transport Management. (2008) doi:10.1016/j.jairtraman.2006.12.002

Logstaff, A.F., Schwartz, E.S.: A simple approach to valuing risky fixed and floating rate debt. Journal of Finance 50(3), 790–819 (1995)

Marichal, J.L.: On Sugeno integral as an aggregation function. Fuzzy Sets and Systems 114, 347–365 (2000)

Merton, R.C.: On the pricing of corporate debt: the risk structure of interest rates. Journal of Finance 28, 87–109 (1974)

Moody's Investors Service, Default report (1999), http://www.moodysqra.com/.

Moody's Japan, K.K.: The Moody's ratings list (1999)

Murofushi, T., Sugeno, M.: Fuzzy t-conorm integral with respect to fuzzy measures: generalization of Sugeno integral and Choquet integral. Fuzzy Sets and Systems 41, 57–71 (1991)

Narukawa, Y., Torra, V.: Generalized transformed t-conorm integral and multifold integral. Fuzzy Sets and Systems 157, 1384–1392 (2006)

Nihon Keizai Shimbun Inc: All listed Japanese company version, Nikkei management index (2000)

Nihon Keizai Shimbun Inc: Nikkei top-rated company ranking (2000)

Nihon Keizai Shimbun Inc: All listed Japanese company version, Nikkei management index (2001)

Nihon Keizai Shimbun Inc: Nikkei top-rated company ranking (2001)

Scott, J.: The probability of bankruptcy: a comparison of empirical predictions and theoretical models. Journal of Banking and Finance 5(3), 317–344 (1981)

Standard & Poor's Corporation, Debt ratings criteria: municipal overview (1986)

Sugeno, M.: Fuzzy measure and fuzzy integral. Journal of the Society of Instrument and Control Engineers 8(2), 218–226 (1972)

Toyo Keizai Inc: Toyo Keizai 1999 data bank (1998)

Vellido, A., Lisboa, P.J.G., Vaughan, J.: Neural networks in business: a survey of applications (1992-1998). Expert Systems with Applications 17, 51–70 (1999)

Wu, J.Q.C., Li, F.: On the restudy of fuzzy complex analysis: Part II. The continuity and differentiation of fuzzy complex functions. Fuzzy Sets and Systems 120(3), 517–521 (2001)

# A Probabilistic Approach to Fuzzy Engineering Economic Analysis*

K. Paul Yoon

Dept. of Information Systems and Decision Sciences, Silberman College of Business,
Fairleigh Dickinson University

**Abstract.** Techniques for ranking simple fuzzy numbers are abundant in the literature. However, we lack efficient methods for comparing complex fuzzy numbers that are generated by fuzzy engineering economic analyses. In this chapter a probabilistic approach is taken instead of the usual fuzzy set manipulations. The Mellin transform is introduced to compute the mean and the variance of a complex fuzzy number. The fuzzy number with the higher mean is to be ranked higher. Two fuzzy cash flow analyses and a fuzzy multiple attribute decision analysis are illustrated in order to demonstrate the suitability of the probabilistic approach. A real beauty of the approach is that it can accommodate all types of fuzzy numbers such as uniform, triangular, or trapezoidal simultaneously, and all computations are easily calculated in the spreadsheet.

## 1 Introduction

Practical applications of fuzzy set in the engineering economy require two laborious tasks: (1) fuzzy arithmetic operations, and (2) comparison of complex fuzzy numbers that are generated by arithmetic operations. Fuzzy algebraic operations can be made based on the extension principle (Dubois and Prade 1980, Zadeh 1975). Even with simple operations such as addition and multiplication, the fuzzy number operation usually tends to be cumbersome. However, financial and engineering economic applications in the fuzzy data environment require higher level fuzzy arithmetics that include the product of several fuzzy numbers and some power of fuzzy numbers (Kaufmann and Gupta 1988). The fuzzy operations of this type may require insurmountable computational effort.

Comparing or ranking fuzzy numbers is very important in applications. For example, the concept of optimum or best choice is completely based on ranking or comparison. The task of comparing fuzzy numbers can be another problem because fuzzy numbers do not always yield a totally ordered set as real numbers do. To resolve this problem, many authors have proposed fuzzy ranking methods that yield a totally ordered set or ranking. These methods range from the trivial to the complex: from one fuzzy number attribute (mode) to many fuzzy number attributes (such as mode, spread, and closeness to a fuzzy ideal). A review and comparison of these existing methods can be found in (Chen and Hwang 1991, Lee and Li 1988, McCahon and Lee 1990, Tseng et al. 1988, Zimmerman 1987).

---

\* This chapter was adapted from the author's article: Yoon KP (1996) A probabilistic approach to rank complex fuzzy numbers. Fuzzy Sets and Systems 80: 167-176.

C. Kahraman (Ed.): Fuzzy Engineering Economics with Appl., STUDFUZZ 233, pp. 217–230, 2008.
springerlink.com                    © Springer-Verlag Berlin Heidelberg 2008

218 K.P. Yoon

The fuzzy ranking methodologies have been widely applied to engineering economic analysis. For example, the analyses with fuzzy cash flow are given in (Chiu and Park 1994, Kahraman et al. 2002, Kahraman et al. 2004, Karsak 1998, Liou and Chen 2006) and with fuzzy multiple attribute decision making are found in (Chen and Hwang 1991, Kickert 1978, Kulak 2005, Kulak et al. 2005, Zimmerman 1985, Zimmerman 1987).

According to the review of Lee and Li (1988), fuzzy ranking methods can be grouped into two classes. One is a possibilistic method based on the possibility theory; the other is a probabilistic method referring to the probability theory. In the probabilistic approach, the mean and variance of fuzzy numbers are computed. The fuzzy number with the higher mean is then ranked higher than the fuzzy number with the lower mean. If the means happen to be equal, the one with the smaller spread (measured by variance) is judged better. Lee and Li conclude that possibilistic methods seem to suffer from a lack of discrimination and occasionally conflict with intuition.

The purpose of this chapter is (1) to present a comparison method for complex fuzzy numbers based on the probability theory, and (2) to demonstrate its applicability in the fuzzy engineering economic environment. The first step in this approach is to generate a corresponding probability function from a membership function. Then the mean and spread of complex fuzzy numbers are calculated by way of the Mellin transform (Giffin 1975, Springer 1979). The fuzzy number with the higher mean is then ranked higher than the fuzzy number with a lower mean. If the means are equal, the one with the smaller variance is judged higher rank.

Section 2 presents how to transform a membership function into a probability density function. Section 3 introduces the Mellin transform to obtain statistical moments of any order of random variables of products and quotients. Useful Mellin transforms for the fuzzy number comparison are derived in Section 4. Two fuzzy cash flow analyses and a fuzzy multi-attribute decision analysis are presented in Section 5. Finally, concluding remarks are given in Section 6.

## 2  Probability Density Function from Membership Function

Fuzzy numbers are a special kind of fuzzy sets which are normal and convex. Although these numbers can be described using many types of shapes, for practical applications it is best to use triangular and trapezoidal shapes. This chapter considers these two fuzzy numbers: triangular fuzzy number (TFN) and trapezoidal fuzzy number (ZFN).

In order to add fuzzy and random data, Kaufmann and Gupta (1985) obtained a membership function from a probability density function by way of a simple linear transformation, which indicates a possible conversion of a membership function into a density function. The conversion of a membership function into a probability density function can be made by one of two linear transformations.

### 2.1  Proportional Probability Distribution

The occurrence probability of fuzzy event A should be proportional to the value of membership function $\mu_A(x)$. That is

$$f_1(x) = c_1 \mu_A(x)$$

A Probabilistic Approach to Fuzzy Engineering Economic Analysis      219

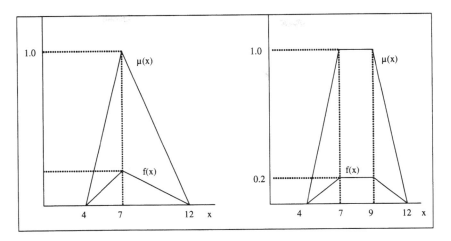

**Fig. 1.** Conversion into proportional probability distribution

where $c_1$ is a proportional constant to satisfy the condition that the area under the continuous probability function is equal to one. Figure 1 illustrates the proportional conversions with triangular and trapezoidal fuzzy numbers. The converted density function retains the domain of variables X, but has lost the original shape of the membership function to some degree. This method corresponds to the approach of Kaufman and Gupta (1985).

## 2.2 Uniform Probability Distribution

The occurrence probability of a fuzzy event should be uniform or consistent to the shape of $\mu_A(x)$. That is

$$f_2(x) = \mu_A(x) + c_2$$

where $c_2$ is a uniform constant to satisfy the probability function requirement. Figure 2 illustrates the uniform conversions with triangular and trapezoidal fuzzy numbers.

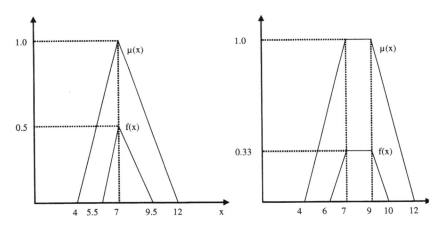

**Fig. 2.** Conversion into uniform probability distribution

The converted density function maintains the original shape, but the domain of X gets contracted (or dilated when the area under the membership function is less than one). Applying geometric similarities of two triangles, we can obtain the analytical relationship between two functions. First the triangular density function from a TFN, which is defined by a triplet ($l$, $m$, $u$), can be determined by ($l'$, $m'$, $u'$) where

$$l' = m - (m - l) \{2/(u - l)\}^{1/2}$$

$$m' = m$$

$$u' = m + (u - m) \{2/(u - l)\}^{1/2}.$$

Similarly, the trapezoidal density function from a ZFN, which is represented by a quadrant ($a$, $b$, $c$, $d$), can be determined by ($a'$, $b'$, $c'$, $d'$) where

$$a' = b - e(b - a)$$

$$b' = b$$

$$c' = c$$

$$d' = c + e(d - c)$$

where

$$e = \frac{(b-c)+\sqrt{(b-c)^2 + 2(-a+b-c+d)}}{(-a+b-c+d)}$$

These formulas are equally applicable when the domain is contracted or dilated.

The choice of the proportional or uniform distribution seems to be rather arbitrary. However, a few observations can be made. In order to make points clear, two distinct fuzzy numbers are converted into density functions: $M = (1, 4, 6)$ and $N = (10, 40, 60)$. Figures 3 and 4 illustrate these conversions. When the domain of fuzzy numbers

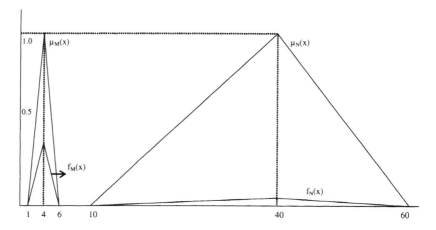

**Fig. 3.** Conversion into proportional probability distribution

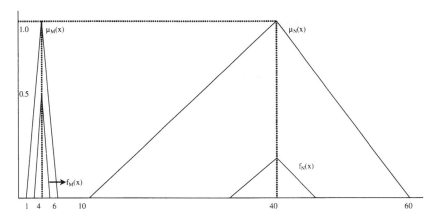

**Fig. 4.** Conversion into uniform probability distribution

is very large (e.g. fuzzy number N), the range of proportional distribution is greatly reduced, where the membership function gives less discrimination power among members in a fuzzy set. When the uniform distribution is used, domain and range of the resulting distribution are reduced simultaneously. But the reduced (or increased) domain indicates the complete ejection (or addition) of some members from (or to) the set. Hence the uniform distribution reveals more undesirable properties. Furthermore, the conversion into a proportional distribution does not require any computation. We, therefore, suggest the use of the proportional distribution when comparing the fuzzy numbers.

## 3 The Mellin Transform

The Mellin transform $M_X(s)$ of probability density function (pdf) $f(x)$, where x is positive, is defined as $M_x(s) = \int_0^\infty x^{s-1} f(x) dx$

Since transforms of probability density functions are of most concern, it is convenient to think the Mellin transform in terms of expected values. Recall that the expected value of any function $g(x)$ of the random variable x, whose distribution is $f(x)$, is given by

$$E[g(x)] = \int_{-\infty}^\infty g(x) f(x) dx$$

Therefore, it follows that

$$M_x(s) = E[X^{s-1}] = \int_0^\infty x^{s-1} f(x) dx$$

or

$$E[X^s] = M_x(s+1)$$

Thus, the first two statistical moments (i.e., mean and variance) of the random variable X can be stated in terms of the Mellin transform as

$$\mu = E[X^1] = M_X(2)$$

$$\sigma^2 = E[X^2] - [E[X]]^2 = M_X(3) - [M_X(2)]^2.$$

Because of the close relationship of Mellin transform and expected value, it is quite simple to establish some important operating properties of the Mellin transform involving products, quotients and powers of random variables. For example, if we try to find the distribution of $Y = X^b$, the expected value argument leads to

$$M_y(s) = E[Y^{s-1}] = E[(X^b)^{s-1}]$$

$$= E[X^{bs-b}] = E[X^{(bs-b+1)-1}] = M_x(bs-b+1)$$

That is, the transform for the distribution of the $b^{th}$ power of X can be obtained by replacing the s argument in the transform of f(x) by the expression (bs –b + 1).

In using the Mellin transform, the most important property is the Mellin convolution of two functions. This convolution is defined by the integral

$$f(z) = \int_0^\infty g(z/y)h(y)dy$$

This equation is precisely the form of the probability density function of the random variable Z=XY, where X and Y are continuously distributed, independent random variables with probability density functions g(x) and h(y), respectively. The transform of this special convolution reduces to a simple product of Mellin transforms. If we define the Mellin transform of f(z) as Mz(s), it follows that

$$M_Z(s) = M_X(s)M_Y(s).$$

The Mellin convolution may be extended to the product of n independent random variables like $Y = X_1 X_2 ... X_n$. The techniques for finding Mellin transforms of the pdfs

**Table 1.** Important Mellin transform operations (adapted from (Giffin 1975, Park 1987))

| Random Variable | pdf | Mellin Transform $M_Z(s)$ |
|---|---|---|
| X | f(x) | $M_X(s)$ |
| aX | f(x) | $a^{s-1} M_X(s)$ |
| $X^b$ | f(x) | $M_X(bs - b + 1)$ |
| 1/X | f(x) | $M_X(2 - s)$ |
| X + Y | f(x) | $M_X(s) + M_Y(s)$ |
| aX +bY | f(x) | $a^{s-1} M_X(s) + b^{s-1} M_Y(s)$ |
| XY | f(x), g(y) | $M_X(s) M_Y(s)$ |
| X/Y | f(x), g(y) | $M_X(s) M_Y(2 - s)$ |
| $aX^bY^cZ^d$ | f(x), g(y), h(z) | $a^{s-1} M_X(bs - b + 1) M_Y(cs - c + 1) M_Z(ds - d + 1)$ |

Note: a, b, c, d are constants; X, Y, Z are random variables.

A Probabilistic Approach to Fuzzy Engineering Economic Analysis    223

for products, quotients, and power of independent random variables are summarized in Table 1.

We have seen that the Mellin transform allows easy calculation of the statistical moments of any order of random variables of products and quotients without going through the inversion process. It is not even necessary to take derivatives. Park (1987) used the Mellin transformation in probabilistic cash flow modeling.

## 4  Mellin Transforms for Selected Probability Functions

In Section 2, we have shown the conversion of triangular and trapezoidal fuzzy membership functions into corresponding probability density function. Mellin transforms of these two density functions will now be derived.

A triangular density function is defined as

$$
f(x) = \begin{cases}
f_1(x) = \dfrac{2(x-l)}{(u-l)(m-l)} & l \leq x \leq m \\[3mm]
f_1(x) = \dfrac{2(u-x)}{(u-l)(u-m)} & m \leq x \leq u
\end{cases}
$$

**Table 2.** Mellin transforms for selected probability functions

| Name | Shape* | Parameters | $M_X(s)$ |
|---|---|---|---|
| Uniform | (a) | u(a ,b) | $\dfrac{b^s - a^s}{s(b-a)}$ |
| Triangular (1) | (b) | $T_1(l, m, u)$ | $\dfrac{2}{(u-l)s(s+1)}\left[\dfrac{u(u^s - m^s)}{(u-m)} - \dfrac{l(m^s - l^s)}{(m-l)}\right]$ |
| Triangular (2) | (c) | $T_2(l, m, \Phi)$ | $\dfrac{2}{(m-l)s(s+1)}\left[sm^s - \dfrac{l(m^s - l^s)}{(m-l)}\right]$ |
| Triangular (3) | (d) | $T_2(\Phi, m, u)$ | $\dfrac{2}{(u-m)s(s+1)}\left[\dfrac{u(u^s - m^s)}{(u-m)} - sm^s\right]$ |
| Trapezoidal (1) | (e) | $TR_1(a, b, c, d)$ | $\dfrac{2}{(c+d-b-a)s(s+1)}\left[\dfrac{(d^{s+1} - c^{s+1})}{(d-c)} - \dfrac{(b^{s+1} - a^{s+1})}{(b-a)}\right]$ |
| Trapezoidal (2) | (f) | $TR_2(a, b, c, \Phi)$ | $\dfrac{2}{(2c-b-a)s(s+1)}\left[(s+1)c^s - \dfrac{(b^{s+1} - a^{s+1})}{(b-a)}\right]$ |
| Trapezoidal (3) | (g) | $TR_3(\Phi, b, c, d)$ | $\dfrac{2}{(c+d-2b)s(s+1)}\left[\dfrac{(d^{s+1} - c^{s+1})}{(d-c)} - (s+1)b^s\right]$ |

*: Shapes are identified in Figure 5.

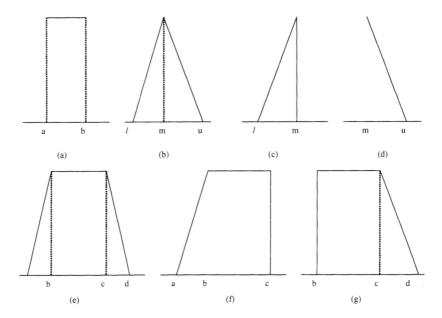

**Fig. 5.** Selected probability density functions

The Mellin transform is then obtained by

$$M_x(s) = \int_0^\infty x^{s-1} f(x) dx$$

$$= \int_l^m x^{s-1} f_1(x) dx + \int_m^u x^{s-1} f_2(x) dx$$

$$= \frac{2}{(u-l)s(s+1)} \left[ \frac{u(u^s - m^s)}{(u-m)} - \frac{l(m^s - l^s)}{(m-l)} \right]$$

The Mellin transform of uniform, triangular and trapezoidal density functions are summarized in Table 2 in conjunction with Figure 5.

Especially, the mean and the variance of a triangular density function can be obtained as

$$\mu = M_X(2) = (l + m + u)/3$$

$$\sigma^2 = M_X(3) - [M_X(2)]^2 = (l^2 + m^2 + u^2 - lu - mu - lm)/18.$$

Lee and Li (1988) obtained the mean and the variance of a fuzzy number based on the probability measure of fuzzy events defined by Zadeh (1968). Their results on mean and variance of TFN are equivalent to the above result by the Mellin transform.

## 5 Engineering Economy Applications

Two fuzzy cash flow analyses and a fuzzy multi-attribute decision problem are presented here to show the applicability of the Mellin transform in the fuzzy engineering economic environment.

### 5.1 Present Worth of Fuzzy Cash Flow Project

Suppose that a lump-sum return (F) is expected 5 years from now. Because of current uncertain economic conditions, the earning interest rate (i) and the lump-sum amount are best estimated by a TFN and a ZFN:

$$i = (l, m, u) = (7, 10, 12) \%$$

$$F = (a, b, c, d) = \$(6000, 10{,}000, 12{,}000, 15{,}000).$$

We want to compute the mean and the variance of the present value of this cash flow. We define the present value as

$$P = F(1+i)^{-5} = FX^{-5} = FY$$

The Mellin transform for P is given as

$$M_P(s) = M_F(s)\, M_Y(s)$$

where

$$M_F(s) = \frac{2}{(c+d-b-a)s(s+1)}\left[\frac{(d^{s+1}-c^{s+1})}{(d-c)} - \frac{(b^{s+1}-a^{s+1})}{(b-a)}\right]$$

$$= \frac{2}{11000 s(s+1)}\left[\frac{(15000^{s+1}-12000^{s+1})}{3000} - \frac{(10000^{s+1}-6000^{s+1})}{4000}\right]$$

$$M_Y(s) = M_X(-5s+6)$$

$$= \left[\frac{2}{0.05(-5s+6)(-5s+7)}\right]$$

$$\times\left[\frac{1.12(1.12^{-5s+6}-1.10^{-5s+6})}{0.02} - \frac{1.07(1.10^{-5s+6}-1.07^{-5s+6})}{0.03}\right]$$

Then, the mean and variance of the present value are determined as

$$\mu_P = M_P(2) = M_F(2)\, M_Y(2)$$

$$= 10696.07 \times 0.6312 = \$\,6752$$

$$\sigma_P{}^2 = M_P(3) - [M_P(2)]^2 = 1.53 \times 10$$

$$\sigma_P = \$1236.$$

226     K.P. Yoon

## 5.2 Comparing Alternatives with Fuzzy Cash Flow

Consider three mutually exclusive investment alternatives: A, B, C. Estimates of cash flow, which are specified by triangular fuzzy number, from three projects for the 6-year planning horizon are given in Table 3. The minimum attractive rate of return for the investment is specified by a triangular fuzzy number: i= (8%, 10%, 11%). The future worth analysis (at the end of year 6) is used to compare the attractiveness of three projects.

**Table 3.** Fuzzy cash flow from three projects

| Year | Projects | | |
|---|---|---|---|
| | A | B | C |
| 0 | $(-750, -700, -650) | $(-700, -600, -550) | $(-750, -700, -650) |
| 1 | (-350, -300, -250) | (-340, -300, -260) | (0, 0, 0) |
| 2 | (170, 200, 250) | (310, 350, 390) | (180, 200, 230) |
| 3 | (200, 250, 320) | (260, 300, 340) | (170, 200, 240) |
| 4 | (150, 300, 400) | (200, 250, 300) | (160, 200, 250) |
| 5 | (200, 350, 450) | (150, 200, 250) | (150, 200, 260) |
| 6 | (250, 400, 500) | (100, 150, 200) | (140, 200, 270) |

The future value (F) of an investment alternative can be determined by

$$F = \sum_{t=0}^{n} A_t (1+i)^{n-t} = \sum_{t=0}^{n} A_t Y^{n-t}$$

where $A_t$ is the cash flow at the end of year t.

Because the Mellin transform of random variable $X^b$ is $M_X(bs - b + 1)$, the Mellin transform for F can be given as

$$M_F(s) = \sum_{t=0}^{n} M_{A_t}(s) M_Y[(n-t)s - (n-t) + 1]$$

The mean future value of each alternatives can be obtained by settling n = 6 and s = 2 in the above equation. That is

$$M_F(2) = \sum M_{A_t}(2) M_Y(-t + 7).$$

Table 4 shows the calculation stages of $M_F(2)$ for project A. For example, when t = 2

$$M_{A2}(2) = (l + m + n)/3 = (170 + 200 + 250)/3 = 206.67$$

$$M_Y(5) = \frac{2}{(1.11 - 1.08)(5)(6)} \left[ \frac{1.11(1.11^5 - 1.10^5)}{(1.11 - 1.10)} - \frac{1.08(1.10^5 - 1.08^5)}{(1.10 - 1.08)} \right]$$

A Probabilistic Approach to Fuzzy Engineering Economic Analysis    227

**Table 4.** Mean future value of alternative A

| t | $M_{At}(2)$ | $M_Y(-t + 7)$ | $M_{At}(2)M_Y(-t + 7)$ |
|---|---|---|---|
| 0 | -$700.00 | 1.74 | -$1218.30 |
| 1 | -300.00 | 1.59 | -476.03 |
| 2 | 206.67 | 1.45 | 298.99 |
| 3 | 256.67 | 1.32 | 338.56 |
| 4 | 283.33 | 1.20 | 240.77 |
| 5 | 333.33 | 1.10 | 365.56 |
| 6 | 383.33 | 1.00 | 383.33 |
| Sum | | | $32.87 |

The sum of six years is $32.87. Similarly the mean future value of other projects can be obtained as [A, B, C] = [$32.87, $22.78, $14.98] which leads to the selection of project A.

### 5.3  Fuzzy Multi-attribute Automobile Selection (Dubois et al. 1988)

A young man wants to buy a used car. Presently seven cars are available at the garage which deals with second-hand cars. He establishes for decision attributes: age ($X_1$), price ($X_2$), gas consumption ($X_3$), and maximum speed ($X_4$) of a car. Table 5 shows his decision matrix which is expressed by fuzzy and crisp data. The fuzzy numbers, which represent the linguistic terms in the decision matrix, are summarized in Table 6.

This is a multiple attribute decision making (MADM) problem where a decision maker is to make preference decisions over the available alternatives which are characterized by multiple, usually conflicting, attributes. The MADM algorithms with crisp data are presented in (Hwang and Yoon 1981, Yoon and Hwang 1995). Reviews on fuzzy MADM methods are presented in (Chen and Hwang 1991, Kickert 1978, Zimmerman 1985, Zimmerman 1987). Most fuzzy MADM methods involve complicated mathematical operations. However, if the Mellin transform is used, a fuzzy MADM problem can be solved with a slight modification of crisp MADM algorithms.

**Table 5.** Decision matrix for used car selection (Dubois et al. 1988)

| Cars | Age | Purchase Price | Consumption | Speed |
|---|---|---|---|---|
| $A_1$ | new | expensive | economical | rather-fast |
| $A_2$ | less-than-3 | around-45000 | rather-economical | 180-200 |
| $A_3$ | very-recent | between-50000-and-60000 | heavy | Fast |
| $A_4$ | around-5 | less-than-20000 | 8-9 | 180 |
| $A_5$ | 5-10 | around-10000 | heavy | rather-fast |
| $A_6$ | old | cheap | economical | not-very-fast |
| $A_7$ | new | 32000-40000 | very-economical | between-140- and-160 |

228     K.P. Yoon

**Table 6.** Fuzzy numbers for linguistic terms

| $X_1$: Age | | $X_2$: Selling Price | |
| --- | --- | --- | --- |
| Term | Number | Term | Number |
| new | $TR_3(\Phi, 0, 1, 2)$ | cheap | $TR_3(\Phi, 5000, 10000, 15000)$ |
| very-recent | $TR_1(0, 1, 2, 3)$ | around-10000 | $TR_1(8000, 9000, 11000, 12000)$ |
| less-than-3 | $U(0, 3)$ | less-than-20000 | $U(0, 20000)$ |
| around-5 | $T_1(4, 5, 6)$ | 32000-40000 | $U(32000, 40000)$ |
| 5-10 | $U(5, 10)$ | around-45000 | $TR_1(43000, 44000, 46000, 47000)$ |
| old | $TR_1(8, 10, 15, 20)$ | moderate | $TR_1(30000, 35000, 60000, 65000)$ |
| | | between-50000- and-60000 | $TR_1(45000, 50000, 60000, 65000)$ |
| | | expensive | $TR_2(75000, 80000, 100000, \Phi)$ |
| $X_3$: Gas Consumption | | $X_4$: Max. Speed | |
| Term | Number | Term | Number |
| Very-economical | $TR_3(\Phi, 5, 6, 6.5)$ | fast | $TR_1(160, 180, 200, 220)$ |
| economical | $TR_1(5.5, 6, 7, 8)$ | rather-fast | $TR_1(130, 150, 180, 200)$ |
| rather-economical | $TR_1(6, 7, 8, 9)$ | not-very-fast | $TR_1(110, 120, 140, 150)$ |
| $8-9$ | $U(8, 9)$ | between-140- and-160 | $TR_1(120, 140, 160, 180)$ |
| heavy | $TR_2(8.5, 9, 15, \Phi)$ | 180-200 | $U(180, 200)$ |

One of the classical MADM methods, the Weighted Product (WP) method, will be used to illustrate the proposed approach. This method was introduced long ago by Bridgman (1922), and has recently been advocated by Starr (1972) due to its super discrimination power.

The value of alternative $A_i$ in WP is determined by

$$V(A_i) = V_i = \prod_{j=1}^{n} X_{ij}^{w_j} \qquad i = 1,...,m$$

where $X_{ij}$ is the performance rating of i-th alternative with respect to j-th attribute, and $W_j$ is the weight assigned to the j-th attribute. Because the Mellin transform of random variable $x^b$ is $M_X(bs - b +1)$, the Mellin transform of $V_i$ is given as

$$M_i(s) = \prod_{j=1}^{n} M_{ij}(w_j s - w_j + 1) \qquad i = 1,...,m$$

Because of the exponent property, this method requires that negative weight (i.e., $-W_j$) should be given to a cost attribute (the more rating value, the less preference).

If the assessment of weights of four attributes be $(W_1, W_2, W_3, W_4) = (.32, .20, .40, .08)$, which car would be chosen? Since X4 is the only benefit attribute, the weight

A Probabilistic Approach to Fuzzy Engineering Economic Analysis 229

can be expressed as $(W_1, W_2, W_3, W_4) = (-.32, -.20, -.40, 0.08)$. The $M_i(s)$ is now rewritten as

$$M_i(s) = M_{i1} (-.32s + 1.32) M_{i2} (-.20s + 1.2) M_{i3} (-.40s + 1.4) M_{i4} (.08s + .92).$$

The mean of each $V_i$ are obtained as

$$(\mu_1, \mu_2, \mu_3, \mu_4, \mu_5, \mu_6, \mu_7)$$

$$= (M_1(2), M_2(2), M_3(2), M_4(2), M_5(2), M_6(2), M_7(2))$$

$$= (.3078, .3291, .2396, .2645, .1881, .1989, .4705)$$

which picks $A_7$ as the first choice and $A_5$ the last. Note that Dubois et al. (1988) solved this problem by weighted fuzzy pattern matching. They also picked $A_7$.

## 6 Concluding Remarks

The literature in evaluating fuzzy alternatives is abundant, but most methods require insurmountable computational effort which may prevents its wide applications. In this chapter a probabilistic approach was chosen instead of fuzzy set theory manipulation.

Membership functions are first converted into probabilistic density functions, then the Mellin transform is used to compute the mean and the variance of the complex fuzzy numbers. The fuzzy number with the higher mean is then ranked higher than the fuzzy number with a lower mean. If the means are equal, the one with the smaller variance is judged higher rank.

Numerical examples reveal the efficiency of the Mellin transform approach in comparing fuzzy alternatives in the engineering economy environment. A real beauty of the approach is that it can accommodate all types of fuzzy numbers such as uniform, triangular, or trapezoidal simultaneously, and all computations are easily calculated in the spreadsheet. Considering that most real-world problems have alternatives ranging from 5, 10, 50, or over 100, this is a great computational advantage.

However, there is one caveat to be raised. The resulting transform may not be analytical for all values of s, especially when quotients are involved. For example, with a random variable Y, $y = 1/x$ where x is a random variable uniformly distributed over [0, 1], the Mellin transform of $f(y)$ is $M_Y(x) = M(2 - s) = 1/(2 - s)$. To find $E[y]$, we need to evaluate $M_Y(2)$, but it is not defined . If the transform is not analytical at $s = 2$ or $s = 3$, Park (1987), Young and Contreras (1975) suggested other techniques such as Laplace transform.

## References

Bridgman, P.W.: Dimensional analysis. Yale University Press, New Haven (1922)

Chen, S.J., Hwang, C.L.: Fuzzy multiple attribute decision making: method and applications. Springer, Berlin (1991)

Chiu, C.Y., Park, C.S.: Fuzzy cash flow analysis using present worth criterion. The Engineering Economist 39, 113–138 (1994)

Dubois, D., Prade, H.: Fuzzy sets and systems: theory and applications. Academic Press, New York (1980)

Dubois, D., Prade, H., Testemale, C.: Weighted fuzzy pattern matching. Fuzzy Sets and Systems 28, 313–331 (1988)

Epstein, B.: Some applications of the Mellin transform in statistics. Annuals of Mathematical Statistics 19, 370–379 (1948)

Giffin, W.C.: Transform techniques for probability modeling. Academic Press, New York (1975)

Hwang, C.L., Yoon, K.: Multiple attribute decision making: methods and applications. Springer, Berlin (1981)

Kahraman, C., Beskese, A., Ruan, D.: Measuring flexibility of computer integrated manufacturing systems using fuzzy cash flow analysis. Information Sciences 168, 77–94 (2004)

Kahraman, C., Ruan, D., Tolga, E.: Capital budgeting techniques using discounted fuzzy versus probabilistic cash flows. Information Sciences 142, 57–76 (2002)

Karsak, E.E.: Measures of liquidity risk supplementing fuzzy discounted cash flow analysis. The Engineering Economist 43, 331–344 (1998)

Kaufmann, A., Gupta, M.M.: Introduction to fuzzy arithmetic theory and applications. Van Nostrand Reinhold, New York (1985)

Kaufmann, A., Gupta, M.M.: Fuzzy mathematical models in engineering and management science. North-Holland, Amsterdam (1988)

Kickert, W.J.M.: Fuzzy theory on decision making, a critical review. In: Martinus Nijhoff Social Science Division, Leiden, Netherlands (1978)

Kulak, O.: A decision support system for fuzzy multi-attribute selection of material handling equipments. Expert Systems with Applications 29, 310–319 (2005)

Kulak, O., Durmusoglu, N.B., Kahraman, C.: Fuzzy multi-attribute equipment selection based on information axiom. Journal of Materials processing Technology 169, 337–345 (2005)

Lee, E.S., Li, R.J.: Comparison of fuzzy numbers based on the probability measure fuzzy events. Computers and Mathematics with Applications 15, 887–896 (1988)

Liou, T.S., Chen, C.W.: Fuzzy decision analysis for alternative selection using a fuzzy annual worth criterion. The Engineering Economist 51, 19–34 (2006)

McCahon, C.S., Lee, E.S.: Comparing fuzzy numbers: the proportion of the optimum method. International Journal of Approximate Reasoning 4, 159–181 (1990)

Park, C.S.: The Mellin transform in probabilistic cash flow modeling. The Engineering Economist 32, 115–134 (1987)

Springer, M.D.: The algebra of random variables. John Wiley, New York (1979)

Starr, M.K.: Production management. Prentice-Hall, Englewood Cliffs (1972)

Tseng, T.Y., Klein, C.M., Leonard, M.S.: A formalism for comparing ranking procedures. In: Proceedings of the 7th Annual Meeting of the North American Fuzzy Information Processing Society (NAFIPS), pp. 231–235 (1988)

Yoon, K.P., Hwang, C.L.: Multiple attribute decision making: an introduction. Sage, Thousand Oaks (1995)

Young, D., Contreras, L.: Expected present worths of cash flows under uncertain timing. The Engineering Economist 20, 257–268 (1975)

Zadeh, L.A.: Probability measures of fuzzy events. Journal of Mathematical Analysis and Applications 23, 421–427 (1968)

Zadeh, L.A.: The concept of linguistic variables and its application to approximate reasoning-I. Information Science 8, 199–249 (1975)

Zimmerman, H.J.: Fuzzy set theory and its applications. Kluwer, Boston (1985)

Zimmerman, H.J.: Fuzzy set, decision making and expert system. Kluwer, Boston (1987)

# Investment Analyses Using Fuzzy Decision Trees

Cengiz Kahraman

Istanbul Technical University, Department of Industrial Engineering,
34367, Macka, İstanbul, Turkey

**Abstract.** A decision tree is a method you can use to help make good choices, especially decisions that involve high costs and risks. Decision trees use a graphic approach to compare competing alternatives and assign values to those alternatives by combining uncertainties, costs, and payoffs into specific numerical values. A fuzzy decision tree is a generalization of the crisp case. Fuzzy decision trees are helpful for representing ill-defined structures in decision analysis. This chapter presents investment analyses using fuzzy decision trees with examples.

## 1 Introduction

In many situations one needs to make a series of decisions. This leads naturally to a structure called a "decision tree." Decision trees provide a geometrical framework for organizing the decisions. A decision tree is a graph of nodes connected by arcs, with each node corresponding to a non-goal attribute and each arc to a possible value of that attribute. A leaf of the tree specifies the expected value of the goal attribute for the records described by the path from root to leaf. *Decision tree* is a classifier in the form of a tree structure, where each node is either:

- a *leaf node* - indicates the value of the target attribute (class) of examples, or
- a *decision node* - specifies some test to be carried out on a single attribute-value, with one branch and sub-tree for each possible outcome of the test.

A decision tree can be used to classify an example by starting at the root of the tree and moving through it until a leaf node, which provides the classification of the instance. The most basic form of decision tree occurs when each alternative can be assumed to result in a single outcome-that is, when certainty is assumed.

## 2 Decision Trees

A decision tree is a diagram of nodes and connecting branches. Nodes indicate decision points, chance events, or branch terminals. Branches correspond to each decision alternative or event outcome emerging from a node.

The diagram is read from left to right. The leftmost node in a decision tree is called the root node. The branches emanating to the right from a decision node represent the set of decision alternatives that are available. One, and only one, of these alternatives can be selected. The small circles in the tree are called chance nodes. The number shown in parentheses on each branch of a chance node is the probability that the outcome shown on that branch will occur at the chance node. The right end of each path

C. Kahraman (Ed.): Fuzzy Engineering Economics with Appl., STUDFUZZ 233, pp. 231–242, 2008.
springerlink.com © Springer-Verlag Berlin Heidelberg 2008

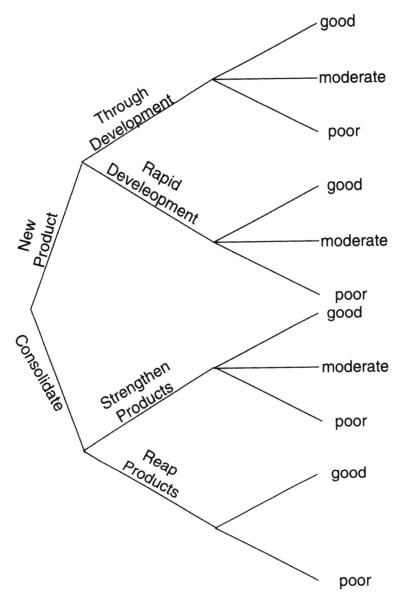

**Fig. 1.** An example decision tree: should we develop new product or consolidate?

through the tree is called an endpoint, and each endpoint represents the final outcome of following a path from the root node of the decision tree to that endpoint. An example for a decision tree problem is given in Figure 1. The problem is if we should develop a new product or consolidate? The decision tree illustrates both the possibilities and their results.

In order to decide which alternative to select in a decision problem, a decision criterion is needed; that is, a rule for making a decision. Expected value is a criterion for making a decision that takes into account both the possible outcomes for each decision alternative and the probability that each outcome will occur.

The expected value for an uncertain alternative is calculated by multiplying each possible outcome of the uncertain alternative by its probability, and summing the results. The expected value decision criterion selects the alternative that has the best expected value. In situations involving profits where "more is better," the alternative with the highest expected value is best, and in situations involving costs, where "less is better," the alternative with the lowest expected value is best.

The process of successively calculating expected values from the endpoints of the decision tree to the root node is called a decision tree rollback. The complete specification of all the preferred decisions in a sequential decision problem is called the decision strategy.

## 3 Fuzzy Decision Trees

The fuzzy rules that were generated were "flat", in the sense they were all applied simultaneously. While this makes more sense with fuzzy rules than with standard bivalent rules, the natural extension is to take a cue from decision trees and attempt to construct fuzzy decision trees. These would be identical to standard decision trees except that the decisions at each branching point would be fuzzy, rather than bivalent. This makes calculating the best split somewhat harder (continuous functions for the information gain must be maximized), but would be more rewarding by allowing smaller trees with less leaves and internal nodes to encapsulate a richer amount of information.

Several papers have been written on this topic. Boyen (1995) describes the application of fuzzy decision trees to assessing the stability of power systems. Heinz (1995) describes a hybrid method of adaptive fuzzy neural trees, combining fuzzy logic with both neural networks and decision trees. Savsek et al. (2006) provide a new definition of the fuzzy relational tree structure and the development of a new comparative method for fuzzy trees and its experimental testing and evaluation. They also provide a new descriptive method of military structures in a fuzzy tree format and the development of a fuzzy decision support system. Scharer et al. (2006) aim at predicting annual phosphorus export from agricultural catchments using a very simple approach that concentrates on the functional behavior of a catchment. Chen et al. (2007) use a fuzzy decision tree to form a search mechanism for vague knowledge in design for outsourcing with index for classifying vague knowledge. Computational experiments are conducted to demonstrate the performance of the proposed search mechanism and knowledge classification. Odejobi et al. (2007) present syllable-based duration modeling in the context of a prosody model for Standard Yoruba (SY) text-to-speech (TTS) synthesis applications. Their prosody model is conceptualized around a modular holistic framework. This framework is implemented using the Relational Tree (R-Tree) techniques. An important feature of their R-Tree framework is its flexibility in that it facilitates the independent implementation of the different dimensions of prosody, i.e. duration, intonation, and intensity, using different techniques and their subsequent integration. They apply the Fuzzy Decision Tree (FDT) technique to model the duration dimension. In order to evaluate the effectiveness of FDT in duration modelling, they also develop a Classification And Regression Tree (CART) based duration model using the same speech data. Beynon et al. (2004) concern with the exposition and application of a fuzzy decision tree approach to a problem involving typical accounting data. More specifically, a set of fuzzy 'if - then' rules is constructed to classify the level of corporate audit costs based on a number of characteristics of the companies and their auditors. The fuzzy rules enable a decision-maker to gain additional insights into the relationship between firm characteristics and audit fees, through human subjective judgments expressed in linguistic terms. They also extend previous research by developing a more objective semi-automated method of constructing the FST related membership functions which mitigates reliance on the input of human expert opinions. Wang et al. (2000) point out the inherent defect of the likes of fuzzy ID3 (ID3 is a typical algorithm for generating decision trees.) and present two optimization principles of fuzzy decision trees and prove that the algorithm complexity of constructing a kind of minimum fuzzy decision tree is NP-hard. They use the following definitions, considering a directed tree of which each edge links two nodes, the initial node and the terminal node. The former is called the fathernode of the latter while the latter is said to be the sonnode of the former. The node having not its fathernode is the root whereas the nodes having not any sonnodes are called leaves. The general matching strategy of fuzzy decision trees is described as follows: (a) Matching starts from the root and ends at a leaf along the branch of the maximum membership. (b) If the maximum membership at the node is not unique, matching proceeds along several branches. (c) The decision with maximum degree of truth is assigned to the matching result.

A comparison between the crisp and fuzzy decision trees is given in Table 1 (Wang et al. 2000).

**Table 1.** Crisp and fuzzy decision trees

| Crisp Decision Trees | Fuzzy Decision Trees |
|---|---|
| Nodes are crisp subsets of X | Nodes are fuzzy subsets of X |
| If N is not a leaf and $\{N_i\}$ is the family off all sonnodes of $N$, then $\bigcup_i N_i = N$ | If N is not a leaf and $\{N_i\}$ is the family off all sonnodes of N, then $\bigcup_i N_i \subset N$ |
| A path from the root to a leaf corresponds to a production rule. | A path from the root to a leaf corresponds to a fuzzy rule with some degree of truth. |
| An example remaining to be classified matches only one path in the tree. | An example remaining to be classified can match several paths in the tree. |
| The intersection of subnodes located on the same layer is empty. | The intersection of subnodes located on the same layer can be nonempty. |

# 4 A Numerical Example

Let us consider the following table (Yuan and Shaw 1995):

**Table 2.** A small training set

| No | Outlook Sunny | Cloudy | Rain | Temperature Hot | Mild | Cool | Humidity Humid | Normal | Humid Normal Windy | Not windy | Class V | S | W |
|---|---|---|---|---|---|---|---|---|---|---|---|---|---|
| 1 | 0.9 | 0.1 | 0.0 | 1.0 | 0.0 | 0.0 | 0.8 | 0.2 | 0.4 | 0.6 | 0.0 | 0.8 | 0.2 |
| 2 | 0.8 | 0.2 | 0.0 | 0.6 | 0.4 | 0.0 | 0.0 | 1.0 | 0.0 | 1.0 | 1.0 | 0.7 | 0.0 |
| 3 | 0.0 | 0.7 | 0.3 | 0.8 | 0.2 | 0.0 | 0.1 | 0.9 | 0.2 | 0.8 | 0.3 | 0.6 | 0.1 |
| 4 | 0.2 | 0.7 | 0.1 | 0.3 | 0.7 | 0.0 | 0.2 | 0.8 | 0.3 | 0.7 | 0.9 | 0.1 | 0.0 |
| 5 | 0.0 | 0.1 | 0.9 | 0.7 | 0.3 | 0.0 | 0.5 | 0.5 | 0.5 | 0.5 | 0.0 | 0.0 | 1.0 |
| 6 | 0.0 | 0.7 | 0.3 | 0.0 | 0.3 | 0.7 | 0.7 | 0.3 | 0.4 | 0.6 | 0.2 | 0.0 | 0.8 |
| 7 | 0.0 | 0.3 | 0.7 | 0.0 | 0.0 | 1.0 | 0.0 | 1.0 | 0.1 | 0.9 | 0.0 | 0.0 | 1.0 |
| 8 | 0.0 | 1.0 | 0.0 | 0.0 | 0.2 | 0.8 | 0.2 | 0.8 | 0.0 | 1.0 | 0.7 | 0.0 | 0.3 |
| 9 | 1.0 | 0.0 | 0.0 | 1.0 | 0.0 | 0.0 | 0.6 | 0.4 | 0.7 | 0.3 | 0.2 | 0.8 | 0.0 |
| 10 | 0.9 | 0.1 | 0.0 | 0.0 | 0.3 | 0.7 | 0.0 | 1.0 | 0.9 | 0.1 | 0.0 | 0.3 | 0.7 |
| 11 | 0.7 | 0.3 | 0.0 | 1.0 | 0.0 | 0.0 | 1.0 | 0.0 | 0.2 | 0.8 | 0.4 | 0.7 | 0.0 |
| 12 | 0.2 | 0.6 | 0.2 | 0.0 | 1.0 | 0.0 | 0.3 | 0.7 | 0.3 | 0.7 | 0.7 | 0.2 | 0.1 |
| 13 | 0.9 | 0.1 | 0.0 | 0.2 | 0.8 | 0.0 | 0.1 | 0.9 | 1.0 | 0.0 | 0.0 | 0.0 | 1.0 |
| 14 | 0.0 | 0.9 | 0.1 | 0.0 | 0.9 | 0.1 | 0.1 | 0.9 | 0.7 | 0.3 | 0.0 | 0.0 | 1.0 |
| 15 | 0.0 | 0.0 | 1.0 | 0.0 | 0.0 | 1.0 | 1.0 | 0.0 | 0.8 | 0.2 | 0.0 | 0.0 | 1.0 |
| 16 | 1.0 | 0.0 | 0.0 | 0.5 | 0.5 | 0.0 | 0.0 | 1.0 | 0.0 | 1.0 | 0.8 | 0.6 | 0.0 |

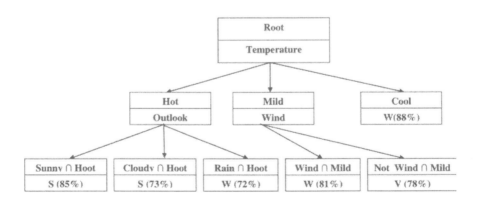

**Fig. 2.** A fuzzy decision tree generated by training Table 1

In Table 2, each column corresponds to a fuzzy subset defined on $X = \{1,2,3,...,16\}$. Four attributes are defined as follows (Wang et al. 2000):

$$\text{Temperature} = \{hot, mild, cool\} \subset F(X),$$

$$\text{Outlook} = \{sunny, cloudy, rain\} \subset F(X),$$

$$\text{Humidity} = \{humid, normal\} \subset F(X),$$

$$\text{Wing} = \{windy, not\_windy\} \subset F(X).$$

In the last column, three symbols, *V, S,* and *W* denote three sports to play on weekends, volleyball, swimming, and weight_lifting, respectively. A fuzzy decision tree for the root *Temperature* is given in Figure 2. The percentage attached to each decision is the degree of truth on the decision.

Consider two examples remaining to be classified $e_1$=(0.0, 0.6, 0.4; 0.3, 0.7, 0.0; 0.5, 0.5; 0.5, 0.5) and $e_2$=(0.9, 0.1, 0.0; 0.8, 0.2, 0.0; 0.5, 0.5; 0.8, 0.2). The process of matching in the fuzzy decision tree (Fig. 2) is shown in Fig. 3.

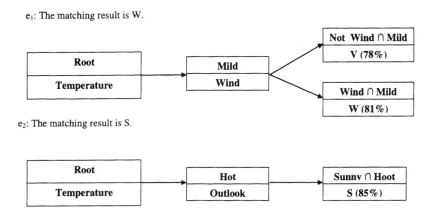

**Fig. 3.** The matching process of two examples

## 5 Investment Analysis Using Fuzzy Decision Trees

In the following two numerical examples for investment analysis will be given. The following approaches are simplistic but effective approaches to decision trees with fuzzy parameters. In these examples, parameters like cash flows and interest rate are fuzzy rather than the tree itself.

The first crisp decision tree problem is about an investment decision. Later, the same problem will be solved using the fuzzy case.

A decision maker may determine that the chance of drilling an oil well that generates $100,000 (outcome) is 25 percent (probability of occurrence). To solve the decision tree, the decision maker begins at the right hand side of the diagram and works toward the initial decision branch on the left. The value of different outcomes is derived by multiplying the probability by the expected outcome; in this example, the value would be $25,000 (0.25 x $100,000). The values of all the outcomes emanating from a chance fork are combined to arrive at a total value for the chance fork. By continuing to work backwards through the chance and decision forks, a value can eventually be assigned to each of the alternatives emanating from the initial decision fork.

In the rudimentary example below, a company is trying to determine whether or not to drill an oil well. If it decides not to drill the well, no money will be made or lost. Therefore, the value of the decision not to drill can immediately be assigned a sum of zero dollars.

If the decision is to drill, there are several potential outcomes, including (1) a 10 percent chance of getting $300,000 in profits from the oil; (2) a 20 percent chance of extracting $200,000 in profits; (3) a 10 percent chance of wresting $100,000 in profits from the well; and (4) a 60 percent chance that the well will be dry and post a loss of $100,000 in drilling costs. Figure 4 shows the decision tree for this data. Multiplying the probability of each outcome by its dollar value, and then combining the results, assigns an expected value to the decision to drill of $20,000 in profits. Thus, the profit maximizing decision would be to drill the well.

For the purposes of demonstration, suppose that the chance of hitting no oil was increased from 60 percent to 70 percent, and the chance of gleaning $300,000 in profits was reduced from ten percent to zero. In that case, the dollar value of the decision to drill would fall to -$20,000. A profit-maximizing decision maker would then elect to not drill the well. The effect of this relatively small change in the probability calculation underscores decision trees' dependence on accurate information, which often may not be available.

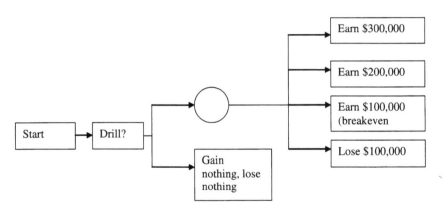

**Fig. 4.** Decision tree for an oil drilling investment

In the fuzzy case, "Earn $300,000 with a probability of 10%" can be expressed by "earn around $30,000". In the same way, the others can be "earn around $40,000", "earn around $10,000", and "lose around $60,000", respectively. "Earn around $30,000" and the others can be represented by the following equations. Let *LSR* and *RSR* be the left and right representations of the fuzzy numbers:

$$\text{earn around } \$30{,}000 = \begin{cases} 25{,}000 + 5{,}000\alpha, & \text{for} \quad LSR \\ 35{,}000 - 5{,}000\alpha, & \text{for} \quad RSR \end{cases}$$

$$\text{earn around } \$40{,}000 = \begin{cases} 35{,}000 + 5{,}000\alpha, & \text{for} \quad LSR \\ 45{,}000 - 5{,}000\alpha, & \text{for} \quad RSR \end{cases}$$

$$\text{earn around } \$10{,}000 = \begin{cases} 8{,}000 + 2{,}000\alpha, & \text{for } LSR \\ 12{,}000 - 2{,}000\alpha, & \text{for } RSR \end{cases}$$

$$\text{lose around } \$60{,}000 = \begin{cases} 55{,}000 + 5{,}000\alpha, & \text{for } LSR \\ 65{,}000 - 5{,}000\alpha, & \text{for } RSR \end{cases}$$

The fuzzy total expected value (*FTEV*) is calculated as in the following:

$$FTEV = \begin{bmatrix} 25{,}000 + 5{,}000\alpha + 35{,}000 + 5{,}000\alpha + 8{,}000 + 2{,}000\alpha - 55{,}000 - 5{,}000\alpha; \\ 35{,}000 - 5{,}000\alpha + 45{,}000 - 5{,}000\alpha + 12{,}000 - 2{,}000\alpha - 65{,}000 + 5{,}000\alpha \end{bmatrix}$$

and

$$FTEV = [13{,}000 + 7{,}000\alpha;\ 27{,}000 - 7{,}000\alpha]$$

FTEV is represented by Fig. 5:

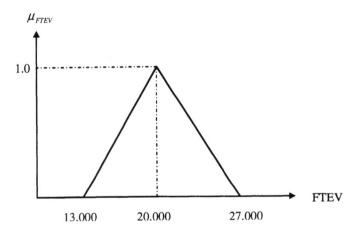

**Fig. 5.** Fuzzy total expected value

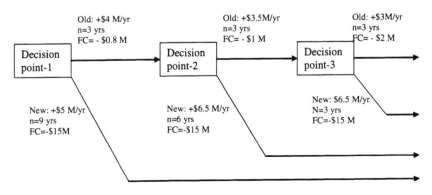

**Fig. 6.** Deterministic decision tree for a replacement decision

According to Figure 5, there is no possibility of having a negative total expected value while it is impossible to obtain a value less than $13,000 and a value more than $27,000. So, this well should be drilled.

As the second example, consider the following deterministic problem illustrated in Figure 6 (Canada and White, 1980). The minimum attractive rate of return is 25% per year.

Table 3 shows the calculations for the deterministic replacement decision:

**Table 3.** Calculations for the deterministic replacement decision

| Decision point | Alternative | PW of monetary outcome | Choice |
|---|---|---|---|
| 3 | Old | $3M(P/A, 25%, 3) - $2M = $3.85M | Old |
|   | New | $6.5M(P/A, 25%, 3) - $15 M = -$2.33M |  |
| 2 | Old | $3.85M(P/F, 25%, 3) + $3.5M(P/A, 25%, 3) - $1M = $7.79 M | Old |
|   | New | $6.5M(P/A, 25% , 6) - $15 M = $4.18 M |  |
| 1 | Old | $7.79M(P/F, 25%, 3) + $4M(P/A, 25%, 3) - $0.8M = $10.98 M | Old |
|   | New | $5.0M(P/A, 25%, 9) - $15M = $2.30M |  |

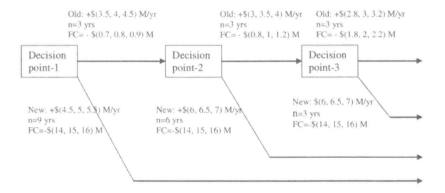

**Fig. 7.** Decision tree with fuzzy cash flows

The decision is to keep the old machine at the decision points 1, 2, and 3.

In the fuzzy case, using a fuzzy discount rate of (23%, 25%, %27) per year, lets decide if the new machine should be invested in or the old machine should be kept. The cash flows are also accepted as fuzzy numbers as indicated in Figure 7.

In this case, the calculations in Table 3 will change to be in Table 4.

240    C. Kahraman

**Table 4.** Calculations for the fuzzy replacement decision

| Decision point | Alternative | PW of monetary outcome | Choice |
|---|---|---|---|
| 3 | Old | $(2.8, 3, 3.2)M(P/A, (23%, 25%, 27%), 3) – | Old |
|   |   | $(1.8, 2, 2.2)M = $(3.12, 3.856, 4.64)M |   |
|   | New | $(6, 6.5, 7)M(P/A, (23%, 25%, 27%), 3) – |   |
|   |   | $(14, 15, 16)M = -$(4.6, 2.312, 0.7798)M |   |
| 2 | Old | $(3.12, 3.856, 4.64)M(P/F, (23%, 25%, 27%), 3) + | Rather old |
|   |   | $(3, 3.5, 4)M(P/A, (23%, 25%, 27%), 3) |   |
|   |   | - $(0.8, 1, 1.2)M = $(6.010, 7.806, 9.739)M |   |
|   | New | $(6, 6.5, 7)M(P/A, (23%, 25%, 27%), 6) |   |
|   |   | - $(14, 15, 16)M = $(0.926, 4.184, 7.644)M |   |
| 1 | Old | $(6.010, 7.806, 9.739)M(P/F, (23%, 25%, 27%), 3) | Old |
|   |   | + $(3.5, 4, 4.5)M(P/A, (23%, 25%, 27%), 3) – |   |
|   |   | $(0.7, 0.8, 0.9)M = $(8.669, 11.005, 11.568)M |   |
|   | New | $(4.5, 5, 5.5)M(P/A, (23%, 25%, 27%), 9) |   |
|   |   | - $(14, 15, 16)M = $(-1.272, 2.315, 6.202)M |   |

On the first line of Table 4, the result of $(3.12, 3.856, 4.64)M is obtained in the following way:

$AW_l(\alpha)=2.8+0.2\alpha,$

$I_l(\alpha)=0.23+0.02\alpha,$

$FC_l(\alpha)= 1.8+0.2\alpha$

$AW_r(\alpha)=3.2-0.2\alpha,$

$I_r(\alpha)=0.27-0.02\alpha,$

$FC_r(\alpha)= 2.2-0.2\alpha$

$PW_l(\alpha)= AW_l(\alpha) (P/A, I_r(\alpha), 3) - FC_r(\alpha)$

$PW_l(\alpha)= (2.8+0.2\alpha) (P/A, 0.27-0.02\alpha, 3) - (2.2-0.2\alpha)$

The least possible value of PW is obtained for $\alpha=0$:

$$PW_l(\alpha)= 2.8(P/A, 0.27, 3) - 2.2 = 2.8\times1.9 - 2.2 = 3.12$$

The most possible value is obtained for $\alpha=1$

$PW_l(\alpha)= 3(P/A, 0.25, 3) - 2 = 3\times1.952 - 2 = 3.856$

$PW_r(\alpha)= AW_r(\alpha) (P/A, I_l(\alpha), 3) - FC_l(\alpha)$

$PW_r(\alpha)= (3.2-0.2\alpha)(P/A, 0.23+0.02\alpha, 3) - (1.8+0.2\alpha)$

The largest possible value is obtained for $\alpha=0$:

$$PW_r(\alpha)= 3.2(P/A, 0.23, 3) - 1.8 = 4.64$$

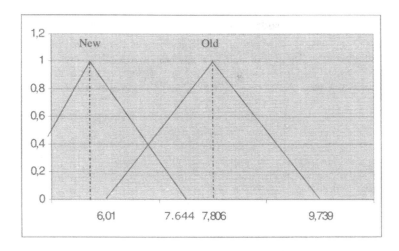

**Fig. 8.** The results of Decision Point 2

For the second decision point, the results have an intersection whereas the others do not.

The intersection level of two results is 0.311. This is the possibility of being that the new machine has a greater present worth than the old machine. But Decision point 1 indicates that the new machine is never superior to the old machine. Then this means that a decision in favor of the new machine in Decision Point 2, considering the possibility of a superiority of 0.311, will be a wrong decision.

## 6 Conclusions

A decision tree is used to identify the strategy most likely to reach a goal. Another use of trees is as a descriptive means for calculating conditional probabilities. The potential of fuzzy decision trees in improving the robustness and generalization in classification is due to the use of fuzzy reasoning. A fuzzy decision tree is the generalization of a decision tree in a fuzzy environment. The knowledge represented by a fuzzy decision tree is more natural to the way of human thinking, but there is the additional work of preprocessing and cost of constructing trees.

## References

Boyen, X.: Design of fuzzy logic-based decision trees applied to power system transient stability assessment. Master's Thesis, University of Liège (1995)
Canada, J.R., White, J.A.: Capital investment decision analysis for management and engineering. Prentice-Hall, Englewood Cliffs (1980)
Chen, R.Y., Sheu, D.D., Liu, C.M.: Vague knowledge search in the design for outsourcing using fuzzy decision tree. Computers & Operations Research 34, 3628–3637 (2007)

Heinz, A.P.: Pipelined neural tree learning by error forward-propagation. ICNN 1995. In: Proceedings of the IEEE International Conference on Neural Networks (1995)

Malcolm, J.B., Michael, J.P., Yu-Cheng, T.: The application of fuzzy decision tree analysis in an exposition of the antecedents of audit fees. Omega 32, 231–244 (2004)

Odejobi, O.A., ShunHa, S.W., Anthony, J.B.: A fuzzy decision tree-based duration model for Standard Yoruba text-to-speech synthesis. Computer Speech and Language 21, 325–349 (2007)

Savsek, T., Vezjak, M., Pavesic, N.: Fuzzy trees in decision support systems. European Journal of Operational Research 174, 293–310 (2006)

Scharer, M., Page, T., Beven, K.: A fuzzy decision tree to predict phosphorus export at the catchment scale. Journal of Hydrology 331, 484–494 (2006)

Wang, X., Chen, B., Qian, G., Ye, F.: On the optimization of fuzzy decision trees. Fuzzy Sets and Systems 112, 117–125 (2000)

Xizhao, W., Bin, C., Guoliang, Q., Feng, Y.: On the optimization of fuzzy decision trees. Fuzzy Sets and Systems 112, 117–125 (2000)

Yuan, Y., Shaw, M.J.: Induction of fuzzy decision trees. Fuzzy Sets and Systems 69, 125–139 (1995)

# Fuzzy Multiobjective Evaluation of Investments with Applications

L. Dymova and P. Sevastjanov

Institute of Comp. & Information Sci., Czestochowa University of Technology,
Dabrowskiego 73, 42-200 Czestochowa, Poland

**Abstract.** The methods for Multiobjective Decision Making in the fuzzy setting in context of Investment Evaluation Problem are analyzed. The problems typical for Multiobjective Decision Making are indicated and new solutions of them are proposed as well. The problem of appropriate common scale for representation of objective and subjective criteria is solved using the simple subsethood measure based on $\alpha$-cut representation of fuzzy values. To elaborate an appropriate method for aggregation of aggregating modes, we use the synthesis of the tools of Type 2 and Level 2 Fuzzy Sets. As the result the final assessments of compared alternatives are presented in form of fuzzy valued membership function defined on the support composed of considered alternatives. To compare obtained fuzzy assessments we use the probabilistic approach to fuzzy values comparison. In is shown that Investment Evaluation Problem is frequently a hierarchical one and a new method for solving such problems, different from commonly used fuzzy $AHP$ method, is proposed.To make the presentation more transparent, it is illustrated throughout the chapter with use of two examples. The first of them is well known Tool Steel Material Selection problem which can be considered as the typical investment problem and relevant test charged by all difficulties concerned with the problems of Multiobjective Decision Making. The next example is a simplified investment project evaluation problem we have used to show that even when project's estimation is based on the budgeting, i.e., only financial parameters are taking into account, we are dealing with multiple-criteria task in the fuzzy setting.

## 1 Introduction

It is well known that evaluation of important investment projects usually can not be successfully carried out using only financial parameters since the possible ecological, social and even political effects of project's implementation should be evaluated as well. The role of these effects rises along with the project's importance. Obviously, such effects as a rule can not be predicted with a high accuracy, moreover their estimations are usually based on the expert's opinions expressed in a verbal form. So the proper mathematical tools are needed to incorporate such ill defined estimations into the general evaluation of investment project. On the other hand, the traditional approaches to the investment project evaluation are usually based on the budgeting, i.e., analysis of the discounted financial parameters of the considered projects, such as Net Present Value ($NPV$), Internal Rate of Return ($IRR$) and so on. It is easy to see that in this case the estimation of the investment efficiency, as well as any forecasting, is rather an uncertain problem and the proper methods for operating in the uncertain

---

C. Kahraman (Ed.): Fuzzy Engineering Economics with Appl., STUDFUZZ 233, pp. 243–287, 2008.
springerlink.com © Springer-Verlag Berlin Heidelberg 2008

setting should be used. Since the applicability of traditional probabilistic methods is often restricted by absence of objective probabilistic information about future events, during last two decades growing interest to the application of interval analysis and fuzzy sets methods in budgeting has being observed, see (Dimova et al. 2000, Kahraman et al. 2002, Ward 1985). On the other hand, when analysing the investment project we consider (sometimes implicitly) some local criteria based on the calculated financial parameters and quantitative evaluations of the project's implementation effects. Therefore, the project estimation is in essence a multiple criteria problem. As the examples of successful systematization the local criteria sets proposed in (Wang 2004, Lopes and Flavel 1998) may be considered. Even skin-deep analysis of these criteria systematization allows us to conclude that investment and project quality estimation is the complicated multiple criteria problem frequently with a certain hierarchical structure. There are a lot of multiple criteria methods proposed in literature for solving economical and financial problems. Steuer (1986) presented the widest review of this problem based on more than 250 literature indices. Nevertheless, we can cite the only few papers devoted to the multiple criteria financial project estimation, see (Wang 2004, Mohamed and McCowan 2001, Li and Sterali 2000).

It is worth noting that in all these works the fuzzy sets theory concepts were used. The method proposed in (Mohamed and McCowan 2001) is based on a representation of the local criteria by membership functions and their aggregation using simple fuzzy summation. The ranks of local criteria and possible hierarchical structure of the problem were not taken into account. The hierarchical structure of the problem is considered in (Weck et al. 1997) and well known AHP method is used for its building, but simple normalization of financial parameters (dividing them by their maximal values) is applied instead of natural local criteria. An interesting example of practical application of the multiple criteria hierarchical analysis is presented in (Li and Sterali 2000). The generalized *AHP* method has been used for estimation of 103 mutually dependent investment projects proposed for the Tumen river region (China) industrial development.

We do not intend to make here the detailed review of these works, but as a result of the analysis we have done in the field of investment project estimation as well as in some other practical applications, we can say that generally the project evaluation is a Multiobjective or Multiple Criteria Decision Making (MCDM) hierarchical problem in the fuzzy setting. It is important that as it has been pointed out in (Bana et al. 1997) "...the theory of MCDM is an open theoretical field and not a closed mathematical theory solving a specific class of problems". Nevertheless, there are a number of methodological problems that are common ones for almost all multiple criteria approaches. They were considered and systemized in (Roy 1996, Stewart 1989, Steuer 1986, Liu and Stewart 2004, Bana and Costa 1990). There are different definitions of Multiple Criteria Decision Making, *MCDM* , in literature, but regardless of what type of *MCDM* task is solving (choice, ranking, sorting, *ets* ) the two pivotal problems arise: how to evaluate an alternative and how to compare them? Last problem is especially important if the results of alternatives evaluations are presented by interval or fuzzy numbers. In the current paper we focus on the class of *MCDM* problems which may be treated as the developing the methods helping the decision maker to choose the best alternative when the several local criteria (sometimes antagonistic ones) effect to the decision. This problem usually is solved in a two phase process (Chen and Klein 1997, Ribeiro 1996, Valls and Torra 2000): the *rating* , i.e., aggregation of the criteria values for comparing alternatives and *ranking* or *ordering* these alternatives. Last phase is not trivial in the fuzzy or interval setting. There are some problems we are faced in the rating phase especially when dealing with fuzzy local criteria and/or their fuzzy weights (ranks). They can be roughly clustered as follows:

### (i) Common representation of different types of local criteria
The local criteria may be constructed on a base of quantitative parameters such as financial ones as well as using expert subjective estimations (verbal assessments of project's scientific, technological level, etc). It is known that experts prefer to provide rather "fuzzy" advises on linguistic level of presentation to avoid possible mistakes caused by the qualitative nature of predictions. However, human experience and intuition play an important role in projects estimation and cannot be ignored, although the specific uncertainty is their inherent property. This uncertainty is of subjective (fuzzy) nature and cannot be described in usual probabilistic way. So a proper methodology is needed to take into account the uncertainty factors, which will allow to build a set of comparable local criteria based on directly measurable quantitative parameters as well as on linguistically formulated assessments. The mathematical tools of fuzzy sets theory elaborated for dealing with subjective kind of uncertainty (Zadeh 1965) may be successfully used for this purpose. Thus, in the real world problems we meet two group of local criteria (Shyi-Ming 1997, Mao-Jiun and Tien-Chien 1995, Chen-Tung 2001): objective criteria based on the numerical parameters and subjective ones based on the subjective expert's opinions. So the problem arises : how to provide an appropriate common scale for representation of objective and subjective criteria?

### (ii) Expert's opinions aggregation
Usually for the evaluation of important investment projects a number of experts in the relevant fields are involved into decision making process. Since in such cases we are dealing with group *MCDM* (Mao-Jiun and Tien-Chien 1995, Herrera et al. 1996, Lee 1996, Shyi-Ming 1997) we face a problem of searching a compromise between expert's opinions available, especially if they are represented by different experts in linguistic form, e.g., as "low importance", " medium importance", " high importance", " large importance" (Mao-Jiun and Tien-Chien 1995).

### (iii) Aggregation of local criteria
A real world decision problem may involve a lot of local criteria to be analyzed simultaneously. Regrettably, the human ability to do this is strongly restricted by the known empirical law of psychology according to which a person can distinguish no more than 7 plus minus 2 classes or grades on some feature scale. If the number of grades is greater, the adjacent grades start to merge and cannot be clustered confidently, see (Miller 1956, Milner 1970). To solve this problem the relevant aggregation of local criteria taking into account their ranking can be used to create some generalized criteria. Therefore, the problem of choice of appropriate aggregation method is of perennial interest, because of its direct relevance to practical decision-making (Yager 1979, Zimmerman and Zysno 1980, Peneva and Popchev

2003). The most popular aggregating mode is the weighted sum. It is used in many well known decision making models such as Analytic Hierarchy Process, *AHP* (Saaty 1977), Multi-attribute Utility Analysis (Pardalos et al. 1995) and so on, but often without any critical analysis. On the other hand, in some fields, e.g., in ecological modelling, the weighted sum is not used for aggregation (Silvert 1997). The reason behind this is that in practice there are the cases when if any of local criteria is totally dissatisfied then considered alternative can be rejected from the consideration at all. Nevertheless, when dealing with a complex task characterized by a great number of local criteria, it seems reasonable to use all types of aggregations relevant to this task. If the results obtained using different aggregation modes are similar, this fact may be considered as a good confirmation of their optimality. In opposite case, an additional analysis of local criteria and their ranking should be advised.

### (iv) Aggregation of aggregating modes

The natural consequence of problem (iii) is a growing interest in the methods for generalizing of the aggregating operators (aggregation of aggregation modes) (Roubens 1997). For this purpose it is proposed to apply the possibility theory (Dubois and Koenig 1991) as well as the weighted sum aggregation (Peneva and Popchev 2003). Also, Yager's $t$-norms are used in (Hauke 1999) and hierarchical aggregation approach is developed in (Dyckhoff 1985), (Migdalas and Pardalos 1996). Nowadays the most popular is so-called $\gamma$-operator (Zimmerman and Zysno 1980):

$$\eta = (\prod_i \mu_i)^{1-\gamma}(1 - \prod_i (1 - \mu_i))^\gamma, i = 1,2,\ldots,n, 0 \le \gamma \le 1 \tag{1}$$

where $\mu_i$ is a membership functions corresponding to the local criteria. Since expression (1) is based only on the multiplicative aggregation, more general approach has been proposed in (Mitra 1988):

$$\eta_{or} = \gamma \max_i (\mu_i) + \frac{(1-\gamma)(\sum_i \mu_i)}{n} \tag{2}$$

$$\eta_{and} = \gamma \min_i (\mu_i) + \frac{(1-\gamma)(\sum_i \mu_i)}{n} \tag{3}$$

These expressions were used in (Shih and Lee 2000) to solve the multiple level decision making problem. As a key issue, the lack of strong rules for choosing the value of $\gamma$ is mentioned. The paper (Choi and Oh 2000) is specifically devoted to this problem, but the method proposed by the authors demands too much additional information to be presented by decision maker in quantitative form. In practice it is hard to get such information since it is not directly related to the decision maker's problems. It is easy to see that local criteria in (2), (3) are considered to be not ranked, whereas their ranking seems to be more important issue than choosing $\gamma$. Finally, the generalizing modes (1)-(3) do not involve all possible approaches to the aggregation.

Of course, the set above mentioned problems of *rating* in *MCDM* is not complete and exhausted. For example, in realm of group *MCDM* the values of membership functions describing the local criteria may be the fuzzy values as well. So the problem of Type 2 Fuzzy representation of local criteria arises. To solve this problem the constructive method based on so called Hyperfuzzy approach (Dymova 2003, Dymova et al. 2002) had been proposed recently. Nontrivial problem in the *ranking* stage of *MCDM* usually arises when as the result of previous rating phase the fuzzy (or interval) valued assessments of alternatives are obtained. Generally, this problem can be formulated as follows.

### (v) Fuzzy values comparison or ordering
It must be emphasized that the problem of interval and fuzzy values comparison plays a pivotal role in *MCDM* and fuzzy optimization (Chanas et al. 1993). There exist numerous definitions of the ordering relation for fuzzy values (as well as crisp intervals). In most cases the authors use some quantitative indices. The values of such indices present the degree to which one value (fuzzy or interval) is greater/smaller than the other value. In some cases, even several indices are used simultaneously. The widest review of this problem based on more than 35 literature indices has been presented in (Wang 2004), where the authors proposed a new interesting classification of methods for fuzzy values comparison. The separate group of methods are based on the so called probabilistic approach to the interval and fuzzy values comparison(Chanas et al. 1993, Krishnapuram et al. 1993, Kundu 1997, Kundu 1998, Nakamura 1986, Sengupta and Pal 2000, Sewastianow et al. 2001, Wadman et al. 1994, Yager et al. 2001). The attractiveness of this approach is caused by the possibility to build interval and fuzzy value relations using the minimum set of preliminary assumptions. In (Sewastianow and Rog 2002, Sewastianow and Rog 2006) as a further development of this approach an effective new method for interval and fuzzy values comparison has been elaborated.

### (vi) Hierarchical structure of local criteria set
The local criteria may compose a multilevel hierarchical structure when given set of local criteria consists of certain subgroups connected logically. Although the *AHP* method and its numerous fuzzy modification nowadays are commonly used to solve this problem, we can say it is still opened one.

The list of problems can be continued. For instance, some problems appear when fuzzy valued financial parameters are used as arguments of functions representing the local criteria (Dimova et al. 2000) or when such functions are fuzzy as well (Dymova 2003, Dymova et al. 2002) but detailed analysis of these issues is out of scope of the current paper. We can say that almost all components for building an efficient method that may be used in the majority cases of real-life project evaluations are already elaborated and described in literature. What is needed is their critical analysis from viewpoint of considered problem, and proper synthesis of them into the integrated method for hierarchical multiple evaluation of investment projects.

The aim of current chapter is to present a synthetical approach to solve the above mentioned problems (i)-(vi) in context of Investment Evaluation. To make the presentation more transparent, it is illustrated throughout the chapter with use of two examples. The first of them is well known Tool Steel Material Selection problem

248     L. Dymova and P. Sevastjanov

(Mao-Jiun and Tien-Chien 1995) which can be considered as a typical investment problem and relevant test charged by all difficulties concerned with the problems of (i)-(v). The next example is a simplified investment project evaluation problem we have used to show that even when project's estimation is based on the budgeting, i.e., only financial parameters such as Net Present Value ( $NPV$ ), Internal Rate of Return ( $IRR$ ) and so on are taking into account, we are dealing with multiple-criteria task in the fuzzy setting.

The rest of the paper is organized as follows. In Section 2, the Tool Steel Material Selection task is recalled and the Expert's opinions aggregation problem (ii) concerned with the linguistic weights of local criteria indicated. Section 3 provides an exposition of the simple and transparent method for evaluation of fuzzy subsethood measure based on the $\alpha$ -cut representation of fuzzy values. This method has been elaborated for solving the above mentioned problems (i) and (ii). Section 4 is devoted to the common representation of different types of local criteria (i). Section 5 describes the probabilistic method for fuzzy values comparison (v). In Section 6, the problem of aggregation of local criteria (iii) is considered and a new approach to aggregation of aggregating modes (iv) based on the synthesis of Type 2 and Level 2 Fuzzy Sets is described. Proposed approach is illustrated with use of Tool Steel Material Selection problem as an example. In Section 7, we present illustrative example of fuzzy multiple criteria investment project evaluation and proposed approach to hierarchical fuzzy $MCDM$ problem solving different from the fuzzy AHP method. Finally, concluding section summarizes the chapter and discusses future research issues.

## 2  Tool Steel Material Selection Problem

This problem has been chosen as the test and illustrative example of investment problem since it is well known and had been discussed in literature earlier (Mao-Jiun and Tien-Chien 1995, Shyi-Ming 1997). The most important is that it is charged with all problems (i)-(v) noted in Section 1. Generally the solution of this $MCDM$ problem is organised as follows. The Decision Maker with a help of analyst considers expert's opinions and aggregates them to obtain a final conclusion. Let us briefly recall the main assumptions and restrictions were made in formulation of Tool Steel Material Selection problem in (Mao-Jiun and Tien-Chien 1995). Suppose there are three experts involved in the decision process. The best among of five tool steel materials $V_1$, $A_2$, $\mathbb{D}_2$, $\gamma_1$, $T_1$ (classification of American Iron and Steel Institute, AISI) must be chosen. The local criteria are clustered into two groups: subjective criteria defined by experts on the base their experience and intuition (the properties of materials in our case) and objective criteria based on the numerical parameters not dependent on the expert's opinions (the cost of material imposed by open market). In (Mao-Jiun and Tien-Chien 1995), the classification of criteria had been proposed as follows. Subjective criteria:

- Non deforming properties for materials - local criterion $C_1$
- Safety in hardening for materials - local criterion $C_2$

- Toughness for materials - local criterion $C_3$
- Resistance to softening effect of heat for materials - local criterion $C_4$
- Wear resistance for materials - local criterion $C_5$
- Machinability for materials - local criterion $C_6$

Objective criterion:

- Cost -local criterion $C_7$

The subjective criteria in (Mao-Jiun and Tien-Chien 1995, Shyi-Ming 1997) were assessed linguistically (see Table 1 and Fig. 1).

**Table 1.** Linguistic values for criteria estimations

| Linguistic terms | Corresponding fuzzy number |
|---|---|
| Worst (W) | (0,0,0.3) |
| Poor (P) | (0,0.3,0.5) |
| Fair (F) | (0.2,0.5,0.8) |
| Good (G) | (0.5,0.7,1) |
| Best (B) | (0.7,1,1) |

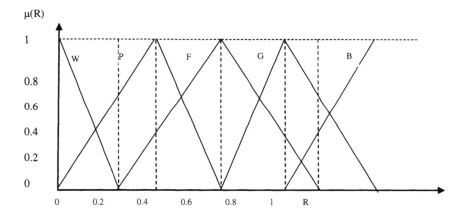

**Fig. 1.** Membership functions for linguistic values

The conventional approach (Delgado et al. 1992, Tong and Bonissone 1980) was used, which assumes that the meaning of each linguistic term (such as "good", "poor", when it is not directly connected with the concrete values) is given by means of a fuzzy subset defined in the [0,1] interval. Of course, there may be different ways to represent the linguistic terms as the triangulars or trapezoidals in the [0,1] interval, but usually the choice of such performance is the prerogative of analyst, since experts as a rule are not familiar enough with Fuzzy Set Theory. In contrast to the subjective criteria, the cost of material is presented directly by fuzzy number. The method for

250     L. Dymova and P. Sevastjanov

obtaining the fuzzy costs is not described in (Mao-Jiun and Tien-Chien 1995, Shyi-Ming 1997, but here we suppose that such fuzzy numbers can be obtained from usual statistic of market prices.Resulting assessments are presented in Table 2. To make our result comparable with those obtained by other authors (Mao-Jiun and Tien-Chien 1995) we used exactly the same initial data and their fuzzy representation as in (Mao-Jiun and Tien-Chien 1995, Shyi-Ming 1997.

**Table 2.** Linguistically and numerically represented local criteria

| Steel quality | $C_1$ | $C_2$ | $C_3$ | $C_4$ | $C_5$ | $C_6$ | $C_7$ |
|---|---|---|---|---|---|---|---|
| $V_1$ | P | F | G | P | F | B | (1.5,1.6,1.7) |
| $A_2$ | B | B | F | F | G | F | (1.8,2.0,2.2) |
| $D_2$ | B | B | F | F | G | P | (1.0,2.0,2.2) |
| $\gamma_1$ | F | G | G | F | F | F | (1.0,1.0,1.0) |
| $T_1$ | G | G | F | B | G | F | (2.5,3.0,3.5) |

Obviously, the problem of common representation of different types of local criteria (i) takes a place.

**Table 3.** Linguistic terms and corresponding triangular fuzzy values

| Linguistic terms | Corresponding fuzzy number |
|---|---|
| Very low (VL) | (0,0,0.3) |
| Low (L) | (0,0.3,0.5) |
| Medium (M) | (0.2,0.5,0.8) |
| High (H) | (0.5,0.7,1) |
| Very high (VH) | (0.7,1,1) |

**Table 4.** Linguistic assessment of importance weight made by three decision makers

| Criteria | Opinions | | |
|---|---|---|---|
| | $E_1$ | $E_2$ | $E_3$ |
| $C_1$ | H | H | VH |
| $C_2$ | M | H | M |
| $C_3$ | VH | VH | H |
| $C_4$ | H | H | M |
| $C_5$ | M | M | M |
| $C_6$ | H | H | VH |
| $C_7$ | VH | VH | VH |

Fuzzy Multiobjective Evaluation of Investments with Applications 251

The importance weights of the local criteria are represented linguistically with use of linguistic term shown in Table 3. There is no consensus among expert's with respect to importance weights of local criteria and their final estimations are presented in Table 4.

For the aggregation of expert's opinions the simple expression was used:

$$W_i = \frac{1}{n} \sum_{j=0}^{n} \oplus W_{ij}, \tag{4}$$

where $W_i$ is the aggregated importance weight of $i$ th local criterion, $W_{ij}$ is the importance weight of $i$ th local criterion given by $j$ th expert, $n$ is the number of participating experts, $\oplus$ is the operation of fuzzy addition. Using expression (4), from the data presented in Tables 3 and 4 we get the aggregated importance weights shown in Table 5.

**Table 5.** Aggregated importance weights

| Criteria | Opinions in fuzzy form |
|---|---|
| $C_1$ | $W_1 = (0.567, 0.800, 1.000)$ |
| $C_2$ | $W_2 = (0.300, 0.567, 0.867)$ |
| $C_3$ | $W_3 = (0.633, 0.800, 1.000)$ |
| $C_4$ | $W_4 = (0.400, 0.633, 0.933)$ |
| $C_5$ | $W_5 = (0.200, 0.500, 0.800)$ |
| $C_6$ | $W_6 = (0.567, 0.800, 1.000)$ |
| $C_7$ | $W_7 = (0.700, 1.000, 1.000)$ |

The fuzzy values from Table 5 may be used as the local criteria weights to calculate the aggregated assessments of compared Tool Steel Materials. Nevertheless, in practice, often the consensus of experts in respect to the obtained weights is needed. To achieve such a consensus the special procedures ,e.g., the Delphi method (Helmer 1966) were elaborated. They are based on correction by the experts their individual opinions taking into account the results of aggregation. In our case, the problem is that the initial experts assessments of weights are represented in linguistic form. Usually experts, e.g., supplies engineers, are not enough familiar with the Fuzzy Sets Theory and know nothing about numerical representation of linguistic terms. Hence, if the problem of consensus arises, an analyst must represent the aggregated weights in the linguistic form to be understood by experts. Moreover, often the experts insist on using the such aggregated linguistic weights (consensus) in further analysis, since initially they presented only the linguistic assessments of weights. It is easy to see that weighs in Table 5 are not coincide with fuzzy numbers representing

252    L. Dymova and P. Sevastjanov

linguistic terms used initially by experts (see Table 3). So the problem of reasonable linguistic interpretation of aggregated importance weights arises. The natural way for its approximate solution is to estimate the degrees to which each of obtained $W_i$ coincides with subsequent fuzzy values from Table 3 and to choose the linguistic term with the most degree of coincidence. In other words, the subsethood measure must be estimated. The simple and transparent method for doing this is presented in the next Section.

## 3 Subsethood Measure for Linguistic Representation of Fuzzy Numbers

As proposed by Kosko (1986), the fuzzy subsethood, $S(A \subset B)$, measures the degree to which $A$ is a subset of $B$ and is given by

$$S(A \subset B) = \frac{\sum\limits_{u \in U} min(\mu_A(u), \mu_B(u))}{\sum\limits_{u \in U} \mu_A(u)} \tag{5}$$

where $U$ is the universe of discuss common for fuzzy subsets $A$ and $B$. Although expression (5) is widely used in applications (Beynon et al. 2004), some of its drawbacks can be noted which prevent from using it for our purposes. Expression (5) is formulated for discrete supports of fuzzy subsets $A$ and $B$, whereas we deal with continuous ones. Besides in asymptotic limit where the support of $A$ is tending to zero (fuzzy value $A$ is reducing to real value) the problem of reasonable interpretation of (5) arises. That is why, in this section the simple and transparent approach free of above mentioned restrictions is described. It is based on the $\alpha$-cuts representation of fuzzy values (Kaufmann and Gupta 1985). So, if $A$ is a fuzzy value then

$$A = \bigcup_{\alpha} \alpha A_{\alpha}$$

where $\alpha A_{\alpha}$ is the fuzzy subset $(x \in U, \mu_A(x) \geq \alpha)$ and $A_{\alpha}$ is the support set of fuzzy subset $\alpha A_{\alpha}$, $U$ is the universe of discourse. It was proved that if $A$ and $B$ are fuzzy numbers (intervals), then all the operations on them may be presented as operations on the set of crisp intervals corresponding to their $\alpha$-cuts: $(A @ B)_{\alpha} = A_{\alpha} @ B_{\alpha}$, $@ \in \{+, , *, /\}$. In a similar way, the subsethood operation can be defined as the set of subsethood operations on corresponding $\alpha$-cuts. So the definition of subsethood measure as the degree to which crisp interval $A_{\alpha}$ is a subset of $B_{\alpha}$, i.e., $S(A_{\alpha} \subset B_{\alpha})$, is needed. Since we deal with the crisp intervals, an intuitively obvious measure of subsethood may be defined as

$$S(A_{\alpha} \subset B_{\alpha}) = \frac{W(A_{\alpha} \cap B_{\alpha})}{W(A_{\alpha})} \tag{6}$$

where $W(A_\alpha)$ is the width of interval $A_\alpha$, $W(A_\alpha \cap B_\alpha)$ is the width of overlapping area of intervals $A_\alpha$ and $B_\alpha$. Expression (6) has the reasonable asymptotic properties. For example, in the case of $W(A_\alpha) \to 0$ from (6) we have an intuitively obvious result

$$\lim_{W(A_\alpha) \to 0} S(A_\alpha \subset B_\alpha) = S(a \in B_\alpha) = \begin{cases} 1 & \text{if } a \in B_\alpha \\ 0 & \text{if } a \notin B_\alpha \end{cases} \quad (7)$$

where $a \in B_\alpha$ is a real value. To get an aggregated estimation of subsethood measure on the base of its $\alpha$-cuts representation (6), we propose to use the next weighted sum:

$$S(A \subset B) = \frac{\sum\limits_\alpha (\alpha S(A_\alpha \subset B_\alpha))}{\sum\limits_\alpha \alpha} \quad (8)$$

Last expression indicates that the contribution of $\alpha$ - cut to the overall subsethood estimation is increasing along with the rise in its number. Proposed method is graphically illustrated in Fig. 2.

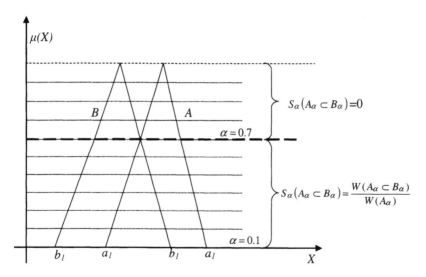

**Fig. 2.** Subsethood degree of fuzzy sets

In asymptotic case, when a degree to which a real value $a$ belongs to the interval $B_\alpha$ must be assessed, expression (8) is reduced to:

$$S(a \subset B) = \frac{\sum\limits_\alpha \alpha S(a \in B_\alpha)}{\sum\limits_\alpha \alpha} \quad (9)$$

This case is shown in Fig. (3). It is clear (even without calculations) that the next inequalities must be verified:

$$S(a^{***} \in B) > S(a^{**} \in B) > S(a^* \in B).$$

Of course, this intuitively obvious result good reflecting the inherent meaning of subsethood is numerically confirmed with use of expression (9).

The other asymptotic case we meet when fuzzy subset $A$ is completely enveloped by fuzzy subset $B$ (see Fig. 4).

In this case, both the common sense and Exp.(8) provide equal results: $S(A \subset B) = 1$. Described approach has been used to estimate the degrees to which

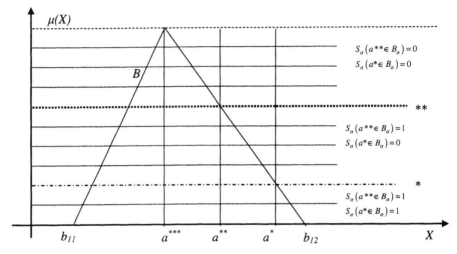

**Fig. 3.** Subsethood degree of real number $a$ in fuzzy set $B$

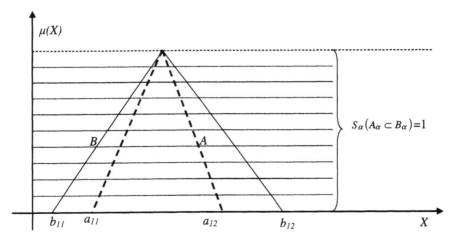

**Fig. 4.** Subsethood degree of fuzzy sets in case when fuzzy set $B$ contains fuzzy set $A$

# Fuzzy Multiobjective Evaluation of Investments with Applications

**Table 6.** The degrees of confidence (subsethood degrees) of importance weight of the criteria with linguistic terms

| Aggregated importance weight | Subsethood degree | | | | |
|---|---|---|---|---|---|
| | $VL$ (0.0,0.0,0.3) | $L$ (0.0,0.3,0.5) | $M$ (0.2,0.5,1.0) | $H$ (0.5,0.7,0.8) | $VH$ (0.7,1.0,1.0) |
| $W_1$ | 0% | 0% | 4.53% | **14.45%** | 12.44% |
| $W_2$ | 0% | 2.83% | **48.04%** | 19.86% | 0.92% |
| $W_3$ | 0% | 0% | 2.44% | **37.45%** | 14.67% |
| $W_4$ | 0% | 0.40% | 24.53% | **35.69%** | 2.82% |
| $W_5$ | 0.18% | 8.98% | **100.00%** | 8.98% | 0.18% |
| $W_6$ | 0% | 0% | 4.53% | **14.45%** | 12.44% |
| $W_7$ | 0% | 0% | 11.51% | 11.51% | **100.00%** |

each of aggregated importance weights, $W_i$, obtained by averaging the expert's opinions (see Table 5) coincides with the linguistic terms used initially by experts (see Table 3). The results are presented in the Table 6, where the bolded values mark the greatest degrees of subsethood.

So we can approximately represent the obtained aggregated importance weights by linguistic terms as shown in Table 7.

**Table 7.** Final linguistic rating of ranks of local criteria

| Criteria | Fuzzy number representation | Linguistic representation |
|---|---|---|
| $C_1$ | $W_1 = (0.567, 0.800, 1.000)$ | H |
| $C_2$ | $W_2 = (0.300, 0.567, 0.867)$ | M |
| $C_3$ | $W_3 = (0.633, 0.800, 1.000)$ | H |
| $C_4$ | $W_4 = (0.400, 0.633, 0.933)$ | H |
| $C_5$ | $W_5 = (0.200, 0.500, 0.800)$ | M |
| $C_6$ | $W_6 = (0.567, 0.800, 1.000)$ | H |
| $C_7$ | $W_7 = (0.700, 1.000, 1.000)$ | VH |

It is easy to see that results obtained with use of described approach (see Table 6, Fig. 5 and Fig. 6 ) are in a good conformity with our intuition.

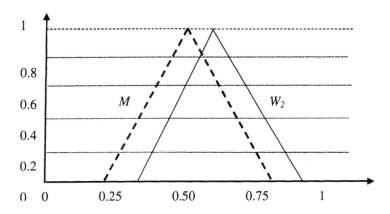

**Fig. 5.** $W_2 \subset M$ ( subsethood degree is 48.04%)

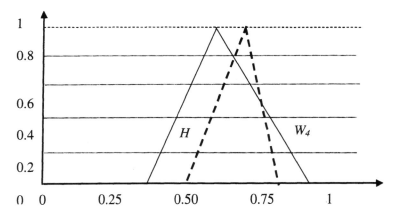

**Fig. 6.** $W_4 \subset H$ (subsethood degree is 35.69%)

## 4 Common Representation of Different Types of Local Criteria

When dealing with real world problem, we can meet two group of local criteria: objective ones, expressed in numerical form directly and subjective criteria represented by experts in form of linguistic terms (Mao-Jiun and Tien-Chien 1995). Linguistic terms are usually described by fuzzy numbers on the conventional common support [0,1], whereas objective criteria may have different supports. So the problem arises in such situations: how to find an appropriate common scale for representation of objective and subjective criteria? The heuristic approaches proposed in (Mao-Jiun and Tien-Chien 1995, Chen-Tung 2001) based on the concept of profit and benefit local criteria are only rough solutions of this problem since they lead to some distortions of initial preferences. To solve this problem the authors of (Mao-Jiun and Tien-Chien 1995) proposed the conversion of initial fuzzy objective criteria

$$RT_i = \{T_i \otimes [T_1^{-1} \oplus T_2^{-1} \oplus \ldots \oplus T_m^{-1}]\}^{-1} \quad (10)$$

Fuzzy Multiobjective Evaluation of Investments with Applications    257

where $T_i$ is the fuzzy value of considered objective criterion assigned to $i$ th alternative. Of course, converted fuzzy values $RT_i$ are defined on the support $[0,1]$. Chen-Tung (2001) proposed to divide all the subjective and objective criteria into two groups: cost criteria (denoted by $C$ ) lowering with rising of the parameter on which the criterion is based (e.g., production cost) and benefit criteria (denoted by $B$ ) (e.g., the quality of goods) with opposite property. In order to ensure the compatibility between objective and subjective criteria in (Chen-Tung 2001) the next method was proposed.

Let $\tilde{x}_{ij}$ is the fuzzy rating of alternative $A_i (i = 1,...,m)$ with respect to criterion $C_j (j = 1,...,n)$. Suppose $\tilde{x}_{ij}$ are represented by triangular fuzzy numbers, i.e., $\tilde{x}_{ij} = (a_{ij}, b_{ij}, c_{ij})$. The next normalizations had been proposed in (Chen-Tung 2001):

for the benefit criteria

$$\tilde{r}_{ij} = \{\frac{a_{ij}}{c_j^*}, \frac{b_{ij}}{c_j^*}, \frac{c_{ij}}{c_j^*}\}, c_j^* = \max_i c_{ij} \tag{11}$$

for cost criteria

$$\tilde{r}_{ij} = \{\frac{\bar{a}_j}{c_{ij}}, \frac{\bar{a}_j}{b_{ij}}, \frac{\bar{a}_j}{a_{ij}}\}, \bar{a}_j = \min_i a_{ij} \tag{12}$$

The normalizations (11),(12) as well as normalization (10) are to preserve the property that supports of normalized fuzzy numbers belong to the interval $[0,1]$. On the other hand, such normalizations lead to some distortions of initial preferences that were explicitly or implicitly taken into account on the stage of local criteria formalization. To clarify, consider an example. Suppose there are two alternatives $A_1$, $A_2$, (the goods to be produced ) and two local criteria for their assessment: investment cost ( $C_1$ ) and expansion possibility ( $C_2$ ) (see Chen-Tung 2001).

Suppose that valuations of considered alternatives $A_1$, $A_2$ with respect to local criteria are represented by triangular fuzzy numbers shown in Table 8. According to

**Table 8.** Fuzzy assessments of alternatives with respect to the local criteria

|       | $C_1$           | $C_2$             |
| ----- | --------------- | ----------------- |
| $A_1$ | (6.0,7.0,8.0)   | (6.3, 8.0, 9.0)   |
| $A_2$ | (3.6,4.0,4.4)   | (9.0, 10.0, 10.0) |

**Table 9.** Fuzzy assessments of alternatives after normalization

|       | $C_1$               | $C_2$               |
| ----- | ------------------- | ------------------- |
| $A_1$ | (0.45, 0.51, 0.60)  | (0.63, 0.80, 0.90)  |
| $A_2$ | (0.81, 0.90, 1.00)  | (0.90, 1.00, 1.00)  |

(Chen-Tung 2001), $C_1$ is the "cost" criterion and $C_2$ is "benefit" criterion. That is why, criterion $C_1$ was normalized using expression (12), whereas $C_2$ was normalized with use of (11). The results of normalization are shown in Table 9.

It could be noted that normalization can not transform the initial fuzzy evaluations of compared alternatives (see Table 8) into local criteria. Factually, in the Table 9 we can see only triangular fuzzy numbers which can not be treated as the criteria since the values of cost criterion must decrease with rising of the parameter on which criterion is based and benefit criterion has opposite property. Conversion (10) brings the similar results. Nevertheless, the data from Table 8 can be used to build the true cost and benefit criteria. For benefit criteria instead of (11) the next function may be used:

$$\mu_B(r_j) = \begin{cases} 0, & \text{if } r_j \leq r_{j\min} \\ \dfrac{(r_j - r_{j\min})}{(r_{j\max} - r_{j\min})}, & \text{if } r_{j\min} < r_j < r_{j\max} \\ 1, & \text{if } r_j \geq r_{j\max} \end{cases} \quad (13)$$

and for the cost criteria:

$$\mu_C(r_j) = \begin{cases} 1, & \text{if } r_j \leq r_{j\min} \\ 1 - \dfrac{(r_j - r_{j\min})}{(r_{j\max} - r_{j\min})}, & \text{if } r_{j\min} < r_j < r_{j\max} \\ 0, & \text{if } r_j \geq r_{j\max} \end{cases} \quad (14)$$

In (13) and (14) $r_{j\min} = \min_i a_{ij}$, $r_{j\max} = \max_i c_{ij}$.

In Fig. 7, the local criteria based on the data presented in Table 8 with use of (13) and (14) are shown. It is easy to see that triangular fuzzy cost and benefits are only fuzzy arguments of membership function representing the corresponding local criteria.

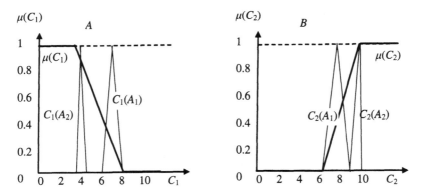

**Fig. 7.** The cost ($A$) and benefit ($B$) criteria based on the data from Table 8

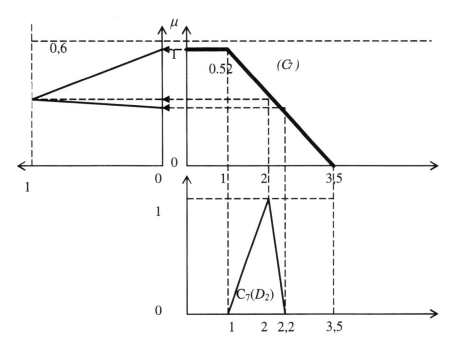

**Fig. 8.** Calculation of fuzzy value of cost criterion $C_7$ for steel $\mathbb{D}_2$

**Table 10.** Linguistic representation of fuzzy values of cost criterion for compared Tool Steel Materials

| Steel quality | Fuzzy cost | Fuzzy value of cost criterion | Subsethood degree | Linguistic approximation of cost criterion's value |
|---|---|---|---|---|
| $V_1$ | (1.50,1.60,1.70) | (0.72,0.76,0.80) | 63.67% | G |
| $A_2$ | (1.80,2.00,2.20) | (0.52,0.60,0.68) | 43.65% | F |
| $\mathbb{D}_2$ | (1.00,2.00,2.20) | (0.52,0.80,1.00) | 43.97% | G |
| $\gamma_1$ | (1.00,1.00,1.00) | (1.00,1.00,1.00) | 100.00% | B |
| $T_1$ | (2.50,3.00,3.50) | (0.00,0.20,0.40) | 39.53% | P |

Let us turn to the Tool Steel Material Selection problem. It is clear that linguistic terms in Table 2 represent final linguistic assessments of criteria and transformations of them are not required. On the other hand, the local criterion based on the cost of Tool Steel Materials is needed. We built it using expression (14). As the result, the membership function representing the local cost criterion has been obtained. The next step is the calculation of fuzzy values of cost criterion as the function of fuzzy argument for all compared Tool Steel Materials. To do this well known procedure

**Table 11.** Fuzzy values of local criteria in linguistic form

| Steel quality | $C_1$ | $C_2$ | $C_3$ | $C_4$ | $C_5$ | $C_6$ | $C_7$ |
|---|---|---|---|---|---|---|---|
| $V_1$ | P | F | G | P | F | B | G |
| $A_2$ | B | B | F | F | G | F | F |
| $D_2$ | B | B | F | F | G | P | G |
| $\gamma_1$ | F | G | G | F | F | F | B |
| $T_1$ | G | G | F | B | G | F | P |

(Kaufmann and Gupta 1985) for evaluation of fuzzy value of function of fuzzy argument illustrated in Fig. 8 has been used.

Obviously, the resulting values of cost criterion are fuzzy values and the problem of common representation of linguistically and numerically defined local criteria arises. For its solution we have used the procedure of calculation of subsethood measure for linguistic representation of fuzzy values described in Section 3.

In Table 10, the calculated maximal degrees to which the fuzzy values of cost criterion coincide with linguistic assessments from Table 1 are presented. The final linguistic estimations of compared alternatives regarding to all local criteria taken into account are shown in Table 11.

## 5 Probabilistic Method For Fuzzy Values Comparison

As we deal with the fuzzy valued local criteria and weights, any their aggregation will be fuzzy as well. Of course, the defuzzification procedure may be used to get real valued final estimations of compared alternatives, but such approach obviously leads to the loss of important information. To avoid this, the method of direct comparison of fuzzy values can be used. As it has been explained in Section 1, we prefer to use the probabilistic approach to fuzzy value comparison. It was already successfully used in some applications (Sewastianow and Jończyk 2003, Sewastianow and Rog 2003).

Let us recall the basics of this approach and elaborated method. We present firstly the probabilistic crisp interval relations and further extend them to the fuzzy values

**Fig. 9.** Examples of interval relations

comparison. There are only two nontrivial situation of intervals setting: the overlapping and inclusion cases (see Fig. 9) are deserved to be considered.

Let $A = [a_1, a_2]$ and $B = [b_1, b_2]$ be independent intervals and $a \in [a_1, a_2], b \in [b_1, b_2]$ be random values distributed on these intervals. As we are dealing with crisp intervals, the natural assumption is that the random values $a$ and $b$ are distributed uniformly. There are some subintervals, which play an important role in our analysis. For example (see Fig. 1), the falling of random $a \in [a_1, a_2], b \in [b_1, b_2]$ into subintervals $[a_1, b_1], [b_1, a_2], [a_2, b_2]$ may be treated as a set of independent random events. Let us define the events $H_k : a \in A_i, b \in B_j$, where $A_i$ and $B_j$ are subintervals formed by the boundaries of compared intervals $A$ and $B$ such that $A = \bigcup A_i, B = \bigcup B_j$. It easy to see that events $H_k$ form the complete set of events, describing all the cases of falling random values $a$ and $b$ into the various subintervals $A_i$ and $B_j$ respectively. Let $P(H_k)$ be the probability of event $H_k$ and $P(B > A/H_k)$ be the conditional probability of $B > A$. Hence, the composite probability can be presented as

$$P(B > A) = \sum_{k=1}^{n} P(H_k) P(B > A/H_k). \tag{15}$$

As we are dealing with uniform distributions of random values $a$ and $b$ in the given subintervals, the probabilities $P(H_k)$ can be easily obtained geometrically. These basic assumptions make it possible to infer the complete set of probabilistic interval relations involving separated equality and inequality relations and comparisons of real numbers with intervals and fuzzy values (Sewastianow and Rog 2006). The complete set of expressions for interval relations is shown in Table 12, the cases without overlapping and inclusion are omitted. In Table 12, only half of cases that may be realized when considering interval overlapping and including are presented since other three cases, e.q., $b_2 > a_2$ for overlapping and so on, can be easily obtained by changing letter $a$ through $b$ and otherwise in the expressions for the probabilities.

It easy to see that in all cases $P(A < B) + P(A = B) + P(A > B) = 1$.

Let $\tilde{A}$ and $\tilde{B}$ be fuzzy values on $X$ with corresponding membership functions $\mu_A(x), \mu_B(x): X \to [0,1]$. We can represent $\tilde{A}$ and $\tilde{B}$ by the sets of $\alpha$-cuts: $\tilde{A} = \bigcup_{\alpha} A_\alpha$, $\tilde{B} = \bigcup_{\alpha} B_\alpha$, where $A_\alpha = \{x \in X : \mu_A(x) \geq \alpha\}, B_\alpha = \{x \in X : \mu_B(x) \geq \alpha\}$ are the crisp intervals. Then all fuzzy value relations $\tilde{A} rel \tilde{B}$, $rel = \{<, =, >\}$, may be presented by the sets of $\alpha$-cut relations $\tilde{A} rel \tilde{B} = \bigcup_{\alpha} A_\alpha rel B_\alpha$.

Since $A_\alpha$ and $B_\alpha$ are crisp intervals, the probability $P_\alpha(B_\alpha > A_\alpha)$ for each pair $A_\alpha$ and $B_\alpha$ can be calculated in the way described above. The set of the probabilities $P_\alpha(\alpha \in (0,1])$ may be treated as the support of the fuzzy subset

$$P(\tilde{B} > \tilde{A}) > \{\alpha/P_\alpha(B_\alpha > A_\alpha)\}$$

## Table 12. The probabilistic interval relations

| $P(B > A)$ | $P(B < A)$ | $P(B = A)$ |
|---|---|---|
| 1. $b_1 > a_1 \wedge b_1 < a_2 \wedge b_1 = b_2$ | | |
| $\dfrac{b_1 - a_1}{a_2 - a_1}$ | $\dfrac{a_2 - b_1}{a_2 - a_1}$ | $0$ |
| 2. $b_1 \geq a_1 \wedge b_2 \leq a_2$ | | |
| $\dfrac{b_1 - a_1}{a_2 - a_1}$ | $\dfrac{a_2 - b_2}{a_2 - a_1}$ | $\dfrac{b_2 - b_1}{a_2 - a_1}$ |
| 3. $a_1 \geq b_1 \wedge a_2 \geq b_2 \wedge a_1 \leq b_2$ | | |
| $0$ | $1 - \dfrac{(b_2 - a_1)^2}{(a_2 - a_1)(b_2 - b_1)}$ | $\dfrac{(b_2 - a_1)^2}{(a_2 - a_1)(b_2 - b_1)}$ |

where the values of $\alpha$ may be considered as grades of membership to fuzzy value $P(\tilde{B} > \tilde{A})$. In this way, the fuzzy subset $P(\tilde{B} = \tilde{A})$ may also be easily created.

Obtained results are simple enough and reflect in some sense the nature of fuzzy arithmetic. The resulting "fuzzy probabilities" can be used directly. For instance, let $\tilde{A}$, $\tilde{B}$, $\tilde{C}$ be fuzzy values and $P(\tilde{A} > \tilde{B})$, $P(\tilde{A} > \tilde{C})$ be fuzzy value expressing the probabilities $A > \tilde{B}$ and $\tilde{A} > \tilde{C}$, respectively. Hence the probability $P(P(\tilde{A} > \tilde{B}) > P(\tilde{A} > \tilde{C}))$ has a sense of probability's comparison and is expressed in the form of fuzzy value as well. Such fuzzy calculations may be useful at the

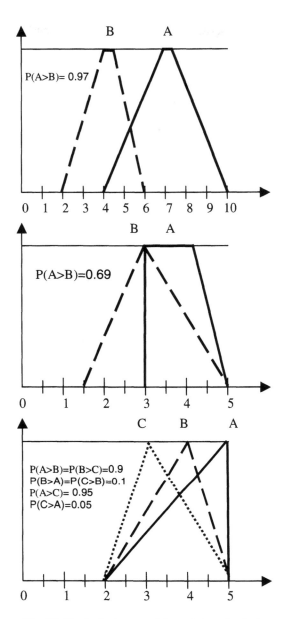

**Fig. 10.** The typical cases of fuzzy interval ordering

intermediate stages of analysis since they preserve the fuzzy information available. Indeed, it can be shown that in any case $P(\tilde{B} > \tilde{A}) + P(\tilde{B} = \tilde{A}) + P(\tilde{B} < \tilde{A}) =$ "near 1", where "near 1" is a symmetrical relative to 1 fuzzy number. It is worth noting here that the main properties of probability are remained in the introduced operations, but in a fuzzy sense. However, a detailed discussion of these issues is out of the scope of this paper.

264     L. Dymova and P. Sevastjanov

Nevertheless, in practice, the real-valued indices are needed for fuzzy value comparison. For this purpose, some characteristic numbers of fuzzy sets could be used. But it seems more natural to use next weighted sum:

$$\overline{P}(\tilde{B} > \tilde{A}) = \sum_{\alpha} \alpha P_{\alpha}(B_{\alpha} > A_{\alpha})/\sum_{\alpha} \alpha \qquad (16)$$

Last expression indicates that contribution of $\alpha$ - cut to the overall probability estimation is increasing along with the rise in its number. Some typical cases of fuzzy values comparison are represented in the Fig. 10. It is easy to see that the resulting quantitative estimations are in a good accordance with our intuition.

## 6   Aggregation of Local Criteria and Aggregating Modes

There are many different methods for aggregation of local criteria proposed in literature, but there is no method proved to be the best in all practical cases. Moreover in (Zimmerman and Zysno 1980) it is stated that choice of aggregation scheme is a context dependent problem. Nevertheless, we contribute here in its consideration from some other point of view.

Firstly, it is possible to represent the membership function of the local criteria as some artificial functions of alternatives. Such a transformation can be carried out formally after calculation of the membership functions values for all alternatives.More strictly, let $A$ and $B$ be local criteria and $\mu_A, \mu_B$ be their membership functions. Then for each $x \in X$, where $X$ is a set of alternatives, the artificial functions $\mu_A(x), \mu_B(x)$ can be formally introduced. Of course, $x$ is not a variable in a common sense. Factually, it is only a label (or number) assigned to the corresponding alternative. Hence, we can say that if for some $x_1 \in X$ we have $\mu_A(x_1) = \mu_B(x_1)$, then the alternative $x_1$ satisfies the local criteria $A$ and $B$ in equal extent, and if for some $x_2 \in X$, we have $\mu_A(x_2) > \mu_B(x_2)$, then alternative $x_2$ satisfies the local criterion $A$ in a greater degree then the criterion $B$. In this way the initially multidimensional problem can be formally transformed into the one-dimensional with the alternative number (label) as the only variable.     To make our consideration more transferable, consider firstly the case of only two local criteria $A$ and $B$ which are equally important for a decision maker and therefore have equal ranks, i.e., $\alpha_A = \alpha_B = 1$. So if $X$ is a set of the alternatives and $\mu_A(x), \mu_B(x), x \in X$ are membership functions representing formally - as it is described above - the local criteria $A$ and $B$ respectively, then the best (optimal) alternative $x_o$ will be such that: a) $\mu_A(x_o) = \mu_B(x_o)$ (since criteria $A$ and $B$ are of equal importance), b) the value $\mu_A(x_o)$ is maximal in comparison with all alternatives for which condition a) is verified.

In this spirit, the theorem useful for our further analysis had been proved in (Sevastianov and Tumanov 1990). It can be formulated as follows:

## Theorem 1

If $A$ and $B$ are equally ranked local criteria represented on a set of the alternatives $X$ by corresponding membership functions $\mu_A(x), \mu_B(x), x \in X$ such that they have unique maximal points $x_A, x_B \in X$ respectively, and

$$\mu_A(x_A) > \mu_B(x_A), \mu_B(x_B) > \mu_A(x_B) \tag{17}$$

then optimal alternative $x_o$ can be found as

$$x_o = \arg \max_{x \in X} (\mu_C(x)) \tag{18}$$

$$\mu_C(x) = \min(\mu_A(x), \mu_B(x)) \tag{19}$$

It is easy to see that function $\mu_C(x)$ can be naturally treated as an aggregation of local criteria $A$ and $B$.

What can be achieved using some other popular aggregation methods is shown in Fig. 11.

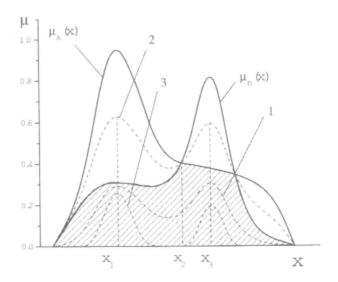

**Fig. 11.** The methods for equally ranked local criteria aggregation:
1. $\mu_C(x) = \mu_A(x) \cdot \mu_B(x)$,
2. $\mu_C(x) = 0.5\mu_A(x) + 0.5\mu_B(x)$,
3. $\mu_C(x) = \max(0, \mu_A(x) + \mu_B(x) - 1)$;

$x_1$ is the optimal alternative for 1, 2 and 3 types of aggregation; $x_2$ is an optimal alternative for the aggregation $\mu_C(x) = min(\mu_A(x), \mu_B(x))$.

Obviously, only the min–type aggregation (18) derives the optimal alternative $x_2$ (see Fig.11) fulfilling the natural restriction a). All other considered aggregation

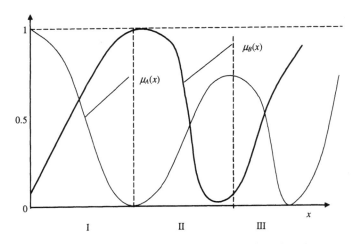

**Fig. 12.** The case of multiple-extreme membership functions

methods get the optimal alternative $x_1$, which is in Pareto region, but far from the actual optimum (see Fig.11). In practice, we can meet more complicated situations than considered above. For example, the membership functions can have several points of extreme. This problem can be resolved by clustering a subset of the alternatives into some Pareto regions as it is shown in Fig.12. Obviously, within such regions (I, II, II in Fig.12) all the conditions of Theorem 1 are verified.

Assume that local criteria $A$ and $B$ are of different importance for a decision maker, i.e., for their ranks we have $\alpha_A \neq \alpha_B$. Since the additive and multiplicative aggregations $\mu_C(x) = \alpha_A \mu_A(x) + \alpha_B \mu_B(x)$ and $\mu_C(x) = (\alpha_A \mu_A(x))(\alpha_B \mu_B(x))$ in the case of $\alpha_A \approx \alpha_B$ result in inappropriate decisions (see Fig. 11), we looked for more correct aggregation rules.

In (Zimmerman 1987), following aggregation has been proposed

$$\mu_C^1(x) = \min(\alpha_A \mu_A(x), \alpha_B \mu_B(x)) \qquad (20)$$

It is easy to see that in asymptotic case $\alpha_A = \alpha_B = 1$, expression (20) reduces to the optimal min-type aggregation (19) and weighting in (20) appears to be logically justified. Nevertheless, in practice such aggregation can produce completely absurd results. For example, let $\alpha_A = 0.8, \alpha_B = 0.2$, i.e., the local criterion $A$ is more important than $B$. As it is shown in Fig. 13, $x_0$ is the optimal alternative for a case of the equally ranked criteria $A$ and $B$, i.e., $x_0 = arg \max_{x \in X} (\min(\mu_A(x), \mu_B(x)))$, and $x_0^1$ is the optimal alternative for a case of the weighted criteria, i.e., $x_0^1 = arg \max_{x \in X} (\min(\alpha_A \mu_A(x), \alpha_B \mu_B(x)))$. It is easy to see that alternative $x_0^1$ satisfies the criterion $B$ in a greater extent then the criterion $A$. This is in contradiction with an initial assumption for $A$ to be more important then $B$.

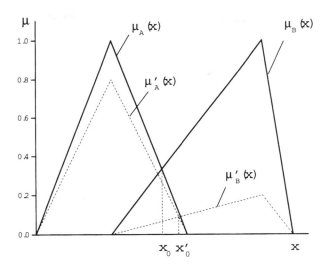

**Fig. 13.** The min-type aggregation of ranked local criteria:
$\mu'_A(x) = 0.8\mu_A(x)$, $\mu'_B(x) = 0.2\mu_B(x)$

Additional drawback of considered aggregation is that the general criterion $\mu^1_C(x) = \min(\alpha_A\mu_A(x), \alpha_B\mu_B(x))$ is not normalized to 1. Hence, it is difficult to assess a closeness of the optimal decision to the global optimum. In (Yager 1979), the aggregation method proposed, which properly reflects the sense of ranking:

$$\mu'_C(x) = \min(\mu_A^{\alpha_A}(x), \mu_B^{\alpha_B}(x)), \quad (\alpha_A + \alpha_B)/2 = 1 \quad (21)$$

Observe that for $\alpha_A = \alpha_B = 1$ expression (21) reduces to (19). It is shown in (Yager 1979) that if $\alpha_A > \alpha_B$ then in optimal point

$$x^1_0 = \arg\max_{x \in X} \min\left(\mu_A^{\alpha_A}(x), \mu_B^{\alpha_B}(x)\right)$$

the natural inequality

$$\mu_A(x^1_0) > \mu_B(x^1_0)$$

always takes place. In a general form expression (20) can be presented as

$$\mu'_C(x) = \mu_1^{\alpha_1}(x) \wedge \mu_2^{\alpha_2}(x) \wedge \ldots \wedge \mu_n^{\alpha_n}(x),$$

$$\alpha_1, \alpha_2, \ldots, \alpha_n > 0, \quad \frac{1}{n}\sum_{i=1}^{n}\alpha_i = 1, \quad (22)$$

where $\wedge$ is a min-operator, $n$ is a number of the local criteria. Therefore, it can be stated that the aggregation (21) is the best one, but it is true only when the conditions of Theorem 1 are verified. In practice, this is not always the case. Moreover, the min-type aggregation sometimes does not comply with intuitive concepts of decision makers about optimality (Dubois and Koenig 1991) since sometimes it does not properly represent the contributions of the local criteria to the overall estimation.

268    L. Dymova and P. Sevastjanov

The detailed analysis of the advantages and drawbacks of aggregating modes can be found in (Dymova et al. 2006), where with a help of two proved theorems and on the base on the author's experience was shown that, in general, the most reliable aggregation approach lies in a use of Yager's (1979) min-type operator (21). The multiplicative mode appears to be somewhat less reliable and, finally, the additive (weighted sum) method may be considered as unreliable and insensitive when choosing an alternative in Pareto-region (see also Fig.15 below). On the other hand, as all known aggregation modes have their own advantages and drawbacks it seems impossible to choose the best one especially when dealing with complicated hierarchical problem. Therefore, when dealing with a complex task characterized by a great number of local criteria, it seems reasonable to use all relevant to the considered problem types of aggregations. Since the different final results may be obtained on the base of different aggregating modes, the problems arises: how to aggregate such results? For this purpose we propose a new method for aggregation of aggregating modes. For further analysis we choose the aggregation modes that are usually used as the atomic ones for building the complex aggregating operations (see expressions (23)-(25)):

$$D_1 = min(\mu_1(C_1)^{W_1}, \mu_2(C_2)^{W_2}, \ldots, \mu_n(C_n)^{W_n}) \qquad (23)$$

$$D_2 = \mu_1(C_1)^{W_1} \otimes \mu_2(C_2)^{W_2}, \ldots, \otimes \mu_n(C_n)^{W_n} \qquad (24)$$

$$D_3 = W_1 \otimes \mu_1(C_1) \oplus W_2 \otimes \mu_2(C_2), \ldots, \oplus W_n \otimes \mu_n(C_n) \qquad (25)$$

where $\oplus$ and $\otimes$ are the operations of fuzzy addition and multiplication respectively.

Let us turn to our example. Since values of all local criteria and their weights are estimated as the linguistic terms (see Table 7 and 11) with corresponding numerical representation (see Tables 1 and 2), the final aggregated assessments of compared Tool Steel Materials obtained using the aggregating modes (23)-(25) must be fuzzy values as well. All arithmetical operations on fuzzy numbers needed to do this are well defined in (Kaufmann and Gupta 1985), and method for fuzzy value comparison has been described in the previous Section. The results of calculations are presented in Table 13.

**Table 13.** The results of aggregations with use of modes (17)-(19)

| Steel quality | $D_1$ | $D_2$ | $D_3$ |
|---|---|---|---|
| $V_1$ | (0.000,0.382,0.675) | (0.020,0.106,0.186) | (0.000,0.191,0.577) |
| $A_2$ | (0.200,0.500,0.855) | (0.077,0.186,0.257) | (0.140,0.500,0.855) |
| $\mathbb{D}_2$ | (0.000,0.382,0.675) | (0.107,0.214,0.286) | (0.350,0.700,1.000) |
| $\gamma_1$ | (0.200,0.574,0.881) | (0.086,0.200,0.257) | (0.140,0.574,0.881) |
| $T_1$ | (0.000,0.300,0.616) | (0.040,0.123,0.214) | (0.000,0.226,0.616) |

Obviously, the different final fuzzy estimations for compared Tool Steel Materials were obtained with use of different aggregating modes. Usually when the results obtained using different aggregation modes are similar, this fact may be considered as a good confirmation of their optimality. In opposite case, an additional analysis of local criteria and their ranking should be advised. It is easy to see that such approach seems as the method based on inexact reasoning,which implicitly takes into account all relevant to the considering task aggregating modes.

To build the method for doing this more rigorously on the base of aggregation of aggregating modes we propose the use of synthesis of Type 2 and Level 2 Fuzzy Sets defined on the support composed of compared alternatives. It is worthy to note that in the framework of proposed approach we avoid the use of *min*, *sum* and *multiplication* operations for aggregation of aggregating modes themselves, since using them leads inevitably to the unlimited sequence of aggregation problems.

Let us recall briefly the basic definitions of Type 2 and Level 2 Fuzzy Sets. As we deal with restricted number of compared alternatives and aggregating modes, it seems reasonable to consider only discrete representation of Type 2 and Level 2 Fuzzy Sets. Type 2 Fuzzy Sets were introduced by L. Zadeh (1975) as the framework for mathematical formalization of linguistic terms. In essence, these sets are the extension of usual Fuzzy Sets (Type 1) to the case when the membership function of fuzzy subset is performed by another fuzzy subset

More strictly, let $A$ be the fuzzy set of type 2 on the support subset $X$. Then, for any $x \in X$ the membership function $\mu_A(x)$ of $A$ is the fuzzy set with the membership function $f_x(y)$. As the result, for the discrete set we get

$$\mu_A(x) = \{\frac{f_x(y_i)}{y_i}\}, i = 1,\ldots,n$$

Further development of the theory of Fuzzy Sets of Type 2 was presented in (Karnik and Mendel 1999, Masaharu and Kokichi 1981, Yager 1980) where the main mathematical operations on such sets were defined. In (Yager 1980), it had been proved that using Zadeh's fuzzy extension principle, it is possible to build the fuzzy sets of types 3,4 and so on. Originally, Level-2 Fuzzy Sets were presented by Zadeh (1971) and were more elaborately studied in (Gottwald 1979, Tre and Caluwe 2003). As proposed by Zadeh (1974), the Level- 2 Fuzzy Set is a Fuzzy Set defined on the support, elements of which are ordinary Fuzzy Sets. So if fuzzy set $A$ is defined on a discrete set of $x_i$, $i = 1,\ldots,N$, and $x_i$ are represented by ordinary fuzzy sets defined on discrete universe set of $z_j$, $j = 1,\ldots,M$, then $A$ is a level- 2 fuzzy subset defined by following expressions:

$$A = \left\{ \frac{\mu_A(x_i)}{x_i} \right\}, x_i = \left\{ \frac{h_i(z_j)}{z_j} \right\} \tag{26}$$

$$A = \left\{ \frac{max[\mu_A(x_i)h_i(z_j)]}{z_j} \right\}, i = 1,\ldots,N, j = 1,\ldots,M \tag{27}$$

It follows from expression (27) that the final degree of membership of $z_j$ in $A$ may be presented as:

$$\mu_A(z_j) = \max_i [\mu_A(x_i) h_i(z_j)], \, j = 1,\ldots,M \tag{28}$$

Suppose there is a set of alternatives $z_j$, $j = 1,\ldots,M$, and $N$ types of aggregating modes $D_i$, $i = 1,\ldots,N$. Since usually in practice it is possible to estimate the relative reliability of aggregating modes at least on a verbal level, it seems natural to introduce membership function $\mu(D_i)$, $i = 1,\ldots,N$, representing expert's opinions about closeness of considered aggregating operator $D_i$ to some perfect types of aggregation, which can be treated as the best one or even "ideal" method of aggregation. Then such "ideal" method, $D_{ideal}$, can be represented by its membership function defined on the set compared aggregation modes as follows

$$D_{ideal} = \left\{ \frac{\mu(D_i)}{D_i} \right\}, i = 1,\ldots,N \tag{29}$$

As for all alternatives $z_j$, $j = 1,\ldots,M$, their estimations, $D_i(z_j)$, with use of aggregation modes $D_i, i = 1,\ldots,N$ can be calculated, each $D_i$ can formally be defined on the set of comparing alternatives $z_j$. As the result, each $D_i$ can be represented by the fuzzy subset

$$D_i = \left\{ \frac{D_i(z_j)}{z_j} \right\}, j = 1,\ldots,M \tag{30}$$

where $D_i(z_j)$ is treated as the degree to which alternative $z_j$ belongs to the set of "good" ones estimated with use of aggregating mode $D_i$. Substituting (30) into (29) with use of definitions (21)-(22) we get

$$D_{ideal} = \left\{ \frac{\mu_{ideal}(D_i)}{D_i} \right\}, i = 1,\ldots,N \tag{31}$$

where

$$\mu_{ideal}(z_j) = \max_i [\mu(D_i) D_i(z_j)] \tag{32}$$

Finally, the best alternative can be found as:

$$z_{best} = \arg \max_i \mu_{ideal}(z_j) \tag{33}$$

In the case when $\mu(D_i)$ or / and $D_i(z_j)$ are fuzzy values, expression (29) represents an object which can be treated simultaneously as level-2 and type 2 fuzzy subset.

To illustrate, let us continue the consideration of the example. At first, we have to calculate the values $\mu(D_i)$, $i = 1,2,3$. As it has been stated above, the $min$ -type

Fuzzy Multiobjective Evaluation of Investments with Applications    271

**Table 14.** Pair comparison of aggregating modes

| Aggregation modes | $D_1$ | $D_2$ | $D_3$ |
|---|---|---|---|
| $D_1$ | 1 | 3 | 9 |
| $D_2$ | $\dfrac{1}{3}$ | 1 | 9 |
| $D_3$ | $\dfrac{1}{9}$ | $\dfrac{1}{9}$ | 1 |

aggregation, $D_1$, is more reliable than the multiplicative aggregation, $D_2$, and both are noticeably more reliable then additive aggregation $D_3$. Such linguistic assessments may be represented in numerical form using the linguistic reciprocal pair comparison matrix (Saaty 1977) which in our case can be presented in Table 14.

The number 3 in this Table indicates that $min$-type aggregation, $D_1$, is more reliable than multiplicative aggregation, $D_2$, an so on. Using the method proposed in (Chu et al. 1979) from this matrix we get

$$\mu(D_1) = 0.7, \mu(D_2) = 0.25, \mu(D_3) = 0.05.$$

Thus, "ideal" method of aggregation in our case can be presented as:

$$D_{ideal} = \left\{ \frac{\mu(D_1)}{D_1}, \frac{\mu(D_2)}{D_2}, \frac{\mu(D_3)}{D_3} \right\}.$$

In our case, compared alternatives are Tool Steel Materials $V_1$, $A_2$, $\mathbb{D}_2$, $\gamma_1$, $T_1$. Hence, expressions (24) can be rewritten as

$$D_i = \left\{ \frac{D_i(V_1)}{V_1}, \frac{D_i(A_2)}{A_2}, \frac{D_i(\mathbb{D}_2)}{\mathbb{D}_2}, \frac{D_i(\gamma_1)}{\gamma_1}, \frac{D_i(T_1)}{T_1} \right\}, i = 1,2,3. \qquad (34)$$

Fuzzy values $D_i(\textit{steel quality})$, i=1,2,3, $steel\ quality \in \{V_1, A_2, \mathbb{D}_2, \gamma_1, T_1\}$ are presented in Table 13. Finally, expression (29) in our example takes a form:

$$D_{ideal} = \left\{ \frac{\mu_{ideal}(V_1)}{V_1}, \frac{\mu_{ideal}(A_2)}{A_2}, \frac{\mu_{ideal}(\mathbb{D}_2)}{\mathbb{D}_2}, \frac{\mu_{ideal}(\gamma_1)}{\gamma_1}, \frac{\mu_{ideal}(T_1)}{T_1} \right\} \qquad (35)$$

where $\mu_{ideal}(\ \textit{steel quality}\ ) = \max_i(\mu(D_i)D_i(\textit{steel quality}))$, i=1,2,3 and $steel\ quality \in \{V_1, A_2, \mathbb{D}_2, \gamma_1, T_1\}$.

After calculations we get

$$\mu_{opt}(V_1) = max((0.000, 0.267, 0.472), (0.005, 0.026, 0.046), (0.000, 0.009, 0.028)),$$

$$\mu_{opt}(A_2) = max((0.140, 0.350, 0.598), (0.019, 0.046, 0.064), (0.007, 0.025, 0.420)),$$

$\mu_{opt}(\mathbb{D}_2) = max((0.000,0.267,0.472),(0.026,0.053,0.071),(0.017,0.035,0.050))$,

$\mu_{opt}(\gamma_1) = max((0.140,0.401,0.616),(0.021,0.050,0.064),(0.007,0.028,0.044))$,

$\mu_{opt}(T_1) = max((0.000,0.210,0.431),(0.008,0.030,0.053),(0.000,0.011,0.030))$.

To find the maximal triangular fuzzy numbers in these expressions the probabilistic approach described in Section 5 has been used. Resulting evaluations of compared Tool Steel Materials are shown in Fig. 14. Comparing them (see Table 15) we conclude that the best one is the steel $\gamma_1$. The final ranking order is $\gamma_1, A_2, V_1, \mathbb{D}_2, T_1$ ($\mathbb{D}_2$ and $V_1$ have the same ranking).

This result is quite opposite to those obtained in (Mao-Jiun and Tien-Chien 1995, Shyi-Ming 1997) where the steel $\gamma_1$ is only on the fourth place in ranking: $A_2, \mathbb{D}_2, T_1, \gamma_1, V_1$. This difference is easy to explain. In the framework of our approach, the final ranking is based on aggregation of aggregating modes producing own and different ranking (see Table 13). Therefore, in general, our final ranking should differ from those obtained using particular aggregating modes. Besides, we have assigned the lowest weight to the addition type of aggregation used in (Mao-Jiun and Tien-Chien 1995, Shyi-Ming 1997). Hence, the contribution of this aggregation to the final ranking is minimal. The reasons for such weighting are partially pointed out at the beginning of this Section. What is worth noting in addition is that addition-type aggregation and so-called bounded difference aggregation,

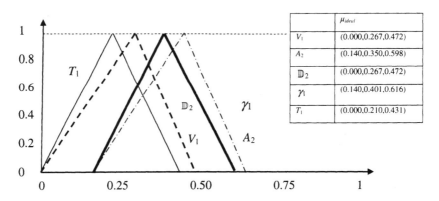

**Fig. 14.** Resulting evaluations of compared Tool Steel Materials

**Table 15.** Final probability relations

| | |
|---|---|
| $P(\gamma_1 > A_2)$ | 77.31% |
| $P(A_2 > V_1) = P(A_2 > \mathbb{D}_2)$ | 92.67% |
| $P(V_1 > T_1)$ | 80.28% |

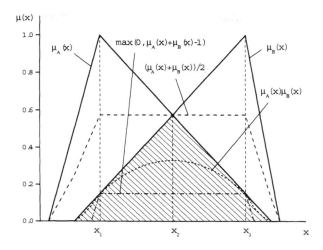

**Fig. 15.** The different types of aggregating operator

$\max(0, \mu_A(x) + \mu_B(X) - 1)$, might be used only with a great prudence, since even in a case of only two local criteria such aggregations frequently cannot reveal any preferences in Pareto-region (see Fig. 15).

We think that just this negative feature of weighted sum type of aggregation was the underlying cause of identical ranking of Tool Steel Materials obtained in (Mao-Jiun and Tien-Chien 1995, Shyi-Ming 1997)in spite of the fact that in (Shyi-Ming 1997) contrary to the work (Mao-Jiun and Tien-Chien 1995) instead of fuzzy values their real valued representations were used (for the same initial data).

## 5 Multiple Criteria Investment Project Evaluation in the Fuzzy Setting

In this section we present the simplified example of investment project evaluation using mainly the financial parameters to stress the multiple criteria nature of this problem and the need for use of Fuzzy Set Theory tools even when we are dealing with the traditional budgeting.

### 5.1 Local Criteria Building

Let us consider such important quantitative financial parameters as Internal Rate of Return ( $IRR$ ) and Net Present Value ( $NPV$ ). Since $IRR$ is measured in percentages whereas $NPV$ accounted in currency units, it seems impossible to compare them while estimating the project. However, it can be done using the functions presenting the local criteria based on these financial parameters. It is worth noting that introducing of such function does not lead to the loss of the original information, quite the contrary. Really, if $x$ is some parameter of analyzed system representing its quality to certain extent, then some scale of preference - presented on numerical or verbal level - for this parameter inevitably exists at least in the decision maker's mind. Indeed, such preference type information is the key of decision-making and local criteria are the mathematical tools for its formalization. Often they can be built using following simple and natural procedure.

For example, when considering $IRR$, it is easy to see that there always exists some lower bound for permissible values of $IRR$, usually being equal to the bank rate, $r$. Further, there is some interval $r \leq IRR \leq IRR_m$, where project's quality is rising gradually along with increasing of $IRR$. Finally, it is expected that if $IRR \geq IRR_m$, the project's efficiency with respect to $IRR$ is so high that it is difficult to make a reasonable choice among such excellent projects.

To transform this description into mathematical form, the membership function, which is the pivotal concept of Fuzzy Sets Theory, may be used. The membership function $\mu(IRR)$ representing the local criterion based on $IRR$ for some analyzed example is shown in (see Fig. 16).

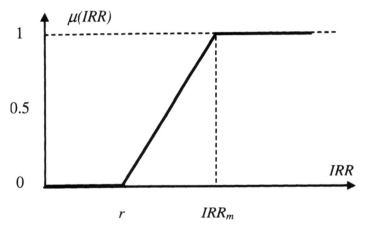

**Fig. 16.** The membership function representing the local criterion based on IRR

The values of membership functions change from zero (for the worst values of quality parameters) to maximum value equal to 1 in the area of best values of analyzed quality parameters. Hence, in a context of the considered problem the values of membership functions may be treated as degrees of quality parameter's preference. The linear form of membership function is not a dogma. However, in the practice what we usually know is only that some value is more preferable than other without any certain quantitative estimation of this preference. In other words, in most cases we deal with so-called "linear ordering" and for such situations the linear form of criterion function is proved to be the best one (Yager 1980). Since the membership function must be convex and normalized to 1, the only few forms of such function, presented in (see Fig. 17), may normally be used. Of course, if the probability distribution of quality parameter is known, the corresponding membership function may have more complex form.

Fuzzy Multiobjective Evaluation of Investments with Applications 275

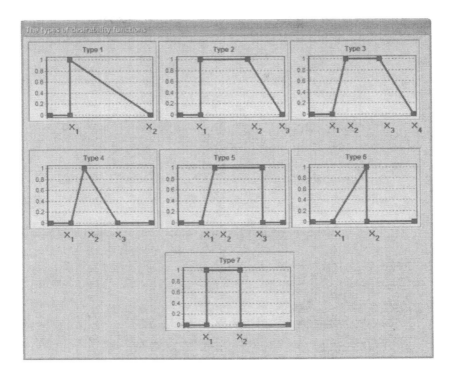

Fig. 17. The typical forms of membership functions

Consider a qualitative parameter presented in a verbal form, e.g., such as "ecological impact on region". It can be described by set of statements that linguistically represent the degrees to which an analyzed project may affect the ecology of region: "not significant", "slightly significant", "noticeably significant" and so on. As it was noted above, no more then 9 such linguistic degrees may be used in practice. The linguistic variables may be translated into mathematical form by presenting them in a form of triangular of trapezoidal fuzzy numbers. This approach is in accordance with a spirit of fuzzy sets theory and undoubtedly is very fruitful in a great number of applications, especially in fuzzy logic. Nevertheless, when dealing with the decision-making problems, we usually do not have enough reliable information to build such fuzzy numbers. Frequently, it is hard to choose the base for such triangular of trapezoidal fuzzy numbers, due to the absence of any evident reasons to prefer some base as the best one. In fact, the set of linguistic terms such as "not significant", "slightly significant", "noticeably significant" and so on in practice often represents only some labels signed on the levels of the decision maker's preference scale.

So the membership function shown in (see Fig. 18) seems to be a quite sufficient level of abstraction for formalization of local criteria in the majority of real-life situations (of course, in some cases the use of more complex descriptions, e.g. on the base of type 2 fussy sets, may be preferable (Dymova 2003, Dymova et al. 2002). Finally, we can see that all the qualitative and quantitative local criteria can be naturally presented within the universal scale of membership function.

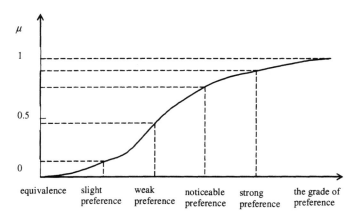

Fig. 18. The membership functions of qualitative local criterion

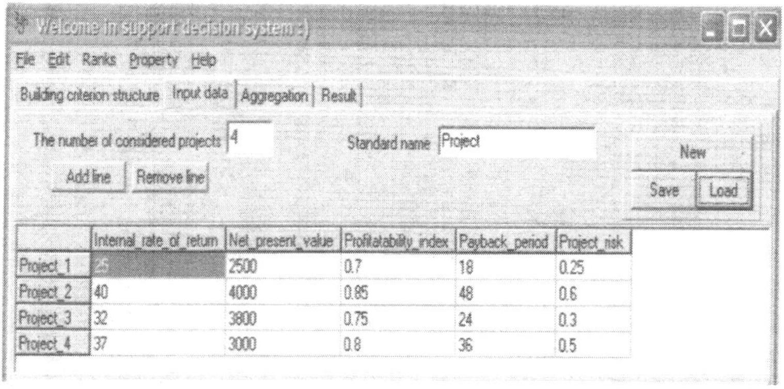

Fig. 19. The typical forms of membership functions

To make our further analysis more transparent, consider simple example which is not the description of any real projects estimation situation but reflects good the main advantages of proposed approach. Assume we have four projects to be compared taking into account five quality parameters (see Fig. 19). Obviously, the number of the parameters can be much greater in a real situation, but four following main financial parameters are almost obligatory for consideration: Internal Rate of Return ($IRR$), Net Present Value ($NPV$), Profitability Index ($P$), Payback period ($PB$). The Project Risk ($R$) is, generally speaking, a complex aggregated characteristic estimated on a basis of the quantitative data as well as expert qualitative estimations. Details can be found in (Doumpos et al. 2002).

Suppose for the sake of simplicity that in our example the risk is estimated in such a way that it ranges in interval from 0 to 1. All numerical values characterizing considered projects are presented in (see Fig. 19). The next step is a building of the membership functions. In fact, the decision maker has to select an appropriate type of the function (see Fig. 17) and choose no more then four reference points on a scale of

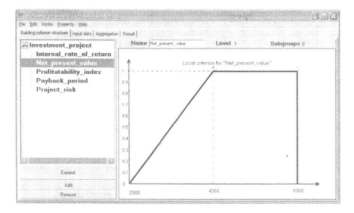

**Fig. 20.** The building of membership function

the analyzing parameter to define completely the corresponding membership function. One may assign the reference points on the basis of expert evaluations, statistical analysis of similar projects or strong (e.g., banking) standards and so on. We will try to use the simpler approach, because in our case we just need to choose the best of four projects. In our example (see Fig. 19) the worst value of *NPV* among all the projects is equal to 2500 and the best one is 4000. Therefore, is seems reasonable to assume for the first reference point $x_1$ =2000, i.e., less then 2500 because we do not want to reject the project No 1 when using some types of local criteria aggregations. Obviously, for the next point it is quite natural to assume $x_2$ =4000, and to obtain a complete description of the function we introduce an auxiliary point $x_3$ =6000 (see Fig. 20).

Other membership functions for considered example are built in a similar way.

## 5.2 Ranking of Local Criteria

Since in the real-world decision problems the local criteria are usually expected to provide the different contribution to the final aggregated estimation of alternatives, the appropriate method for local criteria ranking is needed. It should be emphasized that simple qualitative ordering of local criteria is not enough for practical purposes: usually some quantitative indices representing the local criteria contributions to the overall alternative estimation are necessary. The experience shows that quantitative ranking of the criteria is more difficult task for a decision maker than building of the membership function. Though it is usually hard for decision makers to rank the set of local criteria as whole, they usually can confidently state preference, at least verbally, when comparing only a pair of criteria. Therefore, the proper criteria ranking technique should use this pair comparison in a verbal form. Such technique is based on so-called matrix of linguistic pair comparisons, proposed by Saaty (1977). The procedure of building this matrix for our example is illustrated by (see Fig. 21).

Of course, only 9 basic verbal estimates are in use. Linguistic scales used in such estimations may have different sense, but a number of the scale levels (linguistic

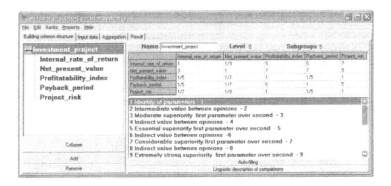

**Fig. 21.** The matrix of pair comparison

granules) cannot be more than 9 in any natural languages: this is an inherent feature of human thinking (Borisov and Korneeva 1980). Nevertheless, to allow the calculations, some natural numbers are assigned to the verbal estimations. Nevertheless, the numbers in Fig. 21 are shown only to illustrate our theoretical considerations. In practice, it is not advised to show any numbers to the experts. More reliable result can be obtained when only the linguistic opinions of an expert are taking into consideration. The matter is that if you propose to a group of experts to estimate some known objects, their verbally expressed opinions usually are quite similar. We cannot expect any other result: these people are learning using the same manuals, reading the same articles, working in the same field. However, if we force them to use some numbers for the estimations (usually it is not easy to do, since nobody loves numbers) we do not obtain any consensus (Zollo et al. 1999). The matter is that "In the beginning was the Word" (The Holy Gospel of Jesus Christ). Numbers had appeared much later and during last few millenniums, which are a minute from historical viewpoint, people are not learnt to use numbers properly yet. So far we are thinking using words, not numbers, and even trying to teach our computers this trick. More strictly, let $C_i$, $i = 1$ to $n$, be local criteria to be ranked. Then a pair comparison matrix $A$ can be created such that any entry $a_{ij} \in A$ represents the relative preference of criterion $C_i$ when it is compared with the criterion $C_j$. A pair comparison matrix is reciprocal meaning that $a_{ij} = 1/a_{ji}$ and $a_{ii} = 1$. If $\alpha_i$ and $\alpha_j$ represent the values of ranks then in the perfect case when we are dealing with a consistent matrix, i.e., $a_{ij} = \dfrac{\alpha_i}{\alpha_j}$, $a_{ij} = a_{ik} a_{kj}$, the ranks $\alpha_i$, $i = 1$ to $n$, can be calculated easily. Unfortunately, in the real-world situations, we usually have only approximate estimations of $a_{ij}$ and actual value of $a_{ij}$ may be unknown. Then the question arises: how to find $\alpha_i$, $i = 1$ to $n$, such that

$$a_{ij} \approx \frac{\alpha_i}{\alpha_j} \qquad (36)$$

Fuzzy Multiobjective Evaluation of Investments with Applications     279

Various approaches have been proposed to obtain these ranks. In (Chang and Lee 1995), they were roughly classified as following: " the eigenvector method; " the least squares method (LSM); " the logarithmic least squares method (LLSM); " the geometric row means method (GRM); " the weighted least squares method (WLSM) and a category of methods that involve only simple arithmetic operations : " the row means of normalized columns approach, " the normalized row sum and the inverted column sum methods (Lootsma 1981). The relative advantages and drawbacks of these methods are reviewed and discussed in literature not only for the case of real valued $a_{ij}$ (Chu et al. 1979), but when the entries of pair comparison matrix $A$ are themselves represented by fuzzy numbers (Chang and Lee 1995, Mikhailov 2003). Nevertheless, since in practice we deal with an approximate equality (36), the natural criterion of pair comparison efficiency can be presented as

$$S = \sum_{i=1}^{n} \sum_{j-1}^{n} (a_{ij} - \frac{\alpha_i}{\alpha_j})^2 \qquad (37)$$

where $\alpha_i$ are the ranks obtained using considered method. For this reason, it is quite natural to use a method based on the minimization of $S$, i.e.,

$$S = \sum_{i=1}^{n} \sum_{j-1}^{n} (a_{ij} - \frac{\alpha_i}{\alpha_j})^2 \rightarrow \min, s.t. \sum_{i=1}^{n} \alpha_i = 1 \qquad (38)$$

(sometimes the restriction $\sum_{i=1}^{n} \alpha_i = n$ is used). Using rich experimental data, (Wagenknecht and Hartmann 1983) have earnestly shown that this method (LSM) is the best one as it provides the least final values of $S$. On the other hand, in some cases the simplest row means of normalized columns method (Saaty 1977)

$$\alpha_i = \frac{\sqrt[n]{\prod_{j=1}^{n} a_{ij}}}{\sum_{i=1}^{n} \sqrt[n]{\prod_{j=1}^{n} a_{ij}}} \qquad (39)$$

may produce good results (Wagenknecht and Hartmann 1983). Of course, the weights of criteria in some special cases may be interpreted differently. Some not exhaustive classification of such possible interpretations and an overview of corresponding ranking methods are presented in (Choo et al. 1999). However, in our case the method based on the expressions (38) seems to be completely corresponding to the nature of the problem and sufficiently justified mathematically.

### 5.3 Numerical Evaluation of the Comparing Investment Projects

As in the Section 6, we will use three most popular aggregation modes:

$$D_1 = \min(\mu_1(C_1)^{\alpha_1}, \mu_2(C_2)^{\alpha_2}, ..., \mu_n(C_n)^{\alpha_n}), \quad D_2 = \prod_{i=1}^{n} \mu_i(C_i)^{\alpha_i}, \quad D_3 = \sum_{i=1}^{n} \alpha_i \mu_i(C_i)$$

The results of multiple criteria evaluation of comparing projects using the data and methods described in previous subsections are presented in (see Fig. 22).

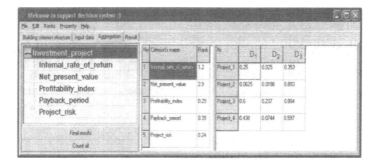

**Fig. 22.** Resulting estimations of projects presented in Fig. 19 (min-type, multiplicative and additive aggregations are denoted by respectively)

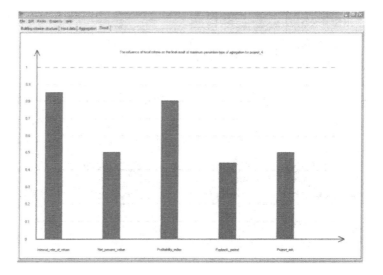

**Fig. 23.** The contributions of the local criteria to the final project's assessment

Observe that the min-type and multiplicative aggregations provide us similar resulting evaluations which are far from those obtained using weighted sum type of aggregations. It is important that proposed approach allows to estimate directly the contribution of each local criterion to the generalized project assessment (see Fig. 23).

As the aggregation modes $D_1$, $D_2$, $D_3$ provide different numerical evaluations of compared four projects, a natural approach to this problem is the aggregation of these modes to get a compromise final evaluation. For this purpose we will use the method presented in Section 6. This method is based of the synthesis of Type 2 and Level 2 Fuzzy Sets defined on the support composed of compared alternatives which in the considered case are the analysed investment projects. The central in this approach is a

Fuzzy Multiobjective Evaluation of Investments with Applications 281

concept of "`ideal'" method of aggregation presented in a form of mathematical object which can be treated as the Level 2 fuzzy set of Type2:

$$D_{ideal} = \left\{ \frac{\mu(D_1)}{D_1}, \frac{\mu(D_2)}{D_2}, \frac{\mu(D_3)}{D_3} \right\}$$

where the values of $\mu(D_i)$ represent expert's opinions about closeness of considering aggregating operators to some perfect type of aggregation which can be treated as the best or "ideal" method of aggregation. As it has been shown in Section 6, the min-type aggregation, $D_1$, is the more reliable one than the multiplicative aggregation, $D_2$, and both are noticeably more reliable than the additive aggregation $D_3$. Therefore, we have used the matrix of pair comparison of these aggregation modes presented in the Table 14. As in the Section 6, using this matrix we get: $\mu(D_1) = 0.7$, $\mu(D_2) = 0.25$, $\mu(D_3) = 0.05$. In the considered case, Exp.(30) takes the form of

$$D_i = \left\{ \frac{D_i(Project_1)}{Project_1}, \frac{D_i(Project_2)}{Project_2}, \frac{D_i(Project_3)}{Project_3}, \frac{D_i(Project_4)}{Project_4} \right\}, i = 1,2,3.$$

Hence, $D_{ideal}$ can be presented as

$$D_{ideal} = \left\{ \frac{\mu_{ideal}(Project_1)}{Project_1}, \frac{\mu_{ideal}(Project_2)}{Project_2}, \frac{\mu_{ideal}(Project_3)}{Project_3}, \frac{\mu_{ideal}(Project_4)}{Project_4} \right\}$$

where $\mu_{ideal}(Project_j) = \max_i(\mu(D_i)D_i(Project_j)), i = 1,2,3$. Finally, after all calculations using the results from Fig. 14 we obtain

$$D_{ideal} = \left\{ \frac{0.18}{Project_1}, \frac{0.05}{Project_2}, \frac{0.42}{Project_3}, \frac{0.31}{Project_4} \right\}$$

Thus, the third project is the best one. Factually, we can infer the same result analyzing data presented in (see Fig. 22), but it would be hard to do for a greater number of projects or/and aggregating operators under consideration.

### 5.4 Hierarchical Structure of Local Criteria

As it has been noticed above, many of real-life problems of investment evaluation are not only multi-criteria but also multi-level (hierarchical) ones. The method presented in previous Sections allows to build in natural way the branched hierarchical structures as in (see Fig. 24).

It can be seen that each criterion of upper $k$ level is built on a basis of local criteria of underlying $(k-1)$ levels using one of the aggregation methods or their generalization as proposed in Section 6. The general expression for calculation of criteria on intermediate levels of hierarchy is as follows:

$$D_{k,i_{n-1},i_{n-2},...,i_k} = f_{k,i_{n-1},i_{n-2},...,i_k}(D_{k-1,i_{n-2},...,i_k}, 1, \alpha_{k-1,i_{n-2},...,i_k}, 1,...,$$

$$D_{k-1,i_{n-1},i_{n-2},...,i_k}, m_{k,i_{n-1},i_{n-2},...,i_k}, \alpha_{k-1,i_{n-1},i_{n-2},...,i_k}, m_{k,i_{n-1},i_{n-2},...,i_k})$$

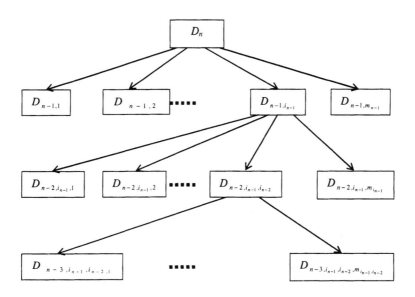

**Fig. 24.** The hierarchical structure of the local criteria

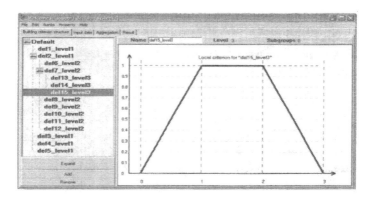

**Fig. 25.** Building of hierarchical local criteria system

where $f_{k,i_{n-1},i_{n-2},...,i_k}$ is an operator of criteria aggregation, $m_{k,i_{n-1},i_{n-2},...,i_k}$ is the number of of local criteria on $(k-1)$ level, aggregating to intermediate local criterion $D_{k,i_{n-1},i_{n-2},...,i_k}$. It is worthy to note that values of $D_{n-1,i_{n-1}}$ are always in interval [0,1] and may be interpreted as some intermediate local criteria assessments. On the lowest level of hierarchy the initial membership functions representing the local criteria based on origin parameters of project's quality are used, i.e.

$$D_{1,i_{n-1},i_{n-2},...,i_1} = f_{1,i_{n-1},i_{n-2},...,i_1}(\mu_{0,i_{n-1},i_{n-2},...,i_1,1}, \alpha_{0,i_{n-1},i_{n-2}...,i_1,1},...,$$
$$\mu_{0,i_{n-1},i_{n-2},...,i_1,m_{1,i_{n-1},i_{n-2},...,i_1}}, \alpha_{0,i_{n-1},i_{n-2},...,i_1,m_{1,i_{n-1},i_{n-2},...,i_1}})$$

where $m_{1,i_{n-1},i_{n-2},\ldots,i_1}$ is the number of initial local criteria on the lowest level. The method for building of hierarchical local criteria systems is based on the multiple criteria approach described in previous Sections and generalizes it. The method is developed in a form of user-friendly software (see Fig. 25).

It is important that presented method not only provide us us some generalized quantitative estimation of a project as a whole, but makes in possible to assess the contribution of each local criterion to this final evaluation.

# 8 Conclusions

The chapter presents the generalized method for multiple criteria hierarchical evaluation of investments in the fuzzy setting. The key issue is the analyzing the familiar approaches to aggregation of local criteria. The problems of ranked local criteria aggregation are analyzed and some new theoretical results which can be useful for proper choice of aggregation method are presented. It is proved that the most popular weighted sum method is the most unreliable one and can provide the wrong results. As all known aggregation modes have their own advantages and drawbacks it seems impossible to choose the best one especially when dealing with complicated hierarchical problem. Therefore, a new approach which makes it possible to generalize the aggregating modes into some "ideal" criterion is elaborated using the mathematical tools of level 2 and type 2 fuzzy sets. It is shown that proposed method allows to build in natural way the branched hierarchical structures of the investment project's local criteria. The theoretical consideration is illustrated using simple examples of investments. The main direction of future research will be generalization of proposed approach using conception of hyperfuzzy sets (Dymova et al. 2002) being the useful particle case of type 2 fuzzy sets for representation of verbally formulated local criteria and parameters of project's quality.

# References

Bana, E., Costa, C.A.: Reading in multiple criteria decision aid. Springer, Berlin (1990)

Bana, E., Costa, C.A., Stewart, T.J., Vansnick, J.C.: Multicriteria decision analysis: Some thoughts based on the tutorial and discussion session of the ESIGMA meetings. European Jurnal of Operational Research 99, 28–37 (1997)

Beynon, M., Peel, M.J., Tang, Y.C.: The application of fuzzy decision tree analysis in an exposition of the antecedents of audit fees. Omega 32, 231–244 (2004)

Borisov, A.N., Korneeva, G.V.: Linguistic approach to decision making model building under uncertainty: Methods of decision making under uncertainty. Riga 7, 4–6 (1980)

Chanas, S., Delgado, M., Verdegay, J.L., Vila, M.A.: Ranking fuzzy interval numbers in the setting of random sets. Information Sciences 69, 201–217 (1993)

Chang, P.T., Lee, E.S.: The estimation of normalized fuzzy weights. Computers and Mathematics with Applications 29, 21–42 (1995)

Chen, C., Klein, C.M.: An efficient approach to solving fuzzy MADM problems. Fuzzy Sets and Systems 88, 51–67 (1997)

Chen-Tung, C.: A fuzzy approach to select the location of distribution center. Fuzzy sets and systems 118, 65–73 (2001)

Choi, D.Y., Oh, K.W.: Asa and its application to multi-criteria decision making. Fuzzy Sets and Systems 114, 89–102 (2000)

Choo, E.U., Schoner, B., Wedley, W.C.: Interpretation of criteria weights in multicriteria decision making. Computers and Industrial Engineering 37, 527–541 (1999)

Chu, A., Kalaba, R., Springarn, R.A.: Comparison of two methods for determining the weights of belonging to fuzzy sets. Journal Of Optimization Theory And Applications 27, 531–538 (1979)

Delgado, M., Verdegay, J.L., Vila, M.A.: Linguistic decision making models. Internat. J. Intell. Systems 7, 479–492 (1992)

Dimova, L., Sevastianov, D., Sevastianov, P.: Application of fuzzy sets theory, methods for the evaluation of investment efficiency parameters. Fuzzy Economic Review 5, 34–48 (2000)

Doumpos, M., Kosmidou, K., Baourakis, G., Zopounidis, C.: Credit risk assessment using a multicriteria hierarchical discrimination approach: A comparative analysis. European Journal of Operational Research 1389, 392–412 (2002)

Dubois, D., Koenig, J.L.: Social choice axioms for fuzzy set aggregation. Fuzzy Sets and Systems 43, 257–274 (1991)

Dyckhoff, H.: Basic concepts for theory of evaluation: hierarchical aggregation via autodistributive connectives in fuzzy set theory. European Journal of Operational Research 20, 221–233 (1985)

Dymova, L.: A constructive approach to managing fuzzy subsets of type 2 in decision making. TASK Quarterly 7, 157–164 (2003)

Dymova, L., Rog, P., Sewastianow, P.: Hyperfuzzy estimations of financial parameters. In: Proceeding of the 2th International Conference on Mathematical Methods in Finance and Econometrics, pp. 78–84 (2002)

Dymova, L., Sewastianow, P., Sewastianow, D.: MCDM in a fuzzy setting: investment projects assessment application. International Journal of Production Economics 100(1), 10–29 (2006)

Gottwald, S.: Set theory for fuzzy sets of higher level. Fuzzy Sets and Systems 2, 125–151 (1979)

Hauke, W.: Using Yager's t-norms for aggregation of fuzzy intervals. Fuzzy Sets and Systems 101, 59–65 (1999)

Helmer, O.H.: The Delphi Method for Systematizing Judgments about the Future. Institute of Government and Public Aairs, University of California (1966)

Herrera, F., Herrera-Vieda, E., Verdegay, J.L.: Direct approach processes in group decision making using linguistic OWA operators. Fuzzy Sets and Systems 79(2), 175–190 (1996)

Kahraman, C., Ruan, D., Tolga, E.: Capital budgeting techniques using discounted fuzzy versus probabilistic cash flows. Information Sciences 142, 57–76 (2002)

Karnik, N.N., Mendel, J.M.: Application of type-2 fuzzy logic systems to forecasting of timeseries. Information Sciences 120, 89–111 (1999)

Kaufmann, A., Gupta, M.: Introduction to fuzzy arithmetic-theory and applications. Van Nostrand Reinhold, New York (1985)

Kosko, B.: Fuzzy entropy and conditioning. Information Science 30, 165–174 (1986)

Krishnapuram, R., Keller, J.M., Ma, Y.: Quantitative analysis of properties and spatial relations of fuzzy image regions. IEEE Trans. Fuzzy Systems 1, 222–233 (1993)

Kuchta, D.: Fuzzy capital budgeting. Fuzzy Sets and Systems 111, 367–385 (2000)

Kundu, S.: Min-transitivity of fuzzy leftness relationship and its application to decision making. Fuzzy Sets and Systems 86(5), 357–367 (1997)

Kundu, S.: Preference relation on fuzzy utilities based on fuzzy leftness relation on interval. Fuzzy Sets and Systems 97, 183–191 (1998)

Lee, H.: Group decision making using fuzzy sets theory for evaluating the rate of aggregative risk in software development. Fuzzy Sets and Systems 80(3), 261–271 (1996)

Li, Q., Sterali, H.D.: An approach for analyzing foreign direct investment projects with application to China's Tumen River Area development. Computers & Operations Research 3, 1467–1485 (2000)

Liu, D., Stewart, T.J.: Object-oriented decision support system modeling for multicriteria decision making in natural resource managment. Computers & Operations Research 31, 985–999 (2004)

Lootsma, F.A.: Performance evaluation of non-linear optimization methods via multi-criteria decision analysis and via linear model analysis. In: Powell, M.J.D. (ed.) Nonlinear Optimization (1981)

Lopes, M., Flavel, R.: Project appraisal-a framework to assess non-financial aspects of projects during the project life cycle. International Journal of Project Management 16, 223–233 (1998)

Mao-Jiun, J.W., Tien-Chien, C.: Tool steel materials selection under fuzzy environment. Fuzzy Sets and Systems 72(3), 263–270 (1995)

Masaharu, M., Kokichi, T.: Fuzzy sets of type II under algebraic product and algebraic sum. Fuzzy Sets and Systems 5, 277–290 (1981)

Migdalas, A., Pardalos, P.M.: Editorial: hierarchical and bilevel programming. J. Global Optimization 8(3), 209–215 (1996)

Mikhailov, L.: Deriving priorities from fuzzy pairwise comparison judgments. Fuzzy Sets and Systems 134, 365–385 (2003)

Miller, G.A.: The magical number seven plus or minus two: some limits on our capacity for processing information. Psychological Review 63, 81–97 (1956)

Milner, P.M.: Physiological psychology. Holt, New York (1970)

Mitra, G.: Mathematical models for decision support. Springer, Berlin (1988)

Mohamed, S., McCowan, A.K.: Modelling project investment decisions under uncertainty using possibility theory. International Journal of Project Management 19, 231–241 (2001)

Nakamura, K.: Preference relations on set of fuzzy utilities as a basis for decision making. Fuzzy Sets and Systems 20, 147–162 (1986)

Pardalos, P.M., Siskos, Y., Zopounidis, C.: Advances in multicriteria analysis. Kluwer Academic Publishers, Dordrecht (1995)

Peneva, V., Popchev, I.: Properties of the aggregation operators related with fuzzy relations. Fuzzy Sets and Systems 139, 615–633 (2003)

Ribeiro, R.A.: Fuzzy multiple attribute decision making: a review and new preference elicitation techniques. Fuzzy Sets and Systems 78, 155–181 (1996)

Roubens, M.: Fuzzy sets and decision analysis. Fuzzy Sets and Systems 90, 199–206 (1997)

Roy, B.: Multicriteria methodology for decision aiding. Kluwer Academic Publlishers, Boston (1996)

Saaty, T.: A scaling method for priorities in hierarchical structures. Journal of Mathematical Psychology 15, 234–281 (1977)

Sengupta, A., Pal, T.K.: On comparing interval numbers. European Journal of Operational Research 127, 28–43 (2000)

Sevastianov, P., Tumanov, N.: Multi-criteria identification and optimization of technological processes. Science and Engineering, Minsk (1990)

Sewastianow, P., Jończyk, M.: Bicriterial fuzzy portfolio selection. Operations Research And Decisions 4, 149–165 (2003)

Sewastianow, P., Rog, P.: A probabilistic approach to fuzzy and interval ordering, Task Quarterly. Special Issue Artificial and Computational Intelligence 7(1), 147–156 (2002)

Sewastianow, P., Rog, P.: Fuzzy modeling of manufacturing and logistic systems. Mathematics and Computers in Simulation 63, 569–585 (2003)

Sewastianow, P., Rog, P.: Two-objective method for crisp and fuzzy interval comparison in Optimization. Computers & Operations Research 33, 115–131 (2006)

Sewastianow, P., Róg, P., Venberg, A.: The constructive numerical method of interval comparison. In: Proceeding of Int. Conf. PPAM 2001, Naleczow (2001)

Shih, H.S., Lee, E.: Compensatory fuzzy multiple level decision making. Fuzzy Sets and Systems 114, 71–87 (2000)

Shyi-Ming, C.: A new method for tool steel materials selection under fuzzy environment. Fuzzy sets and systems 92(3), 265–274 (1997)

Silvert, W.: Ecological impact classification with fuzzy sets. Ecological Moddeling (1997)

Steuer, R.E.: Multiple criteria optimisation: theory, computation and application. Wiley, New York (1986)

Steuer, R.E., Na, P.: Multiple criteria decision making combined with finance. A categorical bibliographic study. European Journal of Operational Research 150, 496–515 (2003)

Stewart, T.J.: A critical survey on the status of multiple criteria decision making. OriON 5, 1–23 (1989)

Tong, M., Bonissone, P.P.: A linguistic approach to decision making with fuzzy sets. IEEE Trans. Systems Man Cybernet 10, 716–723 (1980)

Tre, G., Caluwe, R.: Level-2 fuzzy sets and their usefulness in object-oriented database modeling. Fuzzy Sets and Systems 140, 29–49 (2003)

Valls, A., Torra, V.: Using classification as an aggregation tool in MCDM. Fuzzy Sets and Systems 115, 159–168 (2000)

Wadman, D., Schneider, M., Schnaider, E.: On the use of interval mathematics in fuzzy expert system. International Journal of Intelligent Systems 9, 241–259 (1994)

Wagenknecht, M., Hartmann, K.: On fuzzy rank ordering in polyoptimisation. Fuzzy Sets and Systems 11, 253–264 (1983)

Wang, W.C.: Supporting project cost threshold decisions via a mathematical cost model. International Journal of Project Management 22, 99–108 (2004)

Wang, X., Kerre, E.E.: Reasonable properties for the ordering of fuzzy quantities (I) , (II). Fuzzy Sets and Systems 112, 375–405 (2001)

Ward, T.L.: Discounted fuzzy cash flow analysis. In: Proceeding of the 1985 Fall Industrial Engineering Conference, pp. 476–481 (1985)

Weck, M., Klocke, F., Schell, H., Ruenauver, E.: Evaluating alternative production cycles using the extended fuzzy AHP method. European Journal of Operational Research 100, 351–366 (1997)

Yager, R.R.: Multiple objective decision-making using fuzzy sets. International Journal of Man-Machine Studies 9, 375–382 (1979)

Yager, R.R.: A foundation for a theory of possibility. Journal of Cybernetics 10, 177–209 (1980)

Yager, R.R.: Fuzzy subsets of type II in decisions. Journal of Cybernetics 10, 137–159 (1980)

Yager, R.R.: On ordered weighted averaging aggregation operators in multicriteria decision making. IEEE Trans. Systems Man and Cybern 18(1), 183–190 (1988)

Yager, R.R., Detyniecki, M., Bouchon-Meunier, B.: A context-dependent method for ordering fuzzy numbers using probabilities. Information Sciences 138, 237–255 (2001)

Zadeh, L.A.: Fuzzy sets. Information and Control 8, 338–358 (1965)

Zadeh, L.A.: Quantitative fuzzy semantics. Information Sciences 3, 177–200 (1971)

Zadeh, L.A.: Fuzzy logic and its application to approximate reasoning. Information Processing 74, 591–594 (1974)

Zadeh, L.A.: The Concept of linguistic Variable and its Application to approximate Reasoning-I. Information Sciences 8, 199–249 (1975)

Zimmerman, H.J.: Fuzzy Sets, Decision-Making and Expert Systems. Kluwer Academic Publishers, Dordrecht (1987)

Zimmerman, H.J., Zysno, P.: Latent connectives in human decision making. Fuzzy Sets and Systems 4, 37–51 (1980)

Zimmerman, H.J., Zysno, P.: Decision and evaluations by hierarchical aggregation of information. Fuzzy Sets and Systems 104, 243–260 (1983)

Zollo, G., Iandoli, L., Cannavacciuolo, A.: The performance requirements analysis with fuzzy logic. Fuzzy economic review 4, 35–69 (1999)

# Using Fuzzy Multi-attribute Data Mining in Stock Market Analysis for Supporting Investment Decisions

Francisco Araque[1], Alberto Salguero[1], Ramon Carrasco[1], and Luis Martinez[2]

[1] Department of Software Engineering, University of Granada
[2] Department of Computer Science, University of Jaen

**Abstract.** The stock market analysis is a high demanded task to support investment decisions. The quality of those decisions is the key point in order to obtain profits and obtain new customers and keep old ones. The analysis of stock markets is high complex due to the amount of data analyzed and to the nature of those, in this chapter we propose the use of fuzzy data mining process to support the analysis processes in order to discover useful properties that can help to improve investment decisions.

## 1 Introduction

Usually Market analysts employ a large set of tools to understand the market and forecast its behaviour. How to maximize profits is always the main concern for investors. In the recent literature, various techniques of knowledge discovery have been employed in the stock market to accomplish the analysts' tasks in order to achieve their objectives. However, the highly non-linear, dynamic complicated domain knowledge inherent in the stock market makes it very hard for investors to make the right investment decisions. Moreover, different investors have different preferences about the holding periods of their investments. Financial investment is a knowledge-intensive industry. In recent years, with the advances in electronic transactions, vast amounts of data have been collected. In this context, the emergence of knowledge discovery technology enables the building up financial investment decision support systems. In this chapter we propose the use of analysis and Data Mining (DM) techniques based on a Fuzzy SQL extension to reveal flexible relationships of movements between primary bull and bear markets and help to determine appropriate trading strategies for investors and trend followers. Technical analysis often employs a statistical or artificial intelligence approaches to facilitate the trading strategy-making. It assumes that the market itself contains enough information to forecast its evolution.

The tools used by stock-exchange investors to try to understand the market and forecast its evolution can be arranged in two groups:

- Fundamental analysis: it focuses on key underlying economic and political factors to determine the direction of a currency's value. There are a number of fundamental indicators traders may follow that reflect how an economy is changing and gleam insight into global market prices to come.
- Technical analysis: it is the art of forecasting price movements through the study of chart patterns, indicator signals, sentiment readings, volume and open interest.

C. Kahraman (Ed.): Fuzzy Engineering Economics with Appl., STUDFUZZ 233, pp. 289–306, 2008.
springerlink.com &copy; Springer-Verlag Berlin Heidelberg 2008

Unlike fundamental analysis, technical analysis does not focus on economic data, but rather on interpretation of price data. One of the key tenets of technical analysis is that price patterns repeat themselves, allowing technical traders to make highly probabilistic bets on the direction of the instrument.

Usually investor managers use a large database (DB) that contains information about the market that should be analyzed in order to discover the knowledge that supports the investor's decisions. So the use of data mining processes will help to find out the patterns, features and in general the knowledge that the managers are looking for. In fact to find out the features, patterns, etc. we propose the use of Functional Dependencies (FD) and Gradual Dependencies (GD) (Cubero and Vila 1994) because they reflect immutable properties in a DB hence they facilitate the knowledge discovering which we are interested in.

Functional Dependencies correspond to correlations among data items and are expressed by means of rules showing attribute-value conditions that commonly occur at the same time in some set of data. Another way of considering the connections between data in databases is to specify a relationship between objects in a dataset and reflect monotonicity in the data through gradual dependencies (GDs). GD is a concept closely related to the idea of gradual rules introduced by Dubois and Prade (1992a, 1992b). We shall use as common framework to integrate FD and GD Extended Dependencies (EDs) (Carrasco et al. 2006).

Analysts study quantitative and qualitative features of the markets in their analysis, due to this the fact, the information involved in the DM process can be uncertain and vague because the subjective features usually are qualitative in nature and it is difficult to provide precise information about this type of information (Herrera et al. 2005). The modelling and managing of the uncertainty is a key decision in order to obtain good results from the DM process. The probability theory can be a powerful tool. However, it is not difficult to see that many aspects of uncertainties have a non-probabilistic character. Often the types of uncertainties encountered in market do not fit the axiomatic basis of probability theory, simply because uncertainties in the market are usually caused due to the inherent incompleteness and fuzziness of features rather than randomness. Therefore the use of the fuzzy logic (Zadeh 1965) and linguistic descriptors (Zadeh 1975) may be used to describe subjective features. The linguistic terms are fuzzy judgments and not probabilistic ones. The Fuzzy Linguistic Approach (Herrera and Herrera-Viedma 2000) provides a systematic way to represent linguistic variables in a natural decision-making procedure. It does not require providing a precise point at which a subjective feature exists. So it can be used as a powerful tool complementary to traditional methods to deal with imprecise information, especially linguistic information which is commonly used to represent qualitative information (Delgado et al. 1992, Martínez et al. 2006).

In this chapter we propose to develop a DM process for finding and validating relations between *stock market data* based on the fuzzy logic in order to make it more flexible and on the fuzzy linguistic approach to model uncertain information. To do so, we relax the concept of FD and GD by means of Fuzzy FD (FFD) and Fuzzy GD (FGD) that are quite suitable to model non immutable properties existing in the current manifestation of the data.

The concept of FFD (Cubero and Vila 1994) is a smoothed version of the classical FD. The basic idea consists of replacing the equality used in the FD definition by fuzzy resemblance relations. We can obtain a fuzzy version of GD (FGD) in a similar way. We call Fuzzy Extended Dependencies (FED) to the integration of both FFD and FGD.

The main advantage of FEDs is that they allow us to infer more knowledge from data. Using regular dependencies, only rules that are fulfilled by all of the instances are valid. The use of FEDs supports the discovery of dependencies although there exist instances that do not fulfil them completely. Furthermore, we can obtain the fulfilment degree for each FED stated.

In the last decade many decision-making system which have to deal with multi-criteria decision problems and qualitative information have shown the capability of Fuzzy Decision Analysis (FDA) (Ghotb and Warren 1995, Hussein and Pepe 2002). Liang and Wang (1991) proposed the FDA, which uses fuzzy set representations and utilizes linguistic variables for rating qualitative factors to aggregate decision-makers' assessments, and applied it on facility site selection and personnel selection.

Financial engineering is an emerging research area and also has great commercial potentials. In recent years, with the advances in electronic transactions, vast amounts of data have been collected. The problem of discovering useful knowledge from this data set to support decision-making process has been investigated from many points of views. In (Kuo et al. 2004) the knowledge discovery process is mainly composed of feature representation by visual K-chart technical analysis, feature extraction by discrete wavelet transform, and modelling by two level SOM networks. Lin and Chen (2007) generalize crisp expression trees evolved through genetic programming to fuzzy ones by introducing fuzzy numbers and fuzzy arithmetic operators in the trees. Their objective is to develop a Fuzzy Tree crossover for a multi-valued stock valuation model which improves the convergence phenomenon.

The use of Data Mining techniques has also been proposed in literature for this purpose (Carrasco et al. 1998). In (Boginski et al. 2006) a clustering process for dividing the set of financial instruments into groups of similar objects by computing a clique partition of the market graph is presented. Enke and Thawornwong (2005) introduces an information gain technique used in machine learning for data mining, based on neural networks, to evaluate the predictive relationships of numerous financial and economic variables. They use neural network models for level estimation and classification.

The DM process proposed in this chapter will obtain FEDs by using a flexible query language as the Fuzzy SQL (FSQL) (Carrasco et al. 2006, Carrasco et al. 1998) which will provide the information that will support investors' decisions about which type of investments will be more profitable based on past data and on objective and subjective features. In addition, we have a FSQL Server available to obtain the answers to FSQL queries for Oracle$^{©}$ DBMS (Galindo et al. 1998).

This chapter is organized as follows: in Section 2 different necessary preliminaries to understand the proposal are revised. In Section 3 is introduced the definition of Fuzzy Extended Dependencies (FEDs) based on the FSQL operators. In Section 4 a DM process are shown for finding FEDs (normal ED or ED with a degree of confidence). In section 5 some experimental results are presented and the paper is concluded in section 6.

292     F. Araque et al.

## 2 Preliminaries

In this section we introduce some preliminaries concepts that are necessary to understand the rest of the work.

### 2.1 Basic Concepts

Our work is based on the technical analysis paradigm. Technical analysis usually made use of statistical indicators (Amat and Puig 1992). Through the chapter we shall use the indicators shown below:

- *Moving Averages*: Among the most popular technical indicators, moving averages are used to gauge the direction of the current trend. Every type of moving average is a mathematical result that is calculated by averaging a number of past data points. Once determined, the resulting average allows traders to look at smoothed data rather than focusing on the day-to-day price fluctuations that are inherent in all financial markets. The simplest form of a moving average is appropriately known as a simple moving average (SMA) and is defined like the unweighted mean of the previous n data points. Once in a trend, moving averages will keep us in, but also give late signals. As with most tools of technical analysis, moving averages should not be used on their own, but in conjunction with other tools that complement them. Using moving averages to confirm other indicators and analysis can greatly enhance technical analysis.
- *Oscillators* are an essential group of indicators that futures, options, and stock traders have embraced to reveal turning points in flat markets. An oscillator is a type of momentum indicator that shows the acceleration and deceleration of markets. When a market has been accelerating upward and then begins to slow, the oscillator levels off, indicating the price action as a market top. Likewise, if a market is accelerating downward and that acceleration slows, then a bottom is indicated by the oscillator leveling off and then rising. Moving averages and trends are paramount when studying the direction of an issue. A technician will use oscillators when the charts are not showing a definite trend in either direction. Oscillators are thus most beneficial when a company's stock either is in a horizontal or sideways trading pattern, or has not been able to establish a definite trend in a choppy market.

*William's Oscillator* (also known as %R). It is a momentum indicator measuring overbought and oversold levels, similar to a stochastic oscillator. It was developed by Larry Williams and compares a stock's close to the high-low range over a certain period of time, usually 14 days. It is used to determine market entry and exit points. The Williams %R produces values from 0 to -100, a reading over 80 usually indicates a stock is oversold, while readings below 20 suggests a stock is overbought. The formula to calculate it is shown below:

$$\%R = 100 \times (A - U) / (A - B) \tag{1}$$

where A is the highest value in the period of time, B is the lowest value and U is the latest value.

## 2.2 FSQL: A Language for Flexible Queries

Our proposal of a DM process to knowledge discovering in stock markets dealing with vague and uncertain information needs of a flexible framework to manage the uncertainty of the data stored in our DB. Therefore we shall use a flexible query language (FSQL) to manage uncertainties and imprecise information presented in (Carrasco et al. 2006). It has extended the SQL language to allow flexible queries. Thus, the language can manage fuzzy attributes, from different nature that is necessary in our problem, which are classified by the system in 3 types:

- *Type 1*: These attributes are totally crisp, but they have some linguistic trapezoidal labels defined on them.
- *Type 2*: These attributes admit crisp data as well as possibility distributions over an ordered underlying domain.
- *Type 3*: On these attributes, some labels are defined and on these labels, a similarity relation has yet to be defined. These attributes have no relation of order.
- *Type 4*: There are different kinds of data in a database used in diverse applications (text, XML, etc.) therefore, it would be desirable that a DM system would carry out its work in an effective way. In order to manage these data we have defined these attributes. It is a generic type (fuzzy or crisp), which admits some fuzzy treatment. We permitted this attribute is formed by more than a column of the table (complex attributes).

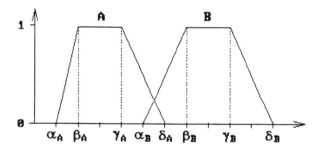

**Fig. 1.** Trapezoidal possibility distributions: A, B

The Fuzzy Meta-knowledge Base (FMB) stores information for the fuzzy treatment of the fuzzy attributes in order to define:

- Representation Functions: these functions are used to show the fuzzy attributes in a comprehensible way for the user and not in the internally used format.
- Fuzzy Comparison Functions: they are utilized to compare the fuzzy values and to calculate the compatibility degrees (CDEG function)

We have extended the SELECT command to express flexible queries and, due to its complex format, we only show an abstract with the main extensions added to this command:
- Fuzzy Comparators: In addition to the common comparators (=, >, etc), FSQL includes fuzzy comparators of two trapezoidal possibility distributions A, B with

A=$[αA,βA,γA,δA] B=$[αB,βB,γB,δB] (see Figure 1) in Table 1. In the same way as in SQL, fuzzy comparators can compare one column with one constant or two columns of the same type. Necessity comparators are more restrictive than possibility comparators, i.e. their fulfillment degree is always lower than the fulfillment degree of their corresponding possibility comparator. More information can be found in (Galindo 1999, Galindo et al. 1998).

- Fulfillment Thresholds $\gamma$: For each simple condition a Fulfillment threshold may be established with the format <condition> THOLD$\gamma$, indicating that the condition must be satisfied with a minimum degree $\gamma$ in [0, 1] fulfilled.

**Table 1.** Fuzzy Comparators for FSQL. Operator NOT can precede to F_Comp.

| F_Comp | Significance | CDEG(A F_Comp B) Possibility operator |
|---|---|---|
| FEQ | Fuzzy Equal | $= \sup_{d \in U} \min(A(d), B(d))$ where U is the domain of A, B. $A(d)$ is the degree of the possibility for $d \in U$ in the distribution A |
| FGT | Fuzzy Greater Than | $= 1$ if $\gamma_A \geq \delta_B$ $= \dfrac{\delta_A - \gamma_B}{(\delta_B - \gamma_B) - (\gamma_A - \delta_A)}$ if $\gamma_A < \delta_B \,\&\, \delta_A > \gamma_B$ $= 0$ otherwise |
| FGEQ | Fuzzy Greater or EQual | $= 1$ if $\gamma_A \geq \beta_B$ $= \dfrac{\delta_A - \alpha_B}{(\beta_B - \alpha_B) - (\gamma_A - \delta_A)}$ if $\gamma_A < \beta_B \,\&\, \delta_A > \alpha_B$ $= 0$ otherwise |
| FLT | Fuzzy Less Than | $= 1$ if $\beta_A \leq \alpha_B$ $= \dfrac{\alpha_A - \beta_B}{(\alpha_B - \beta_B) - (\beta_A - \alpha_A)}$ if $\beta_A > \alpha_B \,\&\, \alpha_A \beta_B$ $= 0$ otherwise |
| FLEQ | Fuzzy Less or EQual | $= 1$ if $\beta_A \leq \gamma_B$ $= \dfrac{\delta_B - \alpha_A}{(\beta_A - \alpha_A) - (\gamma_B - \delta_B)}$ if $\beta_A > \gamma_B \,\&\, \alpha_A < \delta_B$ $= 0$ otherwise |
| MGT | Fuzzy Much Greater Than | $= 1$ if $\gamma_A \geq \delta_B + M$ $= \dfrac{\gamma_B + M - \delta_A}{(\beta_A - \alpha_A) - (\gamma_B - \delta_B)}$ if $\gamma_A < \delta_B + M \,\&\, \delta_A > \gamma_B + M$ $= 0$ otherwise M is the minimum distance to consider two attributes as very separate. M is defined in FMB for each attribute |
| MLT | Fuzzy Much Less Than | $= 1$ if $\beta_A \leq \alpha_B - M$ $= \dfrac{\beta_B - M - \alpha_A}{(\beta_A - \alpha_A) - (\alpha_B - \beta_B)}$ if $\beta_A > \alpha_B - M \,\&\, \alpha_A < \beta_B - M$ $= 0$ otherwise M is the minimum distance to consider two attributes as very separate |

## Table 2. Fuzzy Comparators for FSQL. Operator NOT can precede to F_Comp.

| F_Comp | Significance | CDEG(A F_Comp B) Necessity operator |
|---|---|---|
| NFEQ | Necessarily Fuzzy Equal | $= \inf_{d \in U} \max(1 - A(d), B(d))$ where U is the domain of A, B. $A(d)$ is the degree of the possibility for $d \in U$ in the distribution A |
| NFGT | Necessarily Fuzzy Greater Than | $= 1$ if $\alpha_A \geq \delta_B$ <br> $= \dfrac{\beta_A - \gamma_B}{(\delta_B - \gamma_B) - (\alpha_A - \beta_A)}$ if $\alpha_A < \delta_B \ \& \ \beta_A > \gamma_B$ <br> $= 0$ otherwise |
| NFGEQ | Necessarily Fuzzy Greater or EQual | $= 1$ if $\alpha_A \geq \beta_B$ <br> $= \dfrac{\beta_A - \alpha_B}{(\beta_B - \alpha_B) - (\alpha_A - \beta_A)}$ if $\alpha_A < \beta_B \ \& \ \beta_A > \alpha_B$ <br> $= 0$ otherwise |
| NFLT | Necessarily Fuzzy Less Than | $= 1$ if $\delta_A \leq \alpha_B$ <br> $= \dfrac{\gamma_A - \beta_B}{(\alpha_B - \beta_B) - (\delta_A - \gamma_A)}$ if $\delta_A > \alpha_B \ \& \ \gamma_A < \beta_B$ <br> $= 0$ otherwise |
| NFLEQ | Necessarily Fuzzy Less or EQual | $= 1$ if $\alpha_A \leq \gamma_B$ <br> $= \dfrac{\gamma_A - \delta_B}{(\gamma_B - \delta_B) - (\delta_A - \gamma_A)}$ if $\delta_A > \gamma_B \ \& \ \gamma_A < \delta_B$ <br> $= 0$ otherwise |
| NMGT | Necessarily Fuzzy Much Greater Than | $= 1$ if $\alpha_A \geq \delta_B + M$ <br> $= \dfrac{\gamma_B + M - \beta_A}{(\alpha_A - \beta_A) - (\delta_B - \gamma_B)}$ if $\alpha_A < \delta_B + M \ \& \ \beta_A > \gamma_B + M$ <br> $= 0$ otherwise <br> M is the minimum distance to consider two attributes as very separate |
| NMLT | Necessarily Fuzzy Much Less Than | $= 1$ if $\delta_A \leq \alpha_B - M$ <br> $= \dfrac{\beta_B - M - \gamma_A}{(\delta_A - \gamma_A) - (\alpha_B - \beta_B)}$ if $\delta_A > \alpha_B - M \ \& \ \gamma_A < \beta_B - M$ <br> $= 0$ otherwise <br> M is the minimum distance to consider two attributes as very separate |

- CDEG(<attribute>) function: This function shows a column with the Fulfillment degree of the condition of the query for a specific attribute, which is expressed in brackets as the argument.
- Fuzzy Constants: In FSQL we can use all of the fuzzy constants which appear in Table 2.
- Fuzzy Quantifiers: They can either be relative or absolute with the formats $Quantifier [FUZZY] (<condition>) THOLD $\chi$ or $Quantifier [FUZZY] (<condition_1>) ARE (<condition_2>) THOLD $\chi$, indicating that the quantifier must be satisfied with a minimum degree $\chi$ in [0,1] fulfilled. In the Figure 2 we can see an example of the quantifier most defined in the FMB.

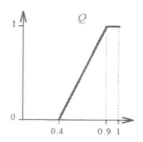

**Fig. 2.** Fuzzy quantifier *most*

**Table 3.** Fuzzy constants of FSQL

| F. Constant | Significance |
|---|---|
| UNKOWN | Unknown value but the attribute is applicable |
| UNDEFINED | The attribute is not applicable or it is meaningless |
| NULL | Total ignorance: We know nothing about it |
| A=$[$\alpha_A,\beta_A, \gamma_A,\delta_A$] | Fuzzy trapezoid ($\alpha_A \leq \beta_A \leq \gamma_A \leq \delta_A$): See Figure 1 |
| $label | Linguistic Label: It may be a trapezoid or a scalar (defined in FMB) |
| [n, m] | Interval "Between n and m" ($\alpha_A=\beta_A$=n and $\gamma_A=\delta_A$=m) |
| #n | Fuzzy value "Approximately n" ($\beta_A=\gamma_A$=n and n-$\alpha_A=\delta_A$=margin) |

We have a FSQL Server available to obtain the answers to FSQL queries for Oracle© DBMS. The FSQL Server maintains a Fuzzy Meta-knowledge Base (FMB) which has all the information about the attributes susceptible to fuzzy treatment.

## 3 Fuzzy Functional Dependencies and Gradual Functional Dependencies

The problem of FD inference has been treated many times in literature (Akutsu and Takasu 1993, Kivinen and Mannila 1992, Mannila and Räihä 1987). Investigated for long years, this issue has been recently addressed in a novel and more efficient way by applying principles of data mining algorithms. This method is useful when we have large sets of materialized data (e.g. Data Warehouse environments...) like in this case. The inference of FD for supporting the decision-making process of market investment is carried out analyzing the data stored in a large data base.

The knowledge that will support the decision investors will reflect relations, correlations, patterns etc., among the analyzed features the use of FDs can help to find them out, but due to the fact that we are dealing with uncertain information it seems suitable to smooth the FDs to facilitate the discovery of such type of information.

There have been several approaches to the problem of defining the concept of FFD but unlike classical FDs one single approach has not dominated. We begin by briefly describing the concept of classical FD, later we give a general definition of FFD and

GFD based on fuzzy functions and then, we shall introduce a more relaxed definition of FFD and GFD in order to manage exceptions.

The relation R with attribute sets X=(x1,...,xn), and Y=(y1,...,ym) in its scheme verifies the FD X→Y if and only if, for every instance r of R it is verified:

$$\forall t_1, t_2 \in r, t_1[X] = t_2[X] \Rightarrow t_1[Y] = t_2[Y] \tag{3}$$

The concept of FFD (Cubero and Vila 1994) is a smoothed version of the classical FD. The basic idea consists in replacing the equality used in the FD definition by fuzzy resemblance relations, in such a way that: The relation R verifies an $\alpha$–ß FFD X→FTY if and only if, for every instance r of R it is verified:

$$\forall t_1, t_2 \in r, F(t_1[X], t_2[X]) \geq \alpha \Rightarrow T(t_1[Y], t_2[Y]) \geq ß \tag{4}$$

where F and T are fuzzy resemblance relations.

The flexibility provided by the combined use of the parameters $\alpha$ and ß and the different kinds of resemblance relation should be noted. If F is a weak resemblance measure and T is a strong one, we get interesting properties for database design (decomposition of relations). A more detailed description of these concepts can be found in (Cubero et al. 1998).

Often just a few tuples in a database can prevent the FFD from being completed. This situation is usual in the stock market are due to the fact that it is a complex and dynamic system with noisy, non-stationary and chaotic data series. To avoid this, we can relax the FFD definition in such a way that all the tuples of the relationship are not forced to fulfill the above condition, therefore we define:

**Definition 1 (confidence of a FFD)**

The relation $R$ verifies an $\alpha$–ß FFD X→$_{FT}$Y with confidence $c$, where $c$ is defined as:

$$c = 0 \text{ if } Card\{(t_1, t_2) \ t_1, t_2 \in r \, / \, F(t_1[X], t_2[X]) \geq \alpha\} = 0$$

$$c = \frac{Card\{(t_1, t_2) \ t_1, t_2 \in r \, / \, F(t_1[X], t_2[X]) \geq \alpha \wedge T(t_1[Y], t_2[Y]) \geq \beta\}}{Card\{(t_1, t_2) \ t_1, t_2 \in r \, / \, F(t_1[X], t_2[X]) \geq \alpha\}} \text{Otherwise} \tag{5}$$

Where $\wedge$ is the logical operator and. The basic idea consists of computing the percentage of tuples which fulfill the antecedent and consequent together with respect to those which only fulfill the consequent.

**Definition 2**

The relation $R$ verifies an $\alpha$–ß FFD X→$_{FT}$Y with support $s$, where $s \in [0, 1]$, is defined as:

$$s = 0 \text{ if } n = 0$$

$$s = \frac{Card\{(t_1, t_2) \ t_1, t_2 \in r \, / \, F(t_1[X], t_2[X]) \geq \alpha \wedge T(t_1[Y], t_2[Y]) \geq \beta\}}{n} \text{Otherwise} \tag{6}$$

where n is the number of tuples of the r instance of the relation R.

The idea is to find the percentage of tuples which fulfill the antecedent and consequent together with respect to the total rows of the relation.

Another way of considering the connections between data in databases is to specify a relationship between objects in a dataset and reflect monotonicity in the data by

298    F. Araque et al.

means of that we have called gradual fuzzy dependencies (GFDs). It is closely related to the idea of gradual rules introduced by Dubois and Prade (1992a, 1992b). An intuitive example of a GFD is "the bigger business is the higher earnings they have" and we assume that the concept of GFD can be considered, in this way, as similar to the FFD one. Therefore we define:

### Definition 3 ($\alpha$–ß gradual functional dependency)

The relation $R$ verifies an $\alpha$–ß GFD $X\mathbf{\Theta}_{FT}Y$ if and only if, for every instance $r$ of $R$ it is verified:

$$\forall t_1, t_2 \in r, \; F'(t_1[X], t_2[X]) \geq \alpha \Rightarrow T'(t_1[Y], t_2[Y]) \geq ß$$

where F' and T' are fuzzy relations of the type: fuzzy greater than, fuzzy greater than or equal to, fuzzy less than, fuzzy less than or equal to, fuzzy not equal, etc. We can define an $\alpha$–ß GFD $X\mathbf{\Theta}F'T'Y$ with confidence c in the same way that we have made it for FFD (see Definition 1).

## 4 Applying FSQL to Obtain Fuzzy Extended Dependencies

Usually, historical market data is represented in form of table. The natural way of managing this kind of information is the relational data model. The problem of this model is that it is not able to manage diffuse information directly. Given the inherent incompleteness and fuzziness of market data it is necessary extend its functionality. In (Medina et al. 1994) a module to allow extending the capacity of a classical Relational Data Base Management Systems (RDBMS) is exposed in order to represent and manipulate "vague" information. This module, called FIRST (Fuzzy Interface Relational Systems), uses GEFRED as theoretical model and the classic relational model resources to be able to represent this type of information. With the intention of completing the FIRST model, to be able to conduct the typical operations (creation of tables, flexible labels, consultations...) on Diffuse Relational Data Base (DRDB) Galindo and Medina (1998) extended language SQL to allow the treatment of the new data. This extension was called Fuzzy SQL (FSQL).

In this section we first introduce a general definition of Fuzzy Extended Dependencies based on FSQL operators and FSQL CDEG function, later we will show how FED can be calculated with FSQL.

### 4.1 Fuzzy Extended Dependencies with FSQL Operators

### Definition 4

The relation $R$ with attribute sets $X=(x_1...x_n)$, and $Y=(y_1...y_m)$ whose attributes are trapezoidal possibility distributions, verifies an $\alpha$–ß FED $X\blacktriangleright_{F*T*}Y$ with $\alpha=(\alpha_1, \alpha_2,...,\alpha_n)$ / $\alpha_i \in [0,1]$ $\forall i=1,...,n$ and $ß=(ß_1, ß_2,...,ß_m)$ / $ß_j \in [0,1]$ $\forall j=1,...,m$, if and only if, for every instance $r$ of $R$ it is verified:

$$\forall t_1, t_2 \in r, \; \wedge_{i=1,2...,n}[F^*_i(t_1[x_i], t_2[x_i]) \geq \alpha_i] \Rightarrow \wedge_{j=1,2...,m}[T^*_j(t_1[y_j], t_2[y_j]) \geq ß_j]$$

where

$$F^*_i : UxU \rightarrow [0,1]/F^*_i(A,B)=CDEG(A \; F\_Comp\_ant_i \; B)$$

$$T^*_j : UxU \rightarrow [0,1]/T^*_j(A,B) = CDEG(A\ F\_Comp\_con_j\ B)$$

$$\forall A = \$[\alpha_A, \beta_A, \gamma_A, \delta_A],\ B = \$[\alpha_B, \beta_B, \gamma_B, \delta_B] \in U\ (\text{see Figure 1})$$

$F\_Comp\_ant_i$, $F\_Comp\_con_j$ defined as any fuzzy comparator in FSQL (any $F\_Comp$ in Table 1, even when preceded by a NOT operator) $\forall i=1,\ldots,n$, $\forall j=1,\ldots,m$

## Definition 5

The relation $R$ with attribute sets $X=(x_1 \ldots x_n)$, and $Y=(y_1 \ldots y_m)$ whose attributes are trapezoidal possibility distributions, verifies an $\alpha$–ß FED $X \blacktriangleright_{F*T*} Y$ with $\alpha \in [0,1]$ and $ß \in [0,1]$, if and only if, for every instance $r$ of $R$ it is verified:

$$\forall t_1, t_2 \in r,\ \wedge_{i=1,2\ldots,n}[F^*_i(t_1[x_i], t_2[x_i]) \geq \alpha] \Rightarrow \wedge_{j=1,2\ldots,m}[T^*_j(t_1[y_j], t_2[y_j]) \geq ß]$$

$$\forall i=1,\ldots,n,\ \forall j=1,\ldots,m$$

Now, we can make a new definition of FFDs and GFDs as a particular case of FEDs.

## Definition 6

If $F\_Comp\_ant_i$, $F\_Comp\_con_j \in \{FEQ,NFEQ\}$ then we say that R verifies an $\alpha_i$–$ß_i$ FFD $X \rightarrow_{F*T*} Y$.

## Definition 7

If $F\_Comp\_ant_i$, $F\_Comp\_con_j \in \{FEQ,NFEQ\}$ then we say that R verifies an $\alpha$–ß FFD $X \rightarrow_{F*T*} Y$.

## Definition 8

If $F\_Comp\_ant$, $F\_Comp\_con$ are any $F\_Comp$ of FSQL such that there exists at least a $k$ from 1 to $n$ which fulfils $F\_Comp\_ant_k \notin \{FEQ,NFEQ\}$ and at least a $s$ from 1 to $m$ which fulfils $F\_Comp\_con_s \notin \{FEQ,NFEQ\}$ then we say that R verifies an $\alpha$–ß GFD $X \bullet_{F*T*} Y$.

## Definition 9

If $F\_Comp\_ant_i$, $F\_Comp\_con_j$ are any $F\_Comp$ of FSQL such that there exists at least a $k$ from 1 to $n$ which fulfils $F\_Comp\_ant_k \notin \{FEQ,NFEQ\}$ and at least a $s$ from 1 to $m$ which fulfils $F\_Comp\_con_s \notin \{FEQ,NFEQ\}$ then we say that R verifies an $\alpha_i$–$ß_i$ GFD $X \bullet_{F*T*} Y$.

Of course we can define an $\alpha$–ß FED $X \blacktriangleright_{F*T*} Y$ with confidence $c$ in the same sense that we have made it for FFD (see Definition 1). To simplify notation, in $X \blacktriangleright_{F*T*} Y$ we will denote $F^*$ as $(F\_Comp\_ant_i)^*$ $\forall i=1,\ldots,n$, and similar notation for $T^*$.

## 4.2 Obtaining Fuzzy Extended Dependencies from a Database by Using FSQL

The definitions given in the previous section should be implemented in a tool, so the investor can make use of them. We have extended the SQL language and developed a server for dealing with this kind of information (Carrasco et al. 2006).

Let R be a relation with attribute sets $X=(x1 \ldots xn)$, $Y=(y1 \ldots ym)$ and $PK=(pk1 \ldots pkS)$ included in its scheme, where PK is the primary key of R. To determine if R verifies an α–ß FED X▶F\*T\*Y for an instance r, we create a FSQL query with the following general format:

```
SELECT count(*) FROM    r A1, r A2
WHERE   (A1.PK <> A2.PK)
AND             (A1.x1 F_Comp_ant1 A2.x1 THOLD α1
     AND ... AND A1.xn F_Comp_antn A2.xn THOLD αn)
AND NOT         (A1.y1 F_Comp_con1 A2.y1 THOLD ß1
     AND ... AND A1.ym F_Comp_conm A2.ym THOLD ßm)
```

The basic idea consists of computing the tuples which fulfill the antecedent and do not fulfill the consequent. Therefore, if the result of the query is 0, we can say that $R$ verifies FED for the instance $r$.

If the result of previous counting is not 0, we can determine if $R$ verifies an α–ß FED X▶$_{F*T*}$Y with confidence $c$ by means of a simple procedure as follows (algorithm 1):

- *Step 1.1:* To obtain the value $a$ as the number of tuples which fulfil the antecedent and consequent together:
```
SELECT count(*) FROM    r A1, r A2
WHERE   (A1.PK <> A2.PK)
  AND             (A1.x₁ F_Comp_ant₁ A2.x₁ THOLD α₁
     AND ... AND A1.xₙ F_Comp_antₙ A2.xₙ THOLD αₙ)
  AND             (A1.y₁ F_Comp_con₁ A2.y₁ THOLD ß₁
     AND ... AND A1.yₘ F_Comp_conₘ A2.yₘ THOLD ßₘ)
```
- *Step 1.2:* To obtain the value $b$ as the number of tuples which fulfil the antecedent:
```
SELECT count(*) FROM    r A1, r A2
WHERE   (A1.PK <> A2.PK)
  AND             (A1.x1 F_Comp_ant1 A2.x1 THOLD α1
     AND ... AND A1.xn F_Comp_antn A2.xn THOLD αn)
```
- *Step 2:* To obtain the degree of confidence $c$ as $c=a/b$.
- *Step 3:* To determine if the computed degree indicates that the FED is good enough, we can compare the value $c$ with some fuzzy quantifier defined in the FMB (by example *most*).

Notice that FSQL also allows us to compare (with fuzzy comparators) crisp attributes. In order to do this, FSQL makes a *fuzzyfication* of the crisp value before the comparison, transforming it into a triangular possibility distribution (according to values stored in the FMB for the attribute). This fuzzyfication can either be implicit or explicit (with the fuzzy constant #). Also, FSQL can work with scalar attributes but with them we can use only use the comparator FEQ (because an order relationship in their domains is not defined).

If the purpose is to search for FFDs in order to discover intentional properties (constraints that exist in every possible manifestation of the database frame) it seems more appropriate to use a weak resemblance measure in the antecedent (FEQ, based on possibility) as a fuzzy comparator and a strong one in the consequent (NFEQ, based on necessity). In this way, we get interesting properties which can help us with the decomposition of relations (Cubero et al. 1998). Searching for FFDs or GFDs to discover extensional properties (those existing in the current manifestation of the data) is a task for DM. In this case, the choice of the fuzzy comparators and the parameters $\alpha$, $\beta$ we will be made according to the specific problem in question.

## 5 Applying a Fuzzy Data Mining Process to Stock Market

In this section we apply the previously outlined process to stock-market sceneries in order to find out behaviour patterns on them. The goal is to identify those patterns which imply earnings of a specific stock. To do so, we are going to apply the previously outlined process which use DM techniques implemented through FSQL.

Let SHARES_ENTERPRISE be a relation defined as:

SHARES_ENTERPRISE (*value_name, date, williams, MA, value*)

which contains data for a specific enterprise, Telefónica S.A., in the Spanish Stock Market. The data corresponds with a period of time between 01/01/2005 and 08/20/2006. The meaning of the attributes is:

- value_name: identifies the enterprise which the stock value belongs to. In this case, all rows shown in table 4 belong to the same enterprise: Telefónica S.A.
- date: instant which data are related to.
- williams: 14-day Williams oscillator value.
- MA: 20-day moving average value.
- value: value in euros for the enterprise at the end of the session.

The step needed to solve the problem are:

1. Identification of the ideal earnings scenaries, taking into account some of the previously detailed indicators used in technical analysis.
2. Expert's theory formulation related to earning scenaries.
3. Expert's theory validation for earning scenaries given the historical data stored.

### 5.1 Earning Sceneries Identification and Expert's Theory Formulation

By means of graphical representations of the data in the table SHARES_ENTERPRISE and the indicators outlined in first section the expert identifies as ideal earning scenery:

- Period: from 06/27/2006 to 07/09/2006.
- Situation: starting from date 06/27/2006, a change in the trend value has been produced with a value of 81.22 (cross with horizontal line 80). There is no significant change until 07/09/2006, when the value line cross the MA line and %R value with horizontal line 20.

So the expert formulates the theory about earning sceneries shown bellow:

*"Greater Williams index and roughly equal moving average implies a greater value for a specific enterprise"*

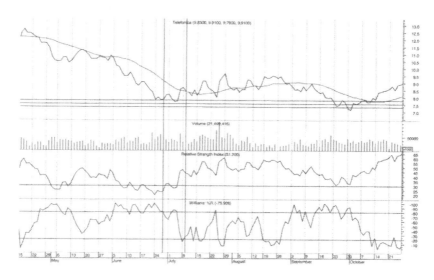

**Fig. 3.** Telefónica S.A. values from 06/27/2006 to 07/09/2006.

### 5.2 Expert's Theory Validation through FEDs Using FSQL

First step consist in define the different attributes of table SHARES_ENTERPRISE in FSQL. The determination that has been taken on the matter is based on the information provided by the stock-exchange technical analysis in the case of the oscillator Williams and for the rest of attributes, not having more information on them, it is going away to determine the use of a similarity function based on the Hamming distance (since it is only necessary to define minimum and maximum values for each attribute). Taking this into account, the attributes are defined in FMB in the following way:

- williams: although it is a crisp value, we decide to define it as diffuse type 1. The value of the margin for approximate values has been determined as 10 and the

**Table 4.** Earning sceneries identified by the expert

| Value_name | Date | Williams | Expert's interpretation | MA(20) | Value |
|---|---|---|---|---|---|
| TELEFONICA | 27-jun-06 | 81,22 | Buy signal | 9,64 | 8,19 |
| TELEFONICA | 28-jun-06 | 68,57 | | 9,50 | 8,50 |
| TELEFONICA | 01-jul-06 | 63,01 | | 9,36 | 8,54 |
| TELEFONICA | 02-jul-06 | 78,54 | | 9,23 | 8,20 |
| TELEFONICA | 03-jul-06 | 84,95 | Buy signal | 9,11 | 8,01 |
| TELEFONICA | 04-jul-06 | 82,26 | Buy signal | 8,98 | 8,06 |
| TELEFONICA | 05-jul-06 | 36,02 | | 8,88 | 8,92 |
| TELEFONICA | 08-jul-06 | 14,57 | Sell signal | 8,83 | 9,02 |
| TELEFONICA | 09-jul-06 | 25,37 | | 8,79 | 8,73 |

minimum range to consider two values as very distanced is 20. Both values are stored in the table of FMB FUZZY_APPROX_MUCH. The objective of this definition is to be able to express linguistic labels for the diffuse management of the Williams oscillator management. The labels are going to be very useful to create a DSS for supporting buying and selling operations. In FSQL the diffuse comparison of the crisps attributes is carried out bluring the attributes, in such a way that they become a triangular possibility distribution (see figure 1) using the margin value defined in FMB. Occasions exist where this oscillator cannot be calculated, e.g. the first day, so the use of the fuzzy constant UNDEFINED turn out useful.

**Fig. 4.** FSQL query for step 1.1

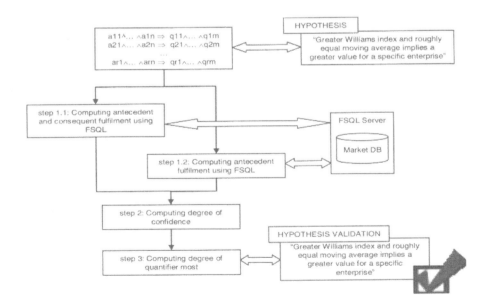

**Fig. 5.** Fuzzy Data Mining Decision process

304    F. Araque et al.

- MA: this is a crisp attribute but we decide define this as Type 1 in the FMB using the fuzzy constants value #n = 2 and margin=5 (approximately n and margin see Table 2)
- value: as well as MA attribute, this is a crisp attribute but we decide define this as Type 1 in the FMB using the fuzzy constants value #n = 2 and margin=5.

Following the scheme showed in figure 5, we show the FSQL query for step 1.1 (fig 4). The FSQL query for step 1.2 is equivalent. After step 2 of algorithm 1 we can say that SHARES_TELEFONICA verifies:

$$(1, 0.7)-(1) \text{ EFD (Williams, MA)} \rightarrow \text{(FGT, FEQ)*(FGT)* (Value)}$$

with confidence c=9/10. Now (Step 3 of Alg. 1) if we compare this value with the fuzzy quantifier *most* we can say that the above FED is verified with fulfilment thresholds 1 for most of the tuples. We can conclude that this FED is fulfilled with a sufficient degree and therefore he will be useful for future earnings sceneries identification.

## 6  Conclusions

The stock market analysis is a complex, time-consuming but highly demanded task nowadays, due to the fact of the huge DBs that could be used in the analysis. The better analysis the higher profits obtained by the investors, so to improve the analysis process we have proposed a Data Mining process dealing with objective and subjective features of the stock markets.

The use of subjective features implies to deal with uncertainty. Therefore, this DM process is based on the use of fuzzy extended dependencies (FEDs) as a common framework to integrate fuzzy functional dependencies and gradual functional dependencies. Also, the process has relaxed the FED definition in order to find FEDs even though there are tuples do not verify them. Therefore, the FSQL language is the natural way to obtain such FEDs. Using possibility in FEDs as a weak resemblance in the antecedent and necessity as a strong one in the consequent, FSQL could be used to find FFDs which portray constraints that exist in every possible manifestation of the frames in a database.

To achieve our aim of discovering useful knowledge for investors a practical application is to search for FFDs or GFDs in order to discover properties which exist in the current manifestation of the data as a task for DM.

## Acknowledgements

This work has been supported by the Spanish Research Program under projects TIN2006-02121 and TIN2005-09098-C05-03, by The Ministry of Education and Science of Spain through Project of Teaching Innovation and the by Andalucía Research Program under projects JA031/06 and GR050/06.

# References

Akutsu, T., Takasu, A.: Inferring approximate functional dependencies from example data. In: Proc. AAAI 1993 Workshop on Knowledge Discovery in Databases. AAAI, New York (1993)

Boginski, V., Butenko, S., Pardalos, P.: Mining market data: a network approach. Computers & Operations Research 33(11), 3171–3184 (2006)

Carrasco, R., Vila, M.A., Araque, F.: dmFSQL: a Language for Data Mining. In: Bressan, S., Küng, J., Wagner, R. (eds.) DEXA 2006. LNCS, vol. 4080. Springer, Heidelberg (2006)

Carrasco, R.A., Galindo, J., Aranda, M.C., Medina, J.M., Vila, M.A.: Classification in Databases using a Fuzzy Query Language. In: 9th International Conference on Management of Data, COMAD 1998, Hyderabad (1998)

Carrasco, R.A., Galindo, J., Vila, M.A., Medina, J.M.: Clustering and Fuzzy Classification in a Financial Data Mining Environment. In: 3rd International ICSC Symposium on Soft Computing, Genova (1998)

Cubero, J.C., Medina, J.M., Pons, O., Vila, M.A.: Fuzzy loss less decompositions in databases. Fuzzy Sets and Systems 97(2), 145–167 (1998)

Cubero, J.C., Vila, M.A.: A new definition of fuzzy functional dependency in fuzzy relational databases. International Journal of Intelligent Systems 9, 441–449 (1994)

Delgado, M., Verdegay, J.L., Vila, M.A.: Linguistic decision making models. International Journal of Intelligent Systems 7, 479–492 (1992)

Dubois, D., Prade, H.: Gradual rules in approximate reasoning. Information Sciences 61(1-2), 103–122 (1992a)

Dubois, D., Prade, H.: Generalized dependencies in fuzzy databases. In: International Conference on Information Processing and Management of Uncertainty in Knowledge Based Systems, Palma de Mallorca (1992b)

Enke, D., Thawornwong, S.: The use of data mining and neural networks for forecasting stock market returns. Expert Systems with Applications 29(4), 927–940 (2005)

Galindo, J.: Tratamiento de la Imprecisión en Bases de Datos Relacionales: Extensión del Modelo y Adaptación de los SGBD Actuales. Ph. Doctoral Thesis University of Granada (1999)

Galindo, J., Medina, J.M., Pons, O., Cubero, J.C.: A Server for Fuzzy SQL Queries. In: Andreasen, T., Christiansen, H., Larsen, H.L. (eds.) Flexible query answering systems. lecture notes in artificial intelligence. Springer, Heidelberg (1998)

Galindo, J., Medina, J.M., Vila, M.A., Pons, O.: Fuzzy comparators for flexible queries to databases. In: Iberoamerican Conference on Artificial Intelligence, Lisbon (1998)

Ghotb, F., Warren, L.: A case study comparison of the analytic hierarchy process and a fuzzy decision methodology. Engineering Economist 40(3), 233–246 (1995)

Herrera, F., Herrera-Viedma, E.: Linguistic decision analysis: steps for solving decision problems under linguistic information. Fuzzy Sets and Systems 115, 67–82 (2000)

Herrera, F., Martínez, L., Sánchez, P.J.: Managing non-homogeneous information in group decision making. European Journal of Operational Research 166(1), 115–132 (2005)

Hussein, D., Pepe, S.: Investment using technical analysis and fuzzy logic. Fuzzy Sets and Systems 127(2), 221–240 (2002)

Medina, J.M., Pons, O., Vila, M.A.: GEFRED: A Generalized Model of Fuzzy Relational Data Bases. Information Sciences 76(1-2), 87–109 (1994)

Kivinen, J., Mannila, H.: Approximate dependency inference from relations, In: Proc. Fourth Internat. In: Conf. on Database Theory (1992)

Kuo, S., Li, S., Cheng, Y., Ho, M.: Knowledge Discovery with SOM Networks in Financial Investment Strategy. HIS 98-103 (2004)

Liang, G., Wang, M.: A fuzzy multi-criteria decision making method for facility site selection. International Journal of Production Research 29(2), 2313–2330 (1991)

Lin, P., Chen, J.: Fuzzy Tree crossover for multi-valued stock valuation. Information Sciences 177(5), 1193–1203 (2007)

Mannila, H., Räihä, K.: Dependency inference, In: Proc. 13th VLDB Conf. (1987)

Martínez, L., Liu, J., Yang, J.B.: A fuzzy model for design evaluation based on multiple-criteria analysis in engineering systems. International Journal of Uncertainty, Fuzziness and Knowledge-Based Systems 14(3), 317–336 (2006)

Amat, O., Puig, X.: Análisis Técnico Bursátil. Ediciones Gestión 2000, S.A., Barcelona (1992)

Zadeh, L.A.: Fuzzy sets. Information and Control 8, 338–353 (1965)

Zadeh, L.A.: The Concept of a Linguistic Variable and Its Applications to Approximate Reasoning. Information Sciences, Part I, II, III 8-8-9: 199–249, 301–357, 43–80 (1975)

# Soft Decision Support Systems for Evaluating Real and Financial Investments

Péter Majlender

Department of Management and Organization,
Swedish School of Economics and Business Administration

**Abstract.** In this chapter, we shall present two main results that we accomplished in the fields of financial management and strategic investment planning. Applying our theoretical results, we will discuss the development of two soft decision support models in detail, which use possibility distributions to describe and characterize the uncertainty about future payoffs.

## 1 Introduction

As practical applications to the normative framework of possibility theory we will formulate two results that we accomplished in the fields of financial management and strategic investment planning. Utilizing our earlier theoretical results we will elaborate the framework of two soft decision support models in detail.

In particular, we will consider the portfolio selection problem under possibility distributions. Originally, the *mean-variance methodology* for the portfolio selection problem was established by Markowitz in 1952, which has been dominating the literature of modern finance since its appearance. Assuming that

- each investor can assign a welfare score, or utility value, to competing investment portfolios, which is based on the expected return and risk of the portfolios, and
- the rates of return on securities are represented by possibility distributions rather than probability distributions,

we will present an algorithm for finding an exact (i.e. *not approximate*) optimal solution to the portfolio selection problem.

Furthermore, we will also show how to work on strategic investment planning problems, where the project can be deferred or postponed for a certain period of time. Utilizing the theories of *Fuzzy Net Present Value Analysis* and *Fuzzy Real Option Valuation*, we will develop an enhanced strategic investment planning model. Incorporating the idea of possibility distributions into the framework of net present value and real option valuation, we are able to formulate and manipulate stochastic as well as non-stochastic uncertainties, which are associated with future cash inflows and outflows. Of particular significance in this development is the methodology to incorporate subjective judgments and statistical uncertainties into an applicable decision support model.

---

C. Kahraman (Ed.): Fuzzy Engineering Economics with Appl., STUDFUZZ 233, pp. 307–338, 2008.
springerlink.com                                    © Springer-Verlag Berlin Heidelberg 2008

308    P. Majlender

## 2  The Basics of Fuzzy Sets and Fuzzy Numbers

### 2.1  Fuzzy Sets

In the following, we shall present the fundamentals of *possibility distributions*.

The new approach to the areas of intelligent systems and artificial intelligence presumes that uncertainty can emerge from sources that are non-stochastic by nature. This type of uncertainty is a result of processes that are *not* random by nature, and thus it cannot be characterized by any statistical methods. We call this particular type of uncertainty *possibilistic* uncertainty. We shall use the theory of fuzzy sets to characterize, manipulate and interpret possibilistic uncertainties. The *risks* associated with possibilistic uncertainties can only be assessed and managed effectively if we take into account that they are implied by non-stochastic sources, and represent them by means of possibility distributions.

**Remark 1.** We note that possibilistic risks usually arise from roots that are neither known nor even describable. For instance, this is the particular case when

- we need to represent human estimations, reasoning and judgments, and
- we encounter a rare or unique situation, where we do not have reliable statistics.

Fuzzy sets were introduced by Zadeh in 1965 as a novel approach to represent and manipulate data with non-statistical uncertainty (Zadeh 1965). Mathematically, a fuzzy set $A$ in a (classical) set $X$ is defined by its membership function

$$\mu_A \colon X \to [0, 1],$$

where value zero represents non-membership, value one represents complete membership and values in between represents intermediate degrees of memberships. Furthermore, we interpret the expression $A(x) = \mu_A(x)$ as the degree of membership of element $x$ in fuzzy set $A$ for each $x \in X$.

In order to embrace a more general view of our framework that we will present in the subsequent sections, we shall briefly go through the basic set theoretic relations and operations on fuzzy sets. Considering the classical set theoretic relations and operations, we can naturally develop the corresponding concepts for fuzzy sets by aptly extending the definitions of ordinary set theory. Formulating in this way, the notions of fuzzy set theory will reduce to their classical set theoretic meaning, when we only consider fuzzy sets with membership values from $\{0, 1\}$.

Let $A$ and $B$ be two fuzzy sets in $X$. We say that $A$ is a *subset* of $B$, denoted by

$$A \subseteq B, \text{ if } A(x) \le B(x) \text{ for all } x \in X.$$

Moreover, we say that $A$ and $B$ are *equal*, denoted by $A = B$, if $A \subseteq B$ and $B \subseteq A$ both hold. We note that $A = B$ if and only if $A(x) = B(x)$ for any $x \in X$.

We call the smallest fuzzy set in $X$ the *empty fuzzy set* in $X$. It is denoted by $\emptyset$, and we formally define $\emptyset(x) = 0$ for all $x \in X$. On the other hand, we call the largest fuzzy set in $X$ the *universal* fuzzy set in $X$. It is denoted by $1_X$, and we define $1_X(x) = 1$ for all $x \in X$. We can easily see that $\emptyset \subseteq A \subseteq 1_X$ holds for any fuzzy set $A$ of $X$.

Let $A$ and $B$ be fuzzy subsets of $X$. The *intersection* of $A$ and $B$ is defined as

$$(A \cap B)(x) = \min\{A(x), B(x)\} = A(x) \wedge B(x), \quad \forall x \in X.$$

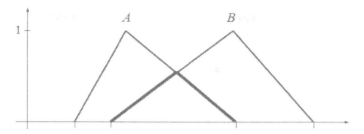

**Fig. 1.** Intersection of two triangular possibility distributions

The *union* of A and B is defined by

$$(A \cup B)(x) = \max\{A(x), B(x)\} = A(x) \vee B(x), \quad \forall x \in X.$$

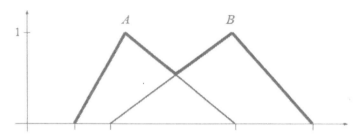

**Fig. 2.** Union of two triangular possibility distributions

Finally, the *complement* of A is defined as

$$(\neg A)(x) = 1 - A(x), \quad \forall x \in X.$$

**Fig. 3.** Complement of a triangular possibility distribution

Using the basic set theoretic operations, we can elaborate the framework of fuzzy set theory with its fundamental relationships and principles[1].

---

[1] We note that *DeMorgan's laws*, the two fundamental identities of classical set theory, also remain valid in fuzzy logic. Namely,

$$\neg(A \wedge B) = \neg A \vee \neg B, \quad \neg(A \vee B) = \neg A \wedge \neg B$$

hold for any fuzzy set A and B.

## 2.2 Fuzzy Numbers

In the following, we shall focus on a special class of fuzzy sets, called *fuzzy numbers*, which plays a significant role in both theory and application. Data and information processed in intelligent systems can almost always be represented or characterized numerically. Introducing fuzzy numbers we are able to represent and manipulate nonstatistical uncertainties on the real line **R**. Furthermore, fuzzy numbers can be considered as *building bricks* of fuzzy sets, since any fuzzy set on **R** can be constructed of fuzzy numbers through fuzzy set theoretic operations (union, intersection and complement).

To introduce the concept of fuzzy number, we need to introduce the following notions:

- Let $A$ be a fuzzy set in $X$. The *support* of $A$ is the crisp (i.e. non-fuzzy) subset of $X$ whose elements all have non-zero membership degree in $A$, that is,

$$\text{supp}(A) = \{x \in X \mid A(x) > 0\}.$$

- A fuzzy subset $A$ of $X$ is called *normal* if there exists $x \in X$ such that $A(x) = 1$. Otherwise $A$ is called *subnormal*.
- A $\gamma$-level set of a fuzzy set $A$ of $X$ is defined by

$$[A]^{\gamma} = \begin{cases} \{x \in X \mid A(x) \geq \gamma\} & \text{if } \gamma > 0, \\ \text{cl supp}(A) & \text{if } \gamma = 0, \end{cases}$$

where cl supp($A$) denotes the *closure* of the support of $A$. A fuzzy set $A$ in $X$ is called *convex* if $[A]^{\gamma}$ is a convex subset of $X$ for all $\gamma \in [0, 1]$.

**Definition 1.** A fuzzy set $A$ in **R** is called a *fuzzy number* if it is normal, fuzzy convex and has an upper semi-continuous membership function of bounded support. The family of all fuzzy numbers will be denoted by $\mathcal{F}$.

For the sake of completeness, we introduce another class of fuzzy sets, which can be considered as a generalization of the concept of fuzzy number.

**Definition 2.** A *quasi fuzzy number* $A$ is a fuzzy set of the real line **R** with a normal, fuzzy convex and upper semi-continuous membership function satisfying the following limit conditions

$$\lim_{x \to -\infty} A(x) = 0, \quad \lim_{x \to \infty} A(x) = 0.$$

It is clear that if $A$ is a fuzzy number then $[A]^{\gamma}$ is a compact and convex subset of **R** for all $\gamma \in [0, 1]$. Furthermore, if $A$ is a quasi fuzzy number, then $[A]^{\gamma}$ is a compact interval for any $0 < \gamma \leq 1$ and a closed interval for $\gamma = 0$.

If $A \in \mathcal{F}$ is a fuzzy number, then let us introduce the following notations

$$a_1(\gamma) = \min[A]^{\gamma}, \quad a_2(\gamma) = \max[A]^{\gamma}.$$

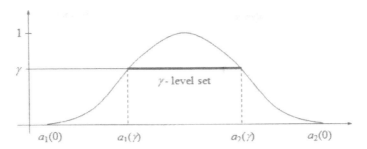

**Fig. 4.** A γ-level set of a fuzzy number

That is, let $a_1(\gamma)$ and $a_2(\gamma)$ denote the left-hand side and the right-hand side of the γ-level set of A, respectively. It is easy to see that if $\gamma \leq \delta$ then $[A]^\delta \subseteq [A]^\gamma$. Furthermore, the left-hand side function $a_1$: [0, 1] → **R** is monotone increasing and lower semi-continuous, and the right-hand side function $a_2$: [0, 1] → **R** is monotone decreasing and upper semi-continuous. We will use the notation

$$[A]^\gamma = [a_1(\gamma), a_2(\gamma)], \quad \gamma \in [0, 1].$$

Here, the support of A is the open interval $(a_1(0), a_2(0))$.

Fuzzy numbers can be considered as possibility distributions (see Zadeh 1965, Zadeh 1978). Let $a, b \in \mathbf{R} \cup \{-\infty, \infty\}$ with $a \leq b$, then the possibility that fuzzy number $A \in \mathscr{F}$ takes its value from the interval [a, b] is defined by (Zadeh 1978)

$$\operatorname{Pos}(A \in [a,b]) = \max_{x \in [a,b]} A(x).$$

In particular,

$$\operatorname{Pos}(A \leq a_1(\gamma)) = \max_{x \leq a_1(\gamma)} A(x) = \gamma,$$

$$\operatorname{Pos}(A \geq a_2(\gamma)) = \max_{x \geq a_2(\gamma)} A(x) = \gamma$$

for any $\gamma \in [0, 1]$ (Alleman and Noam 1999).

In case of side functions $a_1$ and $a_2$ are linear, we get the concept of *triangular* and *trapezoidal* fuzzy numbers.

**Definition 3.** A fuzzy number A is called *triangular* fuzzy number with *center* (or *peak*) a, left width $\alpha \geq 0$ and right width $\beta \geq 0$ if its membership function is of the following form

$$A(x) = \begin{cases} 1 - \dfrac{a-x}{\alpha} & \text{if } a-\alpha < x < a, \\ 1 & \text{if } x = a, \\ 1 - \dfrac{x-a}{\beta} & \text{if } a < x < a+\beta, \\ 0 & \text{otherwise}, \end{cases}$$

and we use the notation $A = (a, \alpha, \beta)$.

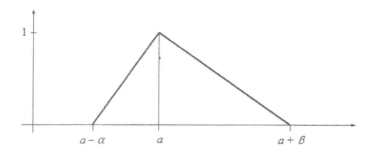

**Fig. 5.** Triangular fuzzy number

We can easily verify that

$$[A]^\gamma = [a - (1 - \gamma)\alpha, a + (1 - \gamma)\beta]$$

holds for all $\gamma \in [0, 1]$, and the support of $A$ is $(a - \alpha, a + \beta)$.

A triangular fuzzy number with center $a$ can be seen as a context-dependent description of the *fuzzy quantity*

"*x* is *approximately* equal to *a*",

where $\alpha$ and $\beta$ define the context.

**Definition 4.** A fuzzy number $A$ is called *trapezoidal* fuzzy number with *tolerance interval* $[a, b]$, left width $\alpha \geq 0$ and right width $\beta \geq 0$, if its membership function is of the following form

$$A(x) = \begin{cases} 1 - \dfrac{a-x}{\alpha} & \text{if } a-\alpha < x < a, \\ 1 & \text{if } a \leq x \leq b, \\ 1 - \dfrac{x-b}{\beta} & \text{if } b < x < b+\beta, \\ 0 & \text{otherwise,} \end{cases} \quad (1)$$

and we use the notation $A = (a, b, \alpha, \beta)$.

We can easily show that

$$[A]^\gamma = [a - (1 - \gamma)\alpha, b + (1 - \gamma)\beta]$$

for any $\gamma \in [0, 1]$, and the support of $A$ is $(a - \alpha, b + \beta)$.

A trapezoidal fuzzy number with tolerance interval $[a, b]$ can be seen as a context-dependent description of the fuzzy quantity

"*x* is *approximately* in the interval $[a, b]$",

where $\alpha$ and $\beta$ define the context.

It is obvious that the concept of trapezoidal fuzzy number extends the notion of triangular fuzzy number with the relationship $(a, \alpha, \beta) = (a, a, \alpha, \beta)$.

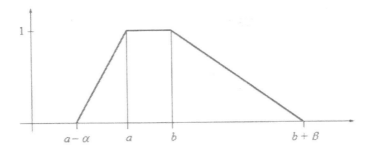

**Fig. 6.** Trapezoidal fuzzy number

From the following definition we can see that the set of real numbers (**R**) actually represent a restriction of the class of fuzzy numbers ($\mathscr{F}$) with no uncertainty or imprecision.

**Definition 5.** Let $A \in \mathscr{F}$ be a fuzzy number. If supp($A$) = $\{a\}$ then we call $A$ a *fuzzy point*, and we use the notation $A = \bar{a}$.

It is clear that if $A$ is a fuzzy point then $[A]^\gamma = [a, a] = \{a\}$ for any $\gamma \in [0, 1]$, and then $A$ defines a *crisp* number.

## 2.3 The Extension Principle

Representing fuzzy sets on the real line implies great advantages from a structural point of view. To model the propagation of possibilistic uncertainty in intelligent systems, we need to carry out mathematical operations with fuzzy numbers. In 1965 Zadeh introduced the *sup-min extension principle*, a cornerstone concept in developing mathematics in possibility theory (Zadeh 1965).

**Definition 6.** Let $A_1, A_2, \ldots, A_n \in \mathscr{F}$ be fuzzy numbers, and let $f : \mathbf{R}^n \to \mathbf{R}$ be a continuous function. Then, the *sup-min extension* of $f$ is defined by

$$f(A_1, A_2, \ldots, A_n)(y) = \sup_{f(x_1, x_2, \ldots, x_n) = y} \min\{A_1(x_1), A_2(x_2), \ldots, A_n(x_n)\}, \quad \forall y \in \mathbf{R}. \tag{2}$$

In reference to the extension principle, in 1978 Nguyen proved that if $A_1, A_2, \ldots, A_n \in \mathscr{F}$ are fuzzy numbers then $f(A_1, A_2, \ldots, A_n)$ is a fuzzy number as well. Furthermore, the following relationship holds for any $\gamma \in [0, 1]$ (Nguyen 1978)

$$[f(A_1, A_2, \ldots, A_n)]^\gamma = f\left([A_1]^\gamma, [A_2]^\gamma, \ldots, [A_n]^\gamma\right)$$

$$= \left\{ f(x_1, x_2, \ldots x_n) \mid x_1 \in [A_1]^\gamma, x_2 \in [A_2]^\gamma, \ldots, x_n \in [A_n]^\gamma \right\}$$

In particular, if $A, B \in \mathscr{F}$ are fuzzy numbers with $\gamma$-level sets $[A]^\gamma = [a_1(\gamma), a_2(\gamma)]$ and $[B]^\gamma = [b_1(\gamma), b_2(\gamma)]$ and $\lambda \in \mathbf{R}$ is a real number, then using the sup-min extension principle Eq. (2) we obtain the following relationships for addition and scalar multiplication of fuzzy numbers

$$(A + B)(z) = \sup_{x+y=z} \min\{A(x), B(y)\}$$

314     P. Majlender

and

$$(\lambda A)(z) = \sup_{xy=z} \min\{A(x), \chi_{\{\lambda\}}(y)\} = \begin{cases} A(z/\lambda) & \text{if } \lambda \neq 0, \\ \chi_{\{0\}}(z) & \text{if } \lambda = 0. \end{cases}$$

Furthermore, we can also verify the following rules for the $\gamma$-level sets

$$[A+B]^{\gamma} = [A]^{\gamma} + [B]^{\gamma} = [a_1(\gamma) + b_1(\gamma), a_2(\gamma) + b_2(\gamma)]$$

and

$$[\lambda A]^{\gamma} = \lambda[A]^{\gamma} = \begin{cases} [\lambda a_1(\gamma), \lambda a_2(\gamma)] & \text{if } \lambda \geq 0, \\ [\lambda a_2(\gamma), \lambda a_1(\gamma)] & \text{if } \lambda < 0 \end{cases}$$

for any $\gamma \in [0, 1]$.

In general, if $A_1 = (a_1, b_1, \alpha_1, \beta_1)$ and $A_2 = (a_2, b_2, \alpha_2, \beta_2)$ are fuzzy numbers of trapezoidal form, and $\lambda \in \mathbf{R}$ is a real number then

$$A_1 + A_2 = (a_1, b_1, \alpha_1, \beta_1) + (a_2, b_2, \alpha_2, \beta_2)$$
$$= (a_1 + a_2, b_1 + b_2, \alpha_1 + \alpha_2, \beta_1 + \beta_2)$$

and

$$\lambda A_1 = \lambda(a_1, b_1, \alpha_1, \beta_1) = \begin{cases} (\lambda a_1, \lambda b_1, \lambda \alpha_1, \lambda \beta_1) & \text{if } \lambda \geq 0, \\ (\lambda b_1, \lambda a_1, -\lambda \beta_1, -\lambda \alpha_1) & \text{if } \lambda < 0. \end{cases}$$

### 2.4 Normative Measures on Fuzzy Numbers

One of the first systematic approaches to define a normative measure of possibility distributions was presented by (Dubois and Prade 1997). They introduced the notion of *interval-valued expectation* of fuzzy numbers by viewing them as *consonant* random sets. They also proved that this expectation, called the *interval-valued probabilistic mean*, is additive in the sense of addition of fuzzy numbers.

Another approach, formulated by Goetschel and Voxman in 1986, defines a measure on fuzzy numbers for a possibilistic interpretation of ordering (Goetschel and Woxman 1986). As they pointed out, their definition was partly motivated by the desire to give less (more) importance to the lower (higher) level sets of fuzzy numbers.

A fundamental approach, based on the view of fuzzy numbers as possibility distributions instead of probability distributions, was presented by Carlsson and Fullér in 2001 (Carlsson and Fullér 2001). In particular, they introduced the concepts of *interval-valued possibilistic mean*, *(crisp) possibilistic mean value* and *(crisp) possibilistic variance* of fuzzy numbers, and showed that those measures are consistent with the extension principle.

In what follows, we shall present the normative measures of *weighted possibilistic mean*, *variance* and *covariance* on fuzzy numbers that we will need to develop our framework. We shall follow the formulation of (Fullér and Majlender 2003) and

# Soft Decision Support Systems for Evaluating Real and Financial Investments

(Majlender 2004). Let $f : [0, 1] \rightarrow \mathbf{R}$ be a non-negative, monotone increasing and normalized function:

$$\inf_{\gamma \in [0,1]} f(\gamma) \geq 0,$$

$$0 \leq \gamma_1 < \gamma_2 \leq 1 \Rightarrow f(\gamma_1) \leq f(\gamma_2),$$

$$\int_0^1 f(\gamma) d\gamma = 1.$$

Then, we call $f$ a *weighting function*. It is clear that by introducing different weighting functions we can magnify the importance of certain $\gamma$-level sets as compared with other ones by overweighing them in the weighted average.

**Definition 7.** Let $A \in \mathcal{F}$ be a fuzzy number with $[A]^\gamma = [a_1(\gamma), a_2(\gamma)]$, $\gamma \in [0, 1]$, and let $f : [0, 1] \rightarrow \mathbf{R}$ be a weighting function. Then the *weighted possibilistic mean value* of $A$ is defined by

$$E_f(A) = \int_0^1 \frac{a_1(\gamma) + a_2(\gamma)}{2} f(\gamma) d\gamma, \tag{3}$$

It can easily be proven that for any weighting function $f$, $E_f : \mathcal{F} \rightarrow \mathbf{R}$ is a linear operator. That is, we have the following result (Fullér and Majlender 2003):

**Theorem 1.** Let $A, B \in \mathcal{F}$ be two fuzzy numbers, and let $\lambda \in \mathbf{R}$ be a real number. Then

$$E_f(A + B) = E_f(A) + E_f(B),$$
$$E_f(\lambda A) = \lambda E_f(A)$$

hold, where addition and multiplication by a scalar of fuzzy numbers is defined by the sup-min extension principle Eq. (2).

**Definition 8.** Let $A \in \mathcal{F}$ be a fuzzy number with $[A]^\gamma = [a_1(\gamma), a_2(\gamma)]$, $\gamma \in [0, 1]$, and let $f : [0, 1] \rightarrow \mathbf{R}$ be a weighting function. Then the *weighted possibilistic variance* of $A$ is defined by

$$\sigma_f^2(A) = \int_0^1 \left[ \frac{a_2(\gamma) - a_1(\gamma)}{2} \right]^2 f(\gamma) d\gamma. \tag{4}$$

**Remark 2.** In particular, if $f(\gamma) = 2\gamma$, $\gamma \in [0, 1]$, is the linear weighting function then we have

$$E_f(A) = \overline{M}(A) = \int_0^1 [a_1(\gamma) + a_2(\gamma)] \gamma d\gamma,$$

$$\sigma_f^2(A) = \text{Var}(A) = \frac{1}{2} \int_0^1 [a_2(\gamma) - a_1(\gamma)]^2 \gamma d\gamma,$$

which gives the notion of crisp possibilistic mean and variance of $A$ introduced in (Carlsson and Fullér 2001).

316    P. Majlender

**Remark 3.** From Eq. (3) we see that the weighted possibilistic mean of $A$ is nothing else but the $f$-weighted average of the *centers* of the $\gamma$-level sets of $A$. Analogously, from Eq. (4) we find that the weighted possibilistic variance of $A$ is nothing else but the $f$-weighted average of the *squared deviations* between the endpoints of the $\gamma$-level sets of $A$ (up to a constant multiplier 1/4).

**Remark 4.** In the particular case of $f(\gamma) = 2\gamma$, $\gamma \in [0, 1]$, the decision maker is assumed to assign a *linear relationship* between the degree of possibility and the degree of *relevance* of values in the fuzzy number.

It is easy to see that if $A = (a, b, \alpha, \beta)$ is a trapezoidal fuzzy number and $f(\gamma) = 2\gamma$, $\gamma \in [0, 1]$, is the linear weighting function then

$$E_f(A) = \int_0^1 [a+b+(\beta-\alpha)(1-\gamma)]\gamma d\gamma = \frac{a+b}{2} + \frac{\beta-\alpha}{6}$$

and

$$\sigma_f^2(A) = \frac{1}{2}\int_0^1 [b-a+(\alpha+\beta)(1-\gamma)]^2 \gamma d\gamma$$

$$= \frac{(b-a)^2}{4} + \frac{(b-a)(\alpha+\beta)}{6} + \frac{(\alpha+\beta)^2}{24} = \left[\frac{b-a}{2} + \frac{\alpha+\beta}{6}\right]^2 + \frac{(\alpha+\beta)^2}{72}.$$

**Remark 5.** Applying the terminology of probability theory, we can call the weighted possibilistic mean as the *first moment* of possibility distributions, and the weighted possibilistic variance as the *second central moment* of possibility distributions.

For the sake of complete presentation of this particular normative aspect of possibility distributions, we present the notion of weighted possibilistic covariance as well as one of its fundamental relationships.

**Definition 9.** Let $A, B \in \mathscr{F}$ be two fuzzy numbers with $[A]^\gamma = [a_1(\gamma), a_2(\gamma)]$ and $[B]^\gamma = [b_1(\gamma), b_2(\gamma)]$, $\gamma \in [0, 1]$, and let $f : [0, 1] \rightarrow \mathbf{R}$ be a weighting function. Then the *weighted possibilistic covariance* of $A$ and $B$ is defined by

$$\mathrm{Cov}_f(A, B) = \int_0^1 \left[\frac{a_2(\gamma)-a_1(\gamma)}{2}\right]\left[\frac{b_2(\gamma)-b_1(\gamma)}{2}\right] f(\gamma)d\gamma$$

Considering fuzzy numbers as possibility distributions we can prove the following relationship that represents a probability-like theorem in a pure possibilistic environment (Fullér and Majlender 2003):

**Theorem 2.** Let $A, B \in \mathscr{F}$ be two fuzzy numbers, and let $\lambda, \mu \in \mathbf{R}$ be real numbers. Then

$$\mathrm{Var}_f(\lambda A + \mu B) = \lambda^2 \mathrm{Var}_f(A) + \mu^2 \mathrm{Var}_f(B) + 2|\lambda\mu|\mathrm{Cov}_f(A, B)$$

holds, where addition and multiplication by a scalar of fuzzy numbers is defined by the sup-min extension principle Eq. (2).

# 3 Soft Decision Support for Financial Investments

## 3.1 Utility Function for Ranking Portfolios

The *mean-variance* methodology for the portfolio selection problem was originally developed by Markowitz (1952). Since its appearance, this approach has been considered as bedrock of the theory of finance, and many important advances in this field have been built on its formulation. By laying down the fundamental principles of the mean-variance methodology, Markowitz actually created one of the most important research fields in modern finance theory (Xia et al. 2000). The key principle of its modeling approach is to use

- the *expected return* of a portfolio as the *investment return*, and
- to use the *variance* of the expected returns of the portfolio as the *investment risk*.

In the following, we will refer to and use the framework of (Alleman and Noam 1999). We shall assume that each investor can assign some value of welfare, or *utility score*, to competing investment portfolios based on the expected return and risk of those portfolios. We will view the utility score as a means of ranking portfolios. Higher (lower) utility values are assigned to portfolios with more (less) attractive risk-return profiles. One reasonable function, which is frequently used in the theory of finance, assigns a risky portfolio $P$ with a risky rate of return $r_P$, an expected rate of return $E(r_P)$, and a variance of the rate of returns $\sigma^2(r_P)$ the following utility score

$$U(P) = E(r_P) - 0.005 \times A \times \sigma^2(r_P) \tag{5}$$

where $A$ stands for a parameter associated with the investor's risk aversion (see Alleman and Noam 1999 for details). This parameter depends only on the attitude of the investor towards future financial risk, and it is essentially determined by the preference relation he or she has in mind when considering different possibilities of financial investments. We note that this parameter was statistically approximated as $A \approx 2.46$ for an average investor in the United States.

**Remark 6.** We can readily estimate risk aversion parameter $A$ of a particular investor by an interview, where we set several appropriately chosen pairs of portfolios with different risk factors against each other, and let the investor select one from each pair. After getting the answers, we only need to carry out a simple analysis to approximate the value of $A$. From mathematical point of view, this method can be formulated as follows. Let $P$ and $Q$ be two portfolios with rates of return $r_P$ and $r_Q$, respectively. Then, an investor with value of risk averseness $A$ *prefers* $r_P$ to $r_Q$, denoted by $r_Q <_A r_P$, if and only if

$$U(Q) = E(r_Q) - 0.005 \times A \times \sigma^2(r_Q) < E(r_P) - 0.005 \times A \times \sigma^2(r_P) = U(P)$$

that is,

$$A < 200 \times \frac{E(r_P) - E(r_Q)}{\sigma^2(r_P) - \sigma^2(r_Q)}$$

318     P. Majlender

In Eq. (5), the factor of 0.005 is a scaling convention that allows us to represent the expected return and standard deviation as percentages rather than decimals or absolute values. Eq. (5) is consistent with the idea of the notion of utility value: utility is enhanced by high expected returns and diminished by high risk.

**Remark 7.** Having a risk-free investment means that $r_P \equiv r_f$ is constant under any circumstances, which implies that $E(r_P) = r_f$ and $\sigma(r_P) = 0$. In this case $U(P) = E(r_P) = r_f$, that is, the utility of the risk-free portfolio equals to its safe rate of return.

Since we can compare utility values with the rate offered on risk-free investments when choosing between a risky portfolio and a safe one, we will interpret a portfolio's utility value as its *certainty equivalent* rate of return to an investor. That is, the certainty equivalent rate of a portfolio is the rate that risk-free investments need to offer with certainty to be considered as equally attractive as the risky portfolio. Now we can say that a portfolio is preferred by an investor with a risk aversion parameter $A$ if its certainty equivalent return exceeds that of the risk-free alternative. That is, from Remark 5 – Remark 6 we have that for a risky portfolio $P$ with rate of return $r_P$ and a risk free portfolio with safe rate of return $r_f$

$$r_f <_A r_P \Leftrightarrow r_f < U(P) \Leftrightarrow A < 200 \times \frac{E(r_P) - r_f}{\sigma^2(r_P)}$$

holds.

In the following, let us formulate the portfolio selection problem in the framework of mean-variance methodology. Let us assume that we have $n$ number of available securities or assets to invest in. Let $x_i$ be the proportion of capital we invest in security $i$, and let $r_i$ denote the risky rate of return we have on security $i$, $i = 1, \ldots, n$. Let us denote the portfolio and its rate of return by $P$ and $r_P$, respectively, and let $r_f$ be the rate of return on a risk-free asset. In these circumstances, the total rate of return on the portfolio equals to the weighted average of the individual rates of return on the individual securities, i.e.

$$r_P = \sum_{i=1}^{n} r_i x_i$$

Furthermore, the portfolio is desirable for the investor if and only if

$$r_f < U\left( \sum_{i=1}^{n} r_i x_i \right)$$

Following the mean-variance methodology, the problem of optimal portfolio selection can be formulated as the following quadratic mathematical programming problem (see Alleman and Noam 1999, Markowitz 1952):

$$\max \ U\left( \sum_{i=1}^{n} r_i x_i \right) = E\left( \sum_{i=1}^{n} r_i x_i \right) - 0.005 \times A \times \sigma^2\left( \sum_{i=1}^{n} r_i x_i \right)$$

$$\text{s.t.} \ \sum_{i=1}^{n} x_i = 1, \quad x_i \geq 0, i = 1, \ldots, n.$$

**Remark 8.** We consider the portfolio selection problem with the *non-negativity* condition. This means that we are not interested in portfolios which involve *short* *selling* of some of its assets. In case of short selling, the investor borrows some securities from its broker with the condition of returning them in some future time. Then, to increase its purchasing power the investor sells those securities, and uses the extra income to buy other assets. In general, the investor holds *short position* in securities that it sold after borrowing them, and the investor holds *long position* in assets that it bought using its own capital. It can readily be seen that by opening a long or short position in a certain asset, the investor speculates for the rise or fall of the price of that asset, respectively.

In the following, we shall assume that the rates of return on the securities are modeled by *possibility distributions* rather than *probability distributions*.

## 3.2 Possibility Theory in Portfolio Selection

In this section, drawing on the results of (Alleman and Noam 1999), we shall describe the significance of considering possibility distributions in the portfolio selection problem, and present the fundamentals of developing the framework of *possibilistic portfolio selection*. Our approach is based on the methodology of employing possibility distributions for estimating future rates of return on financial assets.

**Remark 9.** In classical portfolio theories, the future rates of returns are represented by probability distributions. They are usually defined by *discrete random variables*, where we define the probability of each scenario with the probable future price of each asset. In this case we determine the probability distributions of the random variables by using statistical methods.

In our framework, we represent the rate of return on the $i$-th security by a fuzzy number $r_i$, and we will interpret $r_i(t)$, $t \in \mathbf{R}$, as the *degree of possibility* of the statement

$$\text{"$t$ will be the rate of return on the $i$-th security"}.$$

In our method we will only consider possibility distributions of *trapezoidal form* (see Eq. (1)). Accepting this assumption, we can reduce the complexity of the formulation and calculation we need to carry out when implementing the possibilistic version of the portfolio selection problem. Trapezoidal fuzzy numbers have linear membership functions, which make them easy to represent, compute with and interpret in real-world applications. They can only be characterized by 4 parameters, and yet have the capabilities of defining the quantities associated with future rates of return on financial assets. Nevertheless, we can readily extend our method to include the case of possibility distributions of any type.

The notion of trapezoidal fuzzy number will play fundamental role in our approach. We recall that a trapezoidal fuzzy number with tolerance interval $[a, b]$ can be seen as a context-dependent description of the statement

$$\text{"$t$ is \textit{approximately} in the interval $[a, b]$"},$$

where $\alpha$ and $\beta$ define the context. Hence, by the appropriate choice of the left width $\alpha \geq 0$ and right width $\beta \geq 0$ of a trapezoidal possibility distribution the investor will be able to describe his or her subjective expectations about the rate of returns on each security. Furthermore, those expectations will be incorporated in the framework of possibilistic portfolio selection.

In standard portfolio theories uncertainty is associated with *randomness*, and the financial risk represented by this type of uncertainty is implied by some random process. In this case we have *incomplete* information with respect to the future, and use information from the past to forecast and estimate events in the future. However, we have to be aware of the fact that estimates of future prices or rates of return actually combine both

- objectively observable and
- testable random events

with subjective judgments of the decision maker into probability assessments. As of a pure approach to the fields of finance and statistics, we can accept the use of probability theory to deal with observable random events, but we have to deny the transformation of subjective judgments to probabilities.

The use of probabilities has another drawback. The probabilities assigned to future events give an image of precision that is actually unmerited. As we have found in several cases in the area of portfolio selection, both theorists and practitioners neglect the proper address of the following issue: they usually make the assignment of probabilities through some very rough, subjective estimates, and then carry out the subsequent calculations with a precision of several decimal points. This shows that their routine of using probabilities may not be an appropriate choice. The actual meaning of the results of an analysis obtained in this way might be unclear, and its interpretation can be problematic. Furthermore, since here two types of errors, namely

- errors associated with *incompleteness of information*, which is due to the subjective estimates of the rates of return, and
- errors associated with *imprecision of information*, which is due to computation,

are mixed, results with serious errors can also be accepted at face value.

In standard portfolio theory the decision maker assigns *utility values* to outcomes or consequences, where the outcomes are defined by the combinations of possible actions and random events. Utility theory builds on the decision maker's relative preferences with respect to artificial lotteries. Viewing portfolios as lotteries, where the investor can stake on and against certain financial assets with certain present but uncertain future prices, we can link the issue of portfolio choices to the framework of *von Neumann-Morgenstern* axiomatic utility theory. However, in most of real-world applications, the straightforward use of utility theory turned out to be problematic. In fact, by employing pure probability-based utility theory, we have to face the fact that our model can involve some inconsistencies. From practical point of view having these types of inconsistencies is actually a more serious problem than having axiomatic problems. The main reasons behind the inconsistencies when working with utility scores are the following:

- we cannot validate utility measures inter-subjectively;
- we cannot validate the consistency of utility measures across events or contexts for the same subject;

## Soft Decision Support Systems for Evaluating Real and Financial Investments    321

- in empirical tests, utility measures can show discontinuities (see Nguyen 1978 for details), which should not happen with rational decision makers if the axiomatic foundation is correct; and finally,
- utility measures are artificial, and thus difficult to use on an intuitive basis.

As the combination of utility theory with statistical approaches that are based on pure probabilistic assessments has these unresolved limitations, we shall explore another type of approach. That is, based on recent developments originally presented in (Alleman and Noam 1999), we will consider the use of *possibility theory* as a substituting conceptual framework for the stochastic aspect of utility theory.

Several approaches have been formulated in the literature to enhance classical probability-based portfolio selection procedures through the advancements of fuzzy logic. In 1997, Watada proposed a fuzzy portfolio selection model, where he used fuzzy numbers to characterize the decision maker's aspiration levels with respect to (i) the expected rate of return on an asset and (ii) a certain degree of risk involved in holding that asset (Watada 1997). As of another aspect of this issue, in 2000, Inuiguchi and Tanino introduced a novel mathematical programming approach to the portfolio selection problem that is based on the application of possibility distributions (Inuiguchi and Tanino 2000). That approach is defined by the principle of diversifying portfolios, i.e. it is based on searching distributive investment solutions. Mathematically, it employs a technique, called *minimax regret criterion*, which is about to compute the level of regret that the decision maker is ready to undertake (Inuiguchi and Tanino 2000).

However, in many important cases it might be easier to estimate the possibility distributions of the rates of return on the financial assets than to assess the corresponding probability distributions.

### 3.3  A Possibilistic Approach to the Portfolio Selection Problem

In the following, we shall formulate and algorithmically solve the possibilistic portfolio selection problem. Let us consider the portfolio selection problem with possibility distributions:

$$\max \quad U\left(\sum_{i=1}^{n} r_i x_i\right) = E\left(\sum_{i=1}^{n} r_i x_i\right) - 0.005 \times A \times \sigma^2\left(\sum_{i=1}^{n} r_i x_i\right) \tag{6}$$
$$\text{s.t.} \quad \sum_{i=1}^{n} x_i = 1, \quad x_i \geq 0, i = 1, \ldots, n,$$

where $r_i = (a_i, b_i, \alpha_i, \beta_i)$ are fuzzy numbers of trapezoidal form that represent the risky rate of return on security $i$ (as subjectively assessed by the investor) for $i = 1, \ldots, n$.

**Remark 10.** Here, all $r_i \in \mathcal{F}$, $i = 1, \ldots, n$ can take negative values as well, i.e. their supports *can intersect* into the set of negative numbers. In other words, all financial assets can yield negative rates of return with positive degree of possibility. The only restriction that applies in this representation is that the support of any fuzzy number $r_i$ cannot contain any value that is less than $-100\%$, because we can never

322    P. Majlender

lose more money on the portfolio than the total amount of our original investment. Here we particularly use our earlier assumption that *short positions* in the assets are not allowed.

It is easy to verify that

$$E\left(\sum_{i=1}^{n} r_i x_i\right) = \sum_{i=1}^{n}\left[\frac{a_i + b_i}{2} + \frac{\beta_i - \alpha_i}{6}\right] x_i$$

and

$$\sigma^2\left(\sum_{i=1}^{n} r_i x_i\right) = \left(\sum_{i=1}^{n}\left[\frac{b_i - a_i}{2} + \frac{\alpha_i + \beta_i}{6}\right] x_i\right)^2 + \left(\sum_{i=1}^{n}\frac{\alpha_i + \beta_i}{\sqrt{72}} x_i\right)^2$$

hold. Let us introduce the following notations:

$$\begin{aligned}
u_i &= \frac{a_i + b_i}{2} + \frac{\beta_i - \alpha_i}{6} \\
v_i &= \sqrt{0.005 \times A} \times \left(\frac{b_i - a_i}{2} + \frac{\alpha_i + \beta_i}{6}\right) \\
w_i &= \sqrt{0.005 \times A} \times \frac{\alpha_i + \beta_i}{\sqrt{72}}.
\end{aligned}$$  (7)

Then, we can represent the $i$-th asset by a triplet $(v_i, w_i, u_i) \in \mathbf{R}^3$, where $u_i$ denotes the possibilistic expected value, and $v_i^2 + w_i^2$ stands for the possibilistic variance (multiplied by the constant $0.005 \times A$) of the future rate of return on asset $i$, $i = 1, \ldots, n$. In general, we consider two assets to be *indistinguishable* or *identical* in the framework of mean-variance analysis if they have the same expected value and variance. That is, we assume that for any $1 \leq i, j \leq n$, $i \neq j$ either $u_i \neq u_j$ or $v_i^2 + w_i^2 \neq v_j^2 + w_j^2$ holds.

In these circumstances, we will state the possibilistic portfolio selection problem in Eq. (6), as

$$\max \ \langle \mathbf{u}, \mathbf{x}\rangle - \langle \mathbf{v}, \mathbf{x}\rangle^2 - \langle \mathbf{w}, \mathbf{x}\rangle^2 = \left(\sum_{i=1}^{n} u_i x_i\right) - \left(\sum_{i=1}^{n} v_i x_i\right)^2 - \left(\sum_{i=1}^{n} w_i x_i\right)^2$$  (8)

$$\text{s.t.} \ \sum_{i=1}^{n} x_i = 1, \quad x_i \geq 0, i = 1,\ldots,n.$$

In the following, we will reformulate mathematical programming problem in Eq. (8) and present it of a particular form so that we can readily develop a polynomial algorithm for solving it. Let

$$T = \text{conv}\left\{(v_i, w_i, u_i) : i = 1,\ldots,n\right\}$$

$$= \left\{\left(\sum_{i=1}^{n} v_i x_i, \sum_{i=1}^{n} w_i x_i, \sum_{i=1}^{n} u_i x_i\right) : \sum_{i=1}^{n} x_i = 1, x_i \geq 0, i = 1,\ldots n\right\}$$

denote the *convex hull* of $\{(v_i, w_i, u_i) : i = 1, \ldots, n\}$. It is clear that $T \subset \mathbf{R}^3$ is a convex polyhedron. Then, Eq. (8) turns into the following 3-dimensional nonlinear programming problem

$$\max \quad -(v^2 + w^2 - u)$$
$$\text{s. t.} \quad (v, w, u) \in T,$$

or equivalently,

$$\min \quad f(v, w, u) = v^2 + w^2 - u \qquad (9)$$
$$\text{s. t.} \quad (v, w, u) \in T$$

where $T$ is a *compact* and *convex* subset of $\mathbf{R}^3$, and the implicit function

$$g_c(v, w) = f(v, w, c) = v^2 + w^2 - c$$

is *strictly convex* for any $c \in \mathbf{R}$. This means that any optimal solution to Eq. (9) must be on the *boundary* of $T$.

In what follows, we shall formulate the idea of an algorithm for finding an optimal solution to the constrained optimization problem formulated in Eq. (8). Since $T$ is a compact and convex polyhedron of $\mathbf{R}^3$, and all optimal solutions to Eq. (9) have to be on the boundary of $T$, we have that any optimal solution can be obtained as a *convex combination* of *at most* 3 vertices of $T$. In the algorithm, by lifting the non-negativity conditions for investment proportions $\mathbf{x} \geq \mathbf{0}$ we shall distinguish the following cases and calculate:

- the exact solutions to all conceivable 3-asset problems with non-collinear assets,
- the exact solutions to all conceivable 2-asset problems with distinguishable assets, and
- the utility value of all individual assets.

From these cases we consider the solutions with non-negative weights, i.e. all solutions that represent feasible solutions to the original problem in Eq. (6) as well, and compare their utility values. Portfolios with the highest utility value will be the optimal solutions to the portfolio selection problem in Eq. (8). The algorithm based on this approach will require $O(n^3)$ steps, where $n$ is the number of available securities.

**Remark 11.** Technically, we find the exact solutions to Eq. (9) as follows: Considering cases (i) and (ii) we formulate the problem for 3 and 2 assets *without* the non-negativity condition, respectively, and after solving the particular problem analytically, we verify its optimality by using the Kuhn-Tucker sufficiency conditions for optimality (see Chankong and Haimes 1983) for technical details). As of case (iii), it is only a simple procedure consisting of $n$ number of substitutions.

If we drop the non-negativity condition $\mathbf{x} \geq \mathbf{0}$ in Eq. (8), we call it the *relaxed* version of the possibilistic portfolio selection model.

**Remark 12.** Following our methodology based on (i) – (iii), we solve the possibilistic portfolio selection problem for 2 and 3 assets by dropping the non-negativity condition. However, as we have pointed out in Remark 7, accepting solutions with *not-necessarily non-negative weights*, we actually allow the case of *short selling* of securities.

324     P. Majlender

### 3.4 Algorithm for Solving the Possibilistic Portfolio Selection Problem

In this section, following the framework of (Carlsson et al. 2002), we shall present an algorithm for finding the optimal solutions to the $n$-asset possibilistic portfolio selection problem in Eq. (8). The algorithm will terminate in $O(n^3)$ steps. Furthermore, we note that

- if we are only interested in finding *one* optimal solution then the algorithm needs constant (i.e. $O(1)$) memory, while
- if we seek to find *all* optimal solutions then the algorithm requires $O(n^3)$ memory.

**Step 1**   Let $z = +\infty$ and $\mathbf{x} = [x_1, x_2, ..., x_n] = \mathbf{0} = [0, 0, ..., 0]$.
**Step 2**   Select three points from the set $\{(v_i, w_i, u_i) \in \mathbf{R}^3 : i = 1, ..., n\}$, which have not been considered yet. If there are no such points then go to **Step 9**, otherwise let these three points be $(v_k, w_k, u_k)$, $(v_l, w_l, u_l)$ and $(v_m, w_m, u_m)$.
**Step 3**   Let

$$s_1 = v_k - v_m, \quad s_2 = v_l - v_m$$

and

$$t_1 = w_k - w_m, \quad t_2 = w_l - w_m.$$

If

$$\det \begin{bmatrix} s_1 & t_1 \\ s_2 & t_2 \end{bmatrix} = s_1 t_2 - s_2 t_1 = 0$$

then go to Step 2, otherwise go to Step 4.
**Step 4**   Determine the optimal solution $[x_k^*, x_l^*, x_m^*]$ to the relaxed 3-asset possibilistic portfolio selection problem with assets $k$, $l$ and $m$ by using the formulas

$$\begin{bmatrix} x_k^* \\ x_l^* \end{bmatrix} = \frac{1}{(s_1 t_2 - s_2 t_1)^2} \times \begin{bmatrix} s_2^2 + t_2^2 & -(s_1 s_2 + t_1 t_2) \\ -(s_1 s_2 + t_1 t_2) & s_1^2 + t_1^2 \end{bmatrix}$$

$$\times \begin{bmatrix} 1/2(u_k - u_m) - s_1 v_m - t_1 w_m \\ 1/2(u_l - u_m) - s_2 v_m - t_2 w_m \end{bmatrix},$$

$$x_m^* = 1 - x_k^* - x_l^*.$$

**Step 5**   If $[x_k^*, x_l^*, x_m^*] \in (0, 1)^3$, i.e. $x_j^* > 0$ for $j = k, l, m$, then go to **Step 6**, otherwise go to **Step 2**.
**Step 6**   Let $z_{\{k, l, m\}}$ be the optimal value of the 3-asset portfolio selection problem with assets $k$, $l$ and $m$, i.e.

$$z_{\{k,l,m\}} = (v_k x_k^* + v_l x_l^* + v_m x_m^*)^2 + (w_k x_k^* + w_l x_l^* + w_m x_m^*)^2$$

$$- (u_k x_k^* + u_l x_l^* + u_m x_m^*).$$

If $z_{\{k, l, m\}} < z$ then go to **Step 7**, otherwise go to **Step 2**.

## Soft Decision Support Systems for Evaluating Real and Financial Investments    325

**Step 7**  Let $z = z_{\{k, l, m\}}$, and let

$$x_j = \begin{cases} x_j^* & \text{if } j = k, l, m, \\ 0 & \text{otherwise,} \end{cases}$$

that is,

$$\mathbf{x} = \left[0, \ldots, 0, x_k^*, 0, \ldots, 0, x_l^*, 0, \ldots, 0, x_m^*, 0, \ldots, 0\right]$$

**Step 8**  Go to **Step 2**.

**Step 9**  Select two points from the set $\{(v_i, w_i, u_i) \in \mathbf{R}^3 : i = 1, \ldots, n\}$, which have not been considered yet. If there are no such points then go to **Step 16**, otherwise let these two points be $(v_k, w_k, u_k)$ and $(v_l, w_l, u_l)$.

**Step 10**  If

$$(v_k - v_l)^2 + (w_k - w_l)^2 = 0$$

then go to Step 9, otherwise go to Step 11.

**Step 11**  Calculate the optimal solution $[x_k^*, x_l^*]$ to the relaxed 2-asset possibilistic portfolio selection problem with assets $k$ and $l$ by applying the equations

$$x_k^* = \frac{1}{(v_k - v_l)^2 + (w_k - w_l)^2} \times \left[1/2(u_k - u_l) - (v_k - v_l)v_l - (w_k - w_l)w_l\right],$$

$$x_l^* = 1 - x_k^*.$$

**Step 12**  If $[x_k^*, x_l^*] \in (0, 1)^2$, i.e. $x_j^* > 0$ for $j = k, l$, then go to **Step 13**, otherwise go to **Step 9**.

**Step 13**  Let $z_{\{k, l\}}$ denote the optimal value of the 2-asset portfolio selection problem with assets $k$ and $l$, i.e.

$$z_{\{k,l\}} = (v_k x_k^* + v_l x_l^*)^2 + (w_k x_k^* + w_l x_l^*)^2 - (u_k x_k^* + u_l x_l^*).$$

If $z_{\{k, l\}} < z$ then go to Step 14, otherwise go to Step 9.

**Step 14**  Let $z = z_{\{k, l\}}$, and let

$$x_j = \begin{cases} x_j^* & \text{if } j = k, l, \\ 0 & \text{otherwise,} \end{cases}$$

that is,

$$\mathbf{x} = \left[0, \ldots, 0, x_k^*, 0, \ldots, 0, x_l^*, 0, \ldots, 0\right]$$

**Step 15**  Go to **Step 9**.

**Step 16**  Select a point from the set $\{(v_i, w_i, u_i) \in \mathbf{R}^3 : i = 1, \ldots, n\}$, which have not been considered yet. If there is no such a point then go to **Step 20**, otherwise let this point be $(v_k, w_k, u_k)$.

**Step 17**  Let $z_{\{k\}}$ denote the utility value of the portfolio consisting of the single asset $k$, that is,

$$z_{\{k\}} = v_k^2 + w_k^2 - u_k.$$

If $z_{\{k\}} < z$ then go to Step 18, otherwise go to Step 16.

**Step 18** Let $z = z_{\{k\}}$, and let

$$x_j = \begin{cases} 1 & \text{if } j = k, \\ 0 & \text{otherwise,} \end{cases}$$

that is,

$$\mathbf{x} = \left[0,\dots,0, x_k^*, 0,\dots,0\right]$$

**Step 19** Go to **Step 16**.

**Step 20** Vector $\mathbf{x}$ represents an optimal solution with $-z$ being the optimal value of the possibilistic portfolio selection problem in Eq. (8).

As a conclusion from a financial point of view, we can also find the following result:

**Remark 13.** Concerning an investor whose degree of risk-aversion is equal to $A$, the optimal risky portfolio will be defined by the decision vector x obtained from the algorithm. However, the portfolio will only be preferred and thus constructed, if its expected return $-z$ (which is derived with respect to a utility score incorporating the risk-aversion factor of the investor) is *superior* to the return on the risk-free investment, i.e. $r_f < -z$.

## 4 Soft Decision Support for Strategic Investment Planning

### 4.1 Practical Background

For very large industrial investments we have learned that these kind of industrial activities require a detailed in-depth understanding of the models and methodologies that are applied, since even small deviations in cash flows estimations can cause large capital gains or losses in the prediction. In the following we shall refer to investment opportunities of large scales as *giga-investments*.

To evaluate physical giga-investment opportunities, Carlsson and Fullér (2003) presented a new aspect of real option valuation based on the use of possibility distributions. By considering possibilistic uncertainty as an alternative to probabilistic uncertainty they threw new light upon the classical Black-Scholes formula for real option valuation. Drawing on the principles introduced in (Carlsson and Fullér 2003), we shall generalize its results and propose an extended methodology for the application of fuzzy real options (Majlender 2003).

The motivations for and challenges in managing giga-investments are the following:

- Giga-investments are very large investments with cash flows of size 250 million Euros that usually have long economic lives with 10-35 years.
- The long economic life implies that it becomes increasingly difficult with time to estimate future cash flows in a reliable manner.
- When the lifetime of an investment is long, the fundamentals (markets, products, technologies, etc.) can also change significantly.
- Methods that use precise cash flows estimates become unrealistic and misleading when projects with long economic lives are assessed.

- It is not realistic to assume that managers would remain idle as changes happen – they can act and take actions. The possibility of taking these actions is called *managerial flexibility*.
- It is important to take the right actions at the *optimal time* – the theory of *real options* is a tool that offers solutions for timing.
- Real options take into account the deferral flexibility of investments and provide support with optimal timing. However, it builds on the same precise estimates as the classical capital budgeting methods like *Net Present Value* and *Discounted Cash Flows Analysis*.

Considering the properties and features of giga-investments, it is clear that they essentially differ from ordinary (small or medium size) investment opportunities; furthermore they differ even more from financial investments. Hence, evaluating very large investments requires a different underlying theory than the theory used in classical valuation methods.

There are two main problems arise if we are to use classical investment valuation methods for giga-investment planning:

1. First, the estimates representing the intrinsic value of the investment project are derived in a precise manner, that is, we obtain clear-cut values from the system. Since calculations are carried out on exact values and quantities, there is no uncertainty in the result. However, this can cast doubts on the applicability of the classical models for giga-investments, since there is a large amount of uncertainty that is involved and need to be taken into consideration in future cash flows estimates.

2. Second, and more importantly, classical probability-based investment valuation methods model future uncertainty according to the probabilistic assumption that prices are following stochastic processes. However, it is clear that by implementing a very large investment we can and eventually will shift market and industry conditions on the whole. Changing the *economic equation* in this manner is not included in the financial models which are used in a stochastic environment. Furthermore, since the change of circumstances in industry involves uncertainty, which is non-statistical by nature, it cannot be characterized or modeled by stochastic processes.

In the following, drawing on the results of (Majlender 2003, Majlender 2004), and the capability of our developed Excel-based platform presented in (Majlender 2002), we shall introduce a soft decision support model for giga-investment evaluation, where the expected cash flows and expected costs are estimated by possibility distributions. Proposing an enhanced methodology, we shall extend the traditional investment evaluation approaches by presenting a dynamic model that takes into consideration not only the deferral or cancellation flexibility of the projects, but also the presence of vague information that represents non-statistical uncertainty, which can play a crucial role when long-range investment decisions are made.

## 4.2 Possibility Theory in Strategic Investment Planning

In this section we shall present the fundamentals of our approach to possibilistic investment evaluation (see Majlender 2002, 2003 for practical details). Of particular significance in this view is the development of a methodology to calculate the future return on real asset investments that are based on uncertainty of a non-stochastic nature. We shall enhance the power of decision making with the idea of fuzzy real option valuation by introducing possibilistic uncertainty distributions into the process of strategic investment planning.

Our approach to investment evaluation is drawn on the methodology of employing possibility distributions for estimating future expected cash flows and costs of investments. In our modeling approach we will only consider linear membership functions, that is, possibility distributions of trapezoidal form. We note that there is no critical obstacle to generalize the complete theory of possibilistic investment analysis to the case of possibility distributions of arbitrary types.

**Remark 14.** The possibility distributions of expected cash flows and expected costs could also be represented by nonlinear membership functions. However, from a computational point of view it is easier to use linear membership functions; moreover, our experience with several research projects on very large industrial investments shows that senior managers prefer trapezoidal fuzzy numbers to nonlinear fuzzy numbers when they estimate the uncertainties associated with future cash flows.

Hence, in the following we shall consider trapezoidal possibility distributions to characterize and model the uncertainty involved in future cash flows and costs estimates.

We recall that if $A \in \mathcal{F}$ is a fuzzy number and $t \in \mathbf{R}$ then in a possibilistic setting $A(t)$ denotes the degree of possibility of the statement

$$\text{``}t \text{ is the value of } A\text{''}.$$

Furthermore, a trapezoidal fuzzy number $A = (a, b, \alpha, \beta)$ with core $[a, b]$ can be seen as a context-dependent description of the property

$$\text{``the value of a real variable is } approximately \text{ in the interval } [a, b]\text{''},$$

where left width $\alpha$ and right width $\beta$ define the context.

**Remark 15.** For any $t \in \mathbf{R}$, $A(t)$ represents the degree of possibility of the statement

$$\text{``}t \text{ is in the interval } [a, b]\text{''}.$$

We can see that

- if $t \in [a, b]$, i.e. $t$ belongs to $[a, b]$, then $A(t) = 1$;
- if $t \notin (a - \alpha, b + \beta)$, i.e. $t$ is *too far* from $[a, b]$, then $A(t) = 0$; and otherwise
- $0 < A(t) < 1$ holds, i.e. $t$ is *close enough* to $[a, b]$.

Concerning the representation of the uncertainty associated with future cash flows a trapezoidal fuzzy number $A = (a, b, \alpha, \beta)$ with core $[a, b]$ can be interpreted as a context-dependent description of the property

$$\text{``the value of the expected cash flows is } approximately \text{ in the interval } [a, b]\text{''},$$

where left width $\alpha \geq 0$ and right width $\beta \geq 0$ define the context. Thus, by the appropriate choice of left and right widths of the possibility distribution we can describe our expectations about future uncertainty of cash flows. Furthermore, formulated in this way those expectations will be incorporated into the theory, and we are able to introduce several kinds of qualitative behavior. For instance, by increasing (decreasing) $\beta$ we indicate the positive (negative) potential of beating the estimates, i.e. performing beyond our expectations, while by increasing (decreasing) $\alpha$ we express the positive (negative) potential of underachieving our estimates. In particular, the increase (decrease) in both parameters $\alpha$ and $\beta$ implies the extension (suppression) of future potential in both upward and downward directions, which in turn means the extension (suppression) of future uncertainty.

### 4.3 Fuzzy Net Present Value Analysis

In traditional investment approaches investment activities or projects are often seen as *now or never*, and the main question is whether to undertake an investment – a *yes-or-no* decision (Alleman and Noam 1999). Formulated in this way it is very hard to make a decision when there is uncertainty about the exact outcome of the investment. In order to help solve these tough decision-making problems valuation methods such as *Net Present Value (NPV)* or *Discounted Cash Flow (DCF)* have been developed.

However, these methods ignore the value of flexibility of postponing the investment opportunity as well as the possible effects of learning about the external uncertainties (i.e. potential gains and profits) involved in the investment. In particular, due to the uncertainty that NPV cannot take into consideration, many interesting and innovative investment activities and projects were cancelled. Despite those facts, the method of NPV is widely accepted and employed, since its methodology does not require sophisticated assumptions, and more importantly, its mathematics is straightforward, and the formulas can readily be interpreted in physical investments.

The main question now that an investor or a decision maker faces for a prompt (i.e. non-deferrable) investment opportunity is the following:

*Should we undertake the investment, or do we have to pass it up?*

With the help of the NPV analysis we can evaluate an investment opportunity, which we may immediately enter into or have to abandon. We can compute the exact value of that project according to our forecast and make a decision whether it is worthwhile to undertake it or not. The data needed are the expected cash flows for each year of the project $V_i$, the investment costs $X$ and the required rate of return on the investment $r$. The expected annual cash flows give the annual operational profit, which are actually the difference between operational revenues and operational costs at a certain year of the project. To derive the total operating assets of the investment that we expect to acquire we aggregate the expected annual cash flows with the discounting factor $(1 + r)$ as

$$S_0 = \sum_{i=0}^{L} \frac{V_i}{(1+r)^i}$$

where $L$ denotes the length of the investment activity. The project's internal rate of return $r$ implicitly contains the degree of the investor's or decision maker's risk aversion. Setting $r$ higher means that the underlying investment is thought more risky, because higher future cash flows are needed to reach the same aggregated income. The

330    P. Majlender

investment costs $X$ is a one-time cost, which should be paid at the beginning of the project to be able to enter into it. The value of the underlying investment is

$$NPV = S_0 - X = \sum_{i=0}^{L} \frac{V_i}{(1+r)^i} - X$$

and the decision rule is obvious: That is, if $NPV > 0$ then we should enter the investment, otherwise we have to abandon it.

Usually, the expected cash flows and expected investment costs cannot be characterized by single numbers. We can, however, estimate these quantities by using possibility distributions, i.e. fuzzy numbers. As we have already discussed, we shall only consider trapezoidal possibility distributions. Hence, in each year we estimate the expected cash flows by using a trapezoidal fuzzy number of the form

$$V_i = (a_i, b_i, \alpha_i, \beta_i)$$

for $i = 0, 1, ..., L$. That is, the most possible values of the expected cash flows at year $i$ of the project lie in the interval $[a_i, b_i]$ (which is the core of the trapezoidal fuzzy number $V_i$), and $(b_i + \beta_i)$ is the upward potential and $(a_i - \alpha_i)$ is the downward potential for the expected cash flows at year $i$.

In a similar manner we can estimate the expected investment costs by using a trapezoidal possibility distribution of the form

$$X = (x_1, x_2, \alpha', \beta')$$

i.e. the most possible values of the expected investment costs lie in the interval $[x_1, x_2]$, and $(x_2 + \beta')$ is the upward potential and $(x_1 - \alpha')$ is the downward potential for the expected investment costs.

**Definition 10.** In terms of the concepts we have introduced we suggest the use of the following formula for computing *fuzzy net present values*:

$$FNPV = S_0 - X = \sum_{i=0}^{L} \frac{V_i}{(1+r)^i} - X \tag{10}$$

where the addition and multiplication by scalar of fuzzy numbers are defined by the sup-min extension principle.

Expanding Eq. (10) we have

$$FNPV = \sum_{i=0}^{L} \frac{(a_i, b_i, \alpha_i, \beta_i)}{(1+r)^i} + (-x_2, -x_1, \beta', \alpha')$$
$$= \left( \sum_{i=0}^{L} \frac{a_i}{(1+r)^i} - x_2, \sum_{i=0}^{L} \frac{b_i}{(1+r)^i} - x_1, \sum_{i=0}^{L} \frac{\alpha_i}{(1+r)^i} + \beta', \sum_{i=0}^{L} \frac{\beta_i}{(1+r)^i} + \alpha' \right)$$

Notice that if every possibility distribution has a support containing only one point then it will be a real number, and we get back to the classical method of NPV analysis. Hence, the method of *Fuzzy Net Present Value (FNPV)* implicitly includes the framework of NPV analysis.

We have to make a decision whether to enter into or abandon the underlying project, therefore we need to decide if the trapezoidal possibility distribution *FNPV* is

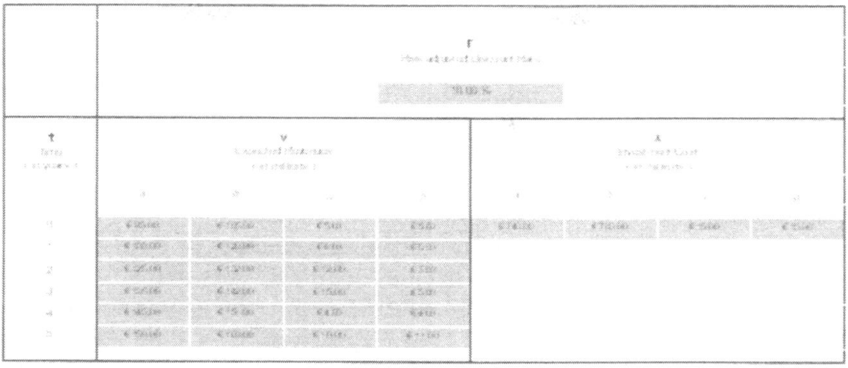

**Fig. 7.** Sample of a Fuzzy Net Present Value table (screenshot)

*greater than zero* or not. However, to define whether a fuzzy number is *greater than zero* is generally impossible (even if it is a trapezoidal one), since it would imply an ordering on fuzzy numbers, but (as it is known) even trapezoidal possibility distributions cannot be ranked universally.

However, we can employ some value function to be able to rank fuzzy net present values of trapezoidal form. In our computerized implementation we used the following value function $u: \mathscr{F} \to \mathbf{R}$: Having $FNPV = (c_1, c_2, \alpha, \beta)$, let the value of $FNPV$ be defined by

$$u(FNPV) = \frac{c_1 + c_2}{2} + r_A \times \frac{\beta - \alpha}{6}$$

where the parameter $r_A \geq 0$ denotes the degree of the investor's risk aversion. If $r_A = 1$ then the investor compares trapezoidal fuzzy numbers by comparing their possibilistic mean values. Furthermore, if $r_A = 0$ then the investor, neglecting the upward and downward potential of the resulting fuzzy net present value, is risk neutral and compares trapezoidal fuzzy numbers by comparing the center of their tolerance intervals.

**Remark 16.** Notice that in this approach the degree of the investor's risk aversion is taken into consideration in two different ways:
1. It is implicitly included in the parameter risk-adjusted discount rate $r$, which is related to future expected payoffs, and
2. it is considered when the final decision is made, i.e. whether to undertake the investment or not, which is decided by the uncertainties associated with the future cash inflows and outflows.

### 4.4 Fuzzy Real Option Valuation

In this section we shall consider investment activities which can be deferred or postponed for a certain period of time. These kinds of investment opportunities are called *real options*, and their idea is an important way of thinking about valuation and strategic decision-making. Increasingly, since very few physical investments are of the

332     P. Majlender

type now-or-never, the power of this approach is starting to change the *economic equation* of many industries.

Real options can be valued using the analogue option theories that have been developed for financial options, which is quite different from the traditional net present value or discounted cash flow investment approaches. In real option valuation the options involve some *real assets* as opposed to financial ones (Alleman and Noam 1999). However, the framework of option pricing theory can be employed for industrial investments in order to built an enhanced and extended valuation method. Since option pricing theories, rooted in and based upon the path-breaking Black-Scholes option pricing model, remain applicable when the uncertainty comes from sources other than the underlying security prices, it will make sense to employ the traditional financial valuation methods as a basis for evaluating investment activities (Hull 1997). Since the 1980s, financial option pricing methods have been utilized more extensively to evaluate the value of flexibility associated with physical investments. In fact, this extension in view has developed the notion of real options. Real options can simply be viewed as the management's opportunities to respond to the changing circumstances and fundamentals (e.g. markets, products and technologies) of an investment (Yeo and Qiu 2003). Since managers do not remain idle as changes happen but do take actions and make choices about the characteristics of the underlying project, this process of adaptability creates *embedded options.* And the presence of these options and opportunities raises the value of the project and in turn invalidate the traditional DCF-based investment approaches. Furthermore, those opportunities to change the behavior in a project and take some appropriate action in the future are rights but not obligations, which necessarily have some value (Dixit and Pindyck 1995).

The exact relationship giving the value of a deferrable investment opportunity is computed by the renown *Black-Scholes pricing formula* for a European call option, extended by Merton to dividend-paying stocks (see Black and Scholes 1973, Leslie and Michaels 1997, Merton 1973 for details):

$$C_0 = S_0 \times e^{-\delta T} N(d_1) - X \times e^{-r_f T} N(d_2)$$

where

$$d_1 = \frac{\ln(S_0 / X) + (r_f - \delta + \sigma^2 / 2)T}{\sigma\sqrt{T}}$$

$$d_2 = \frac{\ln(S_0 / X) + (r_f - \delta - \sigma^2 / 2)T}{\sigma\sqrt{T}} = d_1 - \sigma\sqrt{T}$$

and where $N(d)$ denotes the cumulative normal distribution function, that is, $N(d)$ equals the probability that a random draw from a standard normal distribution will be less than $d$:

$$N(d) = \int_{-\infty}^{d} \frac{1}{\sqrt{2\pi}} e^{-t^2 / 2} dt, \quad d \in \mathbf{R}$$

and the other variables in the frameworks of both the financial call options and the investment opportunities are summarized in the following table (Luehrman 1998):

The real option rule, derived from option pricing theory, is that we should invest today only if the net present value is high enough to compensate for giving up the value of the option to wait, i.e. the deferral flexibility. Since the option to invest loses

| Variable | Call option | Physical investment |
|----------|-------------|---------------------|
| $C_0$ | Price of the European call option | Value of the real option |
| $S_0$ | Stock price | Present value of the expected cash flows |
| $X$ | Exercise price | Nominal value of the investment costs |
| $\sigma$ | Standard deviation of the returns | Uncertainty of the expected cash flows |
| $T$ | Time to expiration | Time to maturity (in years) |
| $\delta$ | Dividends payed out | Value lost over the duration of the option |
| $r_f$ | Rate of return on a safe asset | Rate of return on a safe asset |

**Fig. 8.** Variables of the Black-Scholes formula and their correspondence between the frameworks of financial options and investment opportunities

its value when the investment is irreversibly made, this loss is an opportunity cost of investing.

The main question that an investor or a decision maker must answer for a deferrable investment opportunity is:

*How long do we postpone the investment, if we can postpone it up to T time periods?*

Following the approach of Benaroch and Kauffman (2000), we can apply the following decision rule for an optimal investment strategy: Where the maximum deferral time is $T$, we make the investment (exercise the option) at time $t^*$, $0 \le t^* \le T$, for which the value of the option, $ROV_{t^*}$, is positive and reaches its maximum value:

$$ROV_{t^*} = \max_{0 \le t \le T} ROV_t = S_0^{(t)} \times e^{-\delta t} N(d_1^{(t)}) - X \times e^{-r_f t} N(d_2^{(t)}) > 0 \qquad (11)$$

where

$$d_1^{(t)} = \frac{\ln(S_0^{(t)}/X) + (r_f - \delta + \sigma^2/2)t}{\sigma\sqrt{t}}$$

$$d_2^{(t)} = \frac{\ln(S_0^{(t)}/X) + (r_f - \delta - \sigma^2/2)t}{\sigma\sqrt{t}} = d_1^{(t)} - \sigma\sqrt{t}$$

Here, $S_0^{(t)}$ denotes the present value of the aggregated expected cash flows with deferral time $t$, which can be obtained from the appropriate NPV table that belongs to the case when we enter into the project after waiting $t$ years.

**Remark 17.** The method of real option valuation is a heuristic one, which only presents a benchmark of the value of a deferrable investment opportunity. Thus, we do not have to be precise by setting up all the NPV tables that corresponds to different time periods of undertaking. Instead, we can assume some reasonable relationship between them. Furthermore, since we shall apply the framework of real options to giga-investments (that especially have long economic lives), we can assume that the time scale is stepwise (and not continuous), and the investment activities can only be deferred by whole years.

We note that decision rule formulated in Eq. (11) has to be reapplied every time new information arrives during the deferral period to see how the optimal investment strategy might change in light of the new information.

As in the previous section, we can reason that the present value of expected cash flows cannot be characterized by a single number. However, we can estimate the present value of expected cash flows belonging to a deferral time $t$ by using a trapezoidal possibility distribution of the form

$$S_0^{(t)} = (s_1^{(t)}, s_2^{(t)}, \alpha^{(t)}, \beta^{(t)})$$

for $t = 0, 1, \ldots, T$, i.e. the most possible values of the present value of expected cash flows corresponding to deferral time $t$ lie in the interval $[s_1^{(t)}, s_2^{(t)}]$ (which is the core of the trapezoidal fuzzy number $S_0^{(t)}$), and $(s_2^{(t)} + \beta^{(t)})$ is the upward potential and $(s_1^{(t)} - \alpha^{(t)})$ is the downward potential for the present value of expected cash flows of the case when we wait $t$ years before going ahead with the investment.

In a similar manner, we can estimate the expected investment costs by using a trapezoidal possibility distribution of the form

$$X = (x_1, x_2, \alpha', \beta')$$

i.e. the most possible values of the expected investment costs lie in the interval $[x_1, x_2]$, and $(x_2 + \beta')$ is the upward potential and $(x_1 - \alpha')$ is the downward potential for the expected investment costs.

If the NPV tables of expected cash flows are given by trapezoidal possibility distributions then we can compute the value of $S_0^{(t)}$, $t = 0, 1, \ldots, T$ by aggregating the expected cash flows with the discounting parameter $r$. Considering the formulas for the addition and multiplication by scalar of fuzzy numbers, we can see that the result, $S_0^{(t)}$, will be a trapezoidal fuzzy number as well.

More precisely, let us consider an investment opportunity of $L$ years long with required rate of return $r$, which can be deferred up to $T$ years. In this case, we can apply the following approach:

**Definition 11.** Let us consider a physical investment opportunity, and let its future expected operating assets $\{V_0, V_1, \ldots, V_L\}$ and its required initial capital expenditure $X$ be characterized by trapezoidal possibility distributions. In these circumstances we propose the application of the following possibilistic decision rule for optimal investment strategy: Where the maximum deferral time is $T$, make the investment (exercise the real option) at time $t^*$, $t^* = 0, 1, \ldots, T$, for which the value of the fuzzy real option, $FROV_{t^*}$, is positive and attains its maximum value:

$$FROV_{t^*} = \max_{t=0,1,\ldots,T} FROV_t = S_0^{(t)} \times e^{-\delta t} N(d_1^{(t)}) - X \times e^{-r_f t} N(d_2^{(t)}) > 0$$

where

$$d_1^{(t)} = \frac{\ln(E_f(S_0^{(t)})/E_f(X)) + (r_f - \delta + \sigma_t^2/2)t}{\sigma_t \sqrt{t}}$$

$$d_2^{(t)} = \frac{\ln(E_f(S_0^{(t)})/E_f(X)) + (r_f - \delta - \sigma_t^2/2)t}{\sigma_t \sqrt{t}} = d_1^{(t)} - \sigma_t \sqrt{t}$$

and where $S_0^{(t)}$ denotes the trapezoidal possibility distribution representing the present value of the aggregated expected cash flows of the project that is postponed $t$ years, and

$$\sigma_t = \frac{\sigma(S_0^{(t)})}{E_f(S_0^{(t)})} = \frac{\sqrt{\text{Var}_f(S_0^{(t)})}}{E_f(S_0^{(t)})}$$

is the weighted possibilistic standard deviation about the growth rate of the value of the postponed project's operating cash inflows $S_0^{(t)}$.

Obviously, this decision rule has to be reapplied every time when new information arrives during the deferral period to be able to analyze how the optimal investment strategy should be changed in light of the new information.

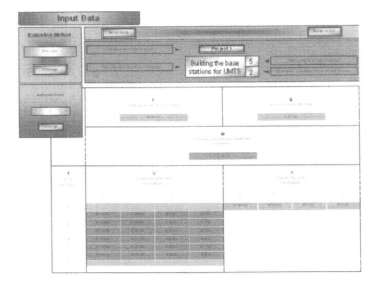

**Fig. 9.** Input data for Fuzzy Real Option Valuation (screenshot)

**Remark 18.** The model of *Fuzzy Real Option Valuation* (*FROV*) represents a hybrid (possibilistic-probabilistic) approach, since considerations of both possibility distributions and probability distributions are incorporated into its theory and methodology. Of particular significance of our representation is the capability to characterize the investment's *external uncertainties* (by means of probability theory with the ability to control and bias the complete real option valuation method by taking into account the investment's *internal uncertainties* (by means of possibility theory) as well.

We have to make a decision how long we should postpone the underlying investment up to $T$ time periods (years) before undertaking it, if the investment is worthwhile to go ahead with at all. Thus, we need to find a maximizing element from the set

$$\{\text{FROV}_0, \text{FROV}_1, \ldots, \text{FROV}_T\}.$$

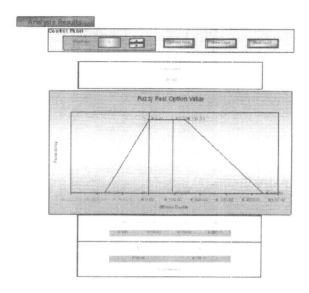

**Fig. 10.** The optimal fuzzy real option value (screenshot)

However, as we have seen in the case of evaluating fuzzy net present values, when we consider arbitrary fuzzy numbers it is generally impossible to decide which one is *the greatest* (even if they are of trapezoidal form), since even on the set of trapezoidal fuzzy numbers there does not exist any well-defined ordering.

However, we can use a value function to define ordering on the set of possibility distributions of trapezoidal form. In our computerized implementation we have employed the following value function $u: \mathscr{F} \to \mathbf{R}$ to define ranking on fuzzy real option values: For $FROV_t = (c_t^L, c_t^R, \alpha_t^*, \beta_t^*)$, $t = 0, 1, \ldots, T$, let

$$u(FROV_t) = \frac{c_t^L + c_t^R}{2} + r_A \times \frac{\beta_t^* - \alpha_t^*}{6}$$

where the parameter $r_A \geq 0$ stands for the degree of the investor's risk aversion. In particular, according to a decision maker with risk aversion parameter $r_A$, $FROV_r$ is preferred to $FROV_s$ if and only if $u(FROV_r)$ is greater than $u(FROV_s)$, i.e.

$$FROV_r \succ FROV_s \Leftrightarrow u(FROV_r) > u(FROV_s), \quad r,s = 0,1,\ldots,T.$$

Especially,

$$r_A = 0 \Rightarrow u(FROV_t) = \text{center}(\text{core}(FROV_t)) = \frac{c_t^L + c_t^R}{2},$$

$$r_A = 1 \Rightarrow u(FROV_t) = E_{2\times\text{id}}(FROV_t) = \frac{c_t^L + c_t^R}{2} + \frac{\beta_t^* - \alpha_t^*}{6}.$$

# References

Alleman, J., Noam, E.: The new investment theory of real options and its implication for telecommunications economics. Kluwer Academic Publishers, Boston (1999)

Balasubramanian, P., Kulatikala, N., Storck, J.: Managing information technology investments using a real-options approach. Journal of Strategic Information Systems 9, 39–62 (2000)

Benaroch, M., Kauffman, R.J.: Justifying electronic banking network expansion using real options analysis. MIS Quarterly 24, 197–225 (2000)

Black, F., Scholes, M.: The pricing of options and corporate liabilities. Journal of Political Economy 81, 637–659 (1973)

Bodie, Z., Kane, A., Marcus, A.J.: Investments. In: Times Mirror Higher Education Group, Irwin, Boston (1996)

Carlsson, C., Fullér, R.: On possibilistic mean value and variance of fuzzy numbers. Fuzzy Sets and Systems 122, 315–326 (2001)

Carlsson, C., Fullér, R., Majlender, P.: A possibilistic approach to selecting portfolios with highest utility score. Fuzzy Sets and Systems 131, 13–21 (2002)

Carlsson, C., Fullér, R.: A fuzzy approach to real option valuation. Fuzzy Sets and Systems 139, 297–312 (2003)

Chankong, V., Haimes, Y.Y.: Multiobjective Decision Making: Theory and Methodology. North-Holland, Amsterdam (1983)

Dixit, A.K., Pindyck, R.S.: The options approach to capital investment. Harvard Business Review 73, 105–115 (1995)

Dubois, D., Prade, H.: The mean value of a fuzzy number. Fuzzy Sets and Systems 24, 279–300 (1997)

Fullér, R., Majlender, P.: On weighted possibilistic mean and variance of fuzzy numbers. Fuzzy Sets and Systems 136, 363–374 (2003)

Goetschel, R., Woxman, W.: Elementary fuzzy calculus. Fuzzy Sets and Systems 18, 31–43 (1986)

Hull, J.C.: Options, futures and other derivatives. Prentice Hall, Englewood Cliffs (1997)

Inuiguchi, M., Tanino, T.: Portfolio selection under independent possibilistic information. Fuzzy Sets and Systems 115, 83–92 (2000)

Leslie, K.J., Michaels, M.P.: The real power of real options. The McKinsey Quarterly 3, 5–22 (1997)

Luehrman, T.A.: Investment opportunities as real options: getting started on the numbers. Harvard Business Review 76, 51–67 (1998)

Majlender, P.: Optimal timing for the exercise of real options. In: Proceedings of the International Real Option Workshop (2002)

Majlender, P.: Strategic investment planning by using dynamic decision trees, In: Proceedings of the 36th Hawaii International Conference on System Sciences (HICSS 1936), Island of Hawaii (2003)

Majlender, P.: A Normative Approach to Possibility Theory and Soft Decision Support, University Doctorate Dissertation, Åbo Akademi University (2004)

Markowitz, H.: Portfolio selection. Journal of Finance 7, 77–91 (1952)

Merton, R.: Theory of rational option pricing. Bell Journal of Economics and Management Science 4, 141–183 (1973)

Nguyen, H.T.: A note on the extension principle for fuzzy sets. Journal of Mathematical Analysis and Applications 64, 369–380 (1978)

Tversky, A.: Intransitivity of preferences. Psychological Review 76, 31–45 (1969)

Watada, J.: Fuzzy portfolio selection and its applications to decision making. Tatra Mountains Math. Publ. 13, 219–248 (1997)

Xia, Y., Liu, B., Wang, S., Lai, K.K.: A model for portfolio selection with order of expected returns. Computers and Operations Research 27, 409–422 (2000)

Yeo, K.T., Qiu, F.: The value of management flexibility – a real option approach to investment evaluation. Journal of Project Management 21, 243–250 (2003)

Zadeh, L.A.: Fuzzy sets. Information and Control 8, 338–353 (1965)

Zadeh, L.A.: Fuzzy sets as a basis for a theory of possibility. Fuzzy Sets and Systems 1, 3–28 (1978)

# Pricing Options, Forwards and Futures Using Fuzzy Set Theory

James J. Buckley[1] and Esfandiar Eslami[2]

[1] University of Alabama at Birmingham, Department of Mathematics Birmingham
[2] Fuzzy Systems and Applications Center of Excellence Shahid Bahonar University of Kerman

**Abstract.** Pricing of options, forwards or futures often requires using uncertain values of parameters in the model. For example future interest rates are usually uncertain. We will use fuzzy numbers for these uncertain parameters to account for this uncertainty. When some of the parameters in the model are fuzzy the price then also becomes fuzzy. We first discuss options: (1) the discrete binomial method; and then (2) the Black-Scholes model. Then we look at pricing futures and forwards.

## 1 Introduction

We first introduce out notation for this chapter. We place a ``bar" over a symbol to denote a fuzzy set. All our fuzzy sets will be fuzzy subsets of the real numbers so $\overline{M}$, $\overline{N}$, $\overline{A}$ are all fuzzy subsets of the real numbers. A triangular fuzzy number $\overline{N}$ is defined by three numbers $a < b < c$ where the base of the triangle is on the interval $[a,c]$ and its vertex is at $x = b$. We write $\overline{N} = (a/b/c)$ for triangular fuzzy number $\overline{N}$. The membership function of fuzzy number $\overline{N}$ evaluated at $x$ is written $\overline{N}(x)$ a number in $[0,1]$. A triangular shaped fuzzy number has base on an interval $[a,c]$, vertex at $x = b$, but the sides are curves not straight lines. We write $\overline{N} \approx (a/b/c)$ for a triangular shaped fuzzy number $\overline{N}$. A trapezoidal fuzzy number $\overline{M}$ is defined by four numbers $a < b < c < d$ where the base of the trapezoid is on $[a,d]$ and the top is over the interval $[b,c]$. We write $\overline{M} = (a/b,c/d)$ for trapezoidal fuzzy number $\overline{M}$. If the sides are curved we write $\overline{M} \approx (a/b,c/d)$ for thr trapezoidal shaped fuzzy number. The $\alpha$-cut of fuzzy number $\overline{N}$ is written $\overline{N}[\alpha]$ and equals $\{x \mid \overline{N}(x) \geq \alpha\}$ for $0 < \alpha \leq 1$. We separately define $\overline{N}[0]$ as the closure of the union of all the $\overline{N}[\alpha]$, $0 < \alpha \leq 1$. We will call the $\alpha = 0$ cut the support of the fuzzy number. Alpha-cuts of fuzzy numbers are always closed, bounded, intervals so we write this as $\overline{N}[\alpha] = [n_1(\alpha), n_2(\alpha)]$, for $0 \leq \alpha \leq 1$. For an introduction to fuzzy sets/logic the reader can consult (Buckley and Eslami 2002).

C. Kahraman (Ed.): Fuzzy Engineering Economics with Appl., STUDFUZZ 233, pp. 339–357, 2008.
springerlink.com                                    © Springer-Verlag Berlin Heidelberg 2008

## 2 Pricing Options

Now we consider pricing options using triangular fuzzy numbers first in the discrete binomial model and then with the Black-Scholes method.

### 2.1 Binomial Model

This section is based on (Buckley and Eslami 2008c). We will be working with the binomial option pricing method. See Figure 1 (One Step) and Figure 3 (Two Step). See (Durbin 2006) for further discussion on this pricing model. The basic assumptions are that we have an European option and the stock pays no dividends during contract period. We purchase the stock option at time $t = 0$ for $\$V$ (the value of the option) per share and we have the option to buy the stock at $\$S$ (strike price) per share, or cancel the contract, at time $t = T$. Time is measured in years. In an American option we may terminate the contract (buy or cancel) at any time $t \in (0,T)$ but in an European option we can terminate (buy or cancel) only at time $t = T$. There is no cash flow in or out of the contract during the time interval $(0,T)$ so the stock pays no dividends in that time interval, there are no transaction costs and no taxes.

The binomial model contains a number of parameters whose values must be given in order that we may price the option (compute a fair value for $V$). Some, or all, of these values may be uncertain and we will model this uncertainty using triangular fuzzy numbers. The first reason for fuzzifying the model is to model uncertainty in the input parameters. A second reason for fuzzifying the model is discussed below.

In the next section we first review the crisp one step binomial price model and then fuzzify it. Then we do the same for the two step model and the n step procedure. See (Buckley and Eslami 2008a, 2008c) for references to other publications about applying fuzzy sets to pricing options.

As we will see in the next three sections the crisp binomial model can consider only $n+1$ possible future stock prices after $n$ steps. This has been a major criticism of the binomial pricing model. Traders wanted to be able to consider all possible future prices for the stock. This lead to taking $n \to \infty$ in the $n$ step binomial pricing model leading to the famous Black-Scholes pricing model discussed below. However in our $n$ step fuzzy binomial pricing model we show that it can consider all possible future stock prices in an interval $[\varepsilon, M]$, for $\varepsilon \approx 0$ and large $M$. This is the second motivation for using fuzzy sets. We do not need to look at $n \to \infty$ in the fuzzy model to be able to consider all possible future stock prices.

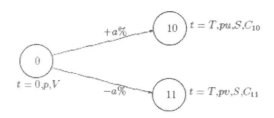

**Fig. 1.** One Step Binomial Pricing Tree

### 2.1.1 One Step Model

The one step model is shown in Figure 1. In Figure 1 $p$ is the share price (spot price) at time zero, $V$ is the value of the option at $t = 0$, and $a$ is the percent increase/decrease of the stock during the time period $T$. We will always write $a$ as a decimal like $a = 0.05$, or $a = 0.10$, etc. If $a = 0.05$, then the stock can rise 5% or decrease 5%. Also $S$ is the strike price, or the amount the option's contract says we will pay per share at time $T$. $C_{10}$ is the value of the stock at time $T$ if it increases $a\%$ and $C_{11}$ is the value if it decreases $a\%$. Let $u = (1+a)$ and $v = (1-a)$. The value of the stock at time $T$ is $pu$ if increases by $a\%$ and $pv$ when it decreases $a\%$. So

$$C_{10} = max(0, pu - S), \tag{1}$$

and

$$C_{11} = max(0, pv - S), \tag{2}$$

Because if $pu \leq S$ the contract is canceled, but if $pu > S$ we buy the stock at $\$S$ per share which is now selling at $pu$ per share. Similar reasoning for $C_{11}$. If the price is going down you can make a profit.

Now we come to determining $V$. Assume we buy $x$ shares of this stock at time zero. The cost is $xp$. To do this we borrow $\$B$ at (risk free) interest rate $r$ per year and make up the difference (if any) with cash $\$V$. So we must have

$$xp = B + V. \tag{3}$$

We determine $x$ and $B$ to be equivalent to our stock option. The two equations are:

$$x(pu) - B\exp(rT) = C_{10}, \tag{4}$$

and

$$x(pv) - B\exp(rT) = C_{11}. \tag{5}$$

In equation (4) we sell our $x$ shares at price $\$pu$ per share and pay back our loan of $\$B$, whose value is now $B\exp(rT)$, and the result must equal the proceeds $C_{10}$. In equation (5) we sell $x$ shares at $\$pv$ per share, pay back our loan and the result equals $C_{11}$.

We will be using continuous interest. If we invest/borrow $\$A$ at interest rate $r$ for time $T$ and interest is compunded continuously, the amount at time $T$ is $A\exp(rT)$ (future value). If we consider $\$A$ in the future at time $T$, its present value today is $A\exp(-rT)$. Reference (Durbin 2006) uses continuous interest and we will do the same.

Equations (4) and (5) have a unique solution for $x$ and $B$ which are

$$x = (C_{10} - C_{11})/(p(u-v)), \tag{6}$$

and

$$B = (xpv - C_{11})\exp(-rT). \tag{7}$$

Using equation (3) we now find $V$

$$V = f_1(p,a,S,r;T) = \exp(-rT)(\theta C_{10} + (1-\theta)C_{11}), \tag{8}$$

where

$$\theta = (\exp(rT)-v)/(u-v), \tag{9}$$

and

$$1-\theta = (u - \exp(rT))/(u-v). \tag{10}$$

Now $\theta$ and $1-\theta$ are like probabilities since they are usually in $(0,1)$ and their sum equals one.

We now let $p$, $a$, $S$ and $r$ all be triangular fuzzy numbers. We may easily consider some of these parameters crisp (not fuzzy). Also, we can extend to all parameters trapeziodal fuzzy numbers but we will use triangular fuzzy numbers in this section. So let $\overline{p} = (p_1/p_2/p_3)$, $\overline{a} = (a_1/a_2/a_3)$, $\overline{S} = (s_1/s_2/s_3)$ and $\overline{r} = (r_1/r_2/r_3)$.

Now we have a choice: (1) fuzzify equations (4) and (5) and solve for $\overline{x}$ and $\overline{B}$ using $\alpha$-cuts and interval arithmetic (Buckley and Eslami 2002) and then compute $\overline{V}$; or (2) fuzzify equation (8) and solve for $\overline{V}$ using the extension principle (Buckley and Eslami 2002). We choose the second option (Buckley and Eslami 2008c).

Fuzzifing equation (8) and using the extension principle gives

$$\overline{V}(\beta) = sup\{min(\overline{p}(z_1),\overline{a}(z_2),\overline{S}(z_3),\overline{r}(z_4)) \mid f_1(z_1,z_2,z_3,z_4;T) = \beta\} \tag{11}$$

However, we know how to find $\alpha$-cuts of $\overline{V}$, Let $\overline{V}[\alpha] = [v_1(\alpha),v_2(\alpha)]$, $\overline{p}[\alpha] = [p_1(\alpha),p_2(\alpha)]$, $\overline{a}[\alpha] = [a_1(\alpha),a_2(\alpha)]$, $\overline{S}[\alpha] = [S_1(\alpha),S_2(\alpha)]$, $\overline{r}[\alpha] = [r_1(\alpha),r_2(\alpha)]$. Then (Buckley and Qu 1990)

$$v_1(\alpha) = min\{f_1(p,a,S,r;T) \mid \mathbf{S}\}, \tag{12}$$

and

$$v_2(\alpha) = max\{f_1(p,a,S,r;T) \mid \mathbf{S}\}, \tag{13}$$

where the statement $\mathbf{S}$ is ``$p \in [p_1(\alpha),p_2(\alpha)], a \in [a_1(\alpha),a_2(\alpha)], S \in [S_1(\alpha),S_2(\alpha)], r \in [r_1(\alpha),r_2(\alpha)]$".

## Example 1

Let $\bar{p} = (57/60/63)$, $\bar{a} = (0.04/0.05/0.06)$, $\bar{S} = (60/62/64)$ and $\bar{r} = (0.05/0.06/0.07)$.
Then $\bar{p}[\alpha] = [57+3\alpha, 63-3\alpha]$, $\bar{a}[\alpha] = [0.04+0.01\alpha, 0.06-0.01\alpha]$,
$\bar{S}[\alpha] = [60+2\alpha, 64-2\alpha]$ and $\bar{r}[\alpha] = [0.05+0.01\alpha, 0.07-0.01\alpha]$. We substitute these into equations (12) and (13) to find the alpha-cuts of $\bar{V}$. Assume that the contract time is six months so $T = 0.5$.

We solved the optimization problem using "Solver" (Frontline Systems, 2008). The "little" Solver is a free add on to Microsoft Excel while the "big" Solver needs to be purchased. The "big" Solver obviously solves larger optimization problems. Since this software is probably not well known we gave the program in (Buckley and Eslami 2008c). The solutions for selected $\alpha$ values are in Table 1. An approximate graph of $\bar{V}$ is in Figure 2. The graph is approximate because we forced the sides only through the end points of the $\alpha = 0, 0.5, 1$ cuts. The numbers on the $x-$axis are the possible values of the option at time zero.

In Figure 1 let us look at $\bar{p}\bar{u}$ and $\bar{p}\bar{v}$, at the end of step one, for the fuzzy values given above, where $\bar{u}[0] = 1+\bar{a}[0]$ and $\bar{v}[0] = 1-\bar{a}[0]$. We get the future stock price in $\bar{p}[0]\bar{u}[0] \cup \bar{p}[0]\bar{v}[0] = [53.58, 66.78]$ using the supports of the fuzzy numbers. So this fuzzy model considers all prices at the end of step one in that interval.

**Table 1.** Alpha-Cuts of Fuzzy Value in Example 1

| $\alpha$ | $\bar{V}[\alpha]$ |
|---|---|
| 0 | [0, 5.22] |
| 0.25 | [0, 4.11] |
| 0.5 | [0, 3.01] |
| 0.75 | [0, 1.90] |
| 1 | 0.78 |

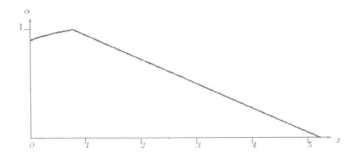

**Fig. 2.** Fuzzy Value $\bar{V}$ in Example 1

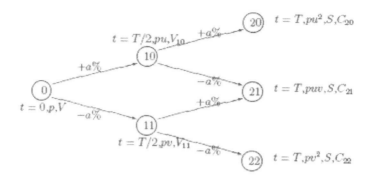

**Fig. 3.** Two Step Binomial Pricing Tree

### 2.1.2 Two Step Model

The two step method is shown in Figure 3. The end of the first step is the same as in Figure 1. The value of the stock at the end of step two is $pu^2$, $puv$ or $pv^2$, with values $C_{20} = max(0, pu^2 - S)$, $C_{21} = max(0, puv - S)$ and $C_{22} = max(0, pv^2 - S)$, respectively. Let $V_{10}$ ($V_{11}$) be the value at node 10 (11). The time for both steps is $T$ so the time for each individual step is $t = T/2$. We find these values $V_{1j}$, $j = 0, 1$, as in the one step case

$$V_{10} = \exp(-rt)(\theta C_{20} + (1-\theta)C_{21}), \qquad (14)$$

and

$$V_{11} = \exp(-rt)(\theta C_{21} + (1-\theta)C_{22}), \qquad (15)$$

for $\theta$ ($(1-\theta)$) defined in equation (9) ((10)). The value of the option at time zero $V$ is then

$$V = \exp(-rt)(\theta V_{10} + (1-\theta)V_{11}), \qquad (16)$$

which equals

$$V = f_2(p, a, S, r; T) = \exp(-rT)(\theta^2 C_{20} + 2\theta(1-\theta)C_{21} + (1-\theta)^2 C_{22}). \qquad (17)$$

Notice that we use $T$ in this equation.

As in the previous section we fuzzify all the parameters except time. The $\alpha$ – cuts of $\overline{V}$ will be found the same way as before

$$v_1(\alpha) = min\{f_2(p, a, S, r; T) | \mathbf{S}\}, \qquad (18)$$

and

$$v_2(\alpha) = max\{f_2(p, a, S, r; T) | \mathbf{S}\}, \qquad (19)$$

where the statement S is
``$p \in [p_1(\alpha), p_2(\alpha)], a \in [a_1(\alpha), a_2(\alpha)], S \in [S_1(\alpha), S_2(\alpha)], r \in [r_1(\alpha), r_2(\alpha)]$''.

**Example 2**
The data for the fuzzy sets is the same as in Example 1. The contract time is six months so $T = 0.5$ and $t = 0.25$. We solved the optimization problem using our "little" Solver. The solutions for selected $\alpha$ values are in Table 2. An approximate graph of $\overline{V}$ (only through three alpha-cuts) is in Figure 4.

In Figure 3 let us look at $\overline{p}\,\overline{u}^2$, $\overline{p}\,\overline{u}\,\overline{v}$ and $\overline{p}\,\overline{v}^2$, at the end of step two, for the fuzzy values in Example 1. We get the future stock price in $\overline{p}[0](\overline{u}[0])^2 \cup \overline{p}[0]\overline{u}[0]\overline{v}[0] \cup \overline{p}[0](\overline{v}[0])^2 = [50.36, 70.79]$ using the supports of the fuzzy numbers. So this fuzzy model considers all prices at the end of step two in that interval.

**Table 2.** Alpha-Cuts of Fuzzy Value in Example 2

| $\alpha$ | $\overline{V}[\alpha]$ |
|---|---|
| 0 | [0, 5.58] |
| 0.25 | [0, 4.38] |
| 0.5 | [0.37, 3.18] |
| 0.75 | [1.04, 2.37] |
| 1 | 1.71 |

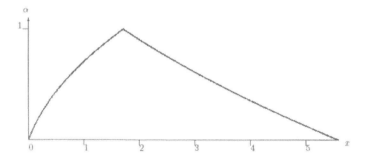

**Fig. 4.** Fuzzy Value $\overline{V}$ in Example 2

### 2.1.3 n Step Model

We see the pattern starting in the previous section. Now $t = T/n$. At the end of $n$ steps we have nodes labeled $n0$, $n1$,..., $nn$ with the stock priced at $pu^{n-i}v^i$,

$i = 0, 1, ..., n$ with values $C_{ni}$, $i = 0, ..., n$. Now $C_{ni} = max(0, pu^{n-i}v^i - S)$, $i = 0, ..., n$. Then

$$V = F(p, a, S, r; T) = \exp(-rT) \sum_{i=0}^{n} \binom{n}{i} \theta^i (1-\theta)^i C_{ni}. \tag{20}$$

We fuzzify all the parameters except time. The $\alpha$-cuts of $\overline{V}$ will be found the same way as before

$$v_1(\alpha) = min\{F(p, a, S, r; T) \mid \mathbf{S}\}, \tag{21}$$

and

$$v_2(\alpha) = max\{F(p, a, S, r; T) \mid \mathbf{S}\}, \tag{22}$$

where the statement $\mathbf{S}$ is
"$p \in [p_1(\alpha), p_2(\alpha)], a \in [a_1(\alpha), a_2(\alpha)], S \in [S_1(\alpha), S_2(\alpha)], r \in [r_1(\alpha), r_2(\alpha)]$".

### Example 3

The data for the fuzzy sets is the same as in Examples 1 and 2. Contract $T = 0.5$ so $t = T/10 = 0.05$ using $n = 10$. For large $n$, say $n \geq 100$, this problem becomes too much for our "little" solver. So let us use $n = 10$. The results are in Table 3. Approximate graph (only through three $\alpha$-cuts) is in Figure 5.

**Table 3.** Alpha-Cuts of Fuzzy Value in Example 3 Using $n = 10$

| $\alpha$ | $\overline{V}[\alpha]$ |
|---|---|
| 0 | [1.07, 7.58] |
| 0.25 | [1.55, 6.47] |
| 0.5 | [2.12, 5.46] |
| 0.75 | [2.95, 4.62] |
| 1 | 3.78 |

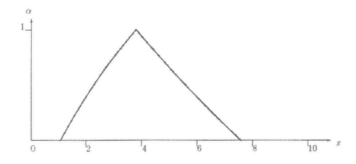

**Fig. 5.** Fuzzy Value $\overline{V}$ in Example 3

Let us look at $\overline{pu}^{n-i}\overline{v}^i$, $i = 0,...,n$ at the end of step n, for the fuzzy values in Examples 1 and 2. We get the price in

$$\bigcup_{i=0}^{n} \overline{p}[0](\overline{u}[0])^{n-i}(\overline{v}[0])^i, \tag{23}$$

using the supports of the fuzzy numbers. For the data in the previous examples this interval is shown in Table 4. So, for $n = 1000$ and this fuzzy model we get essentially all possible prices. The same is probably true for $n = 100$. So for $n = 100$ the fuzzy model probably considers all possible future stock prices.

**Table 4.** Values for the Stock Price, n Steps Fuzzy Model

| $n$ | Price in Interval |
|---|---|
| 10 | $[30.7, 112.8]$ |
| 100 | $[0.12, 21(10)^3]$ |
| 1000 | $[7.6(10)^{-26}, 1.27(10)^{27}]$ |

### 2.1.4 Discussion

Let $\overline{V}_n$ be the fuzzy value for a $n$ step model. From the Tables and Figures, $\overline{V}_n$ becomes more fuzzy as $n$ grows. Uncertainty (fuzziness) grows as you perform more and more computations. And we see that the fuzzy sets $\overline{V}_n$ shift to the right as $n$ grows. What is

$$\lim_{n\to\infty} \overline{V}_n \tag{24}$$

and how is it related to the non-fuzzy (and fuzzy ) Black-Scholes model? We do not know what the above limit is but the Black-Scholes model is the next topic. In the crisp case the limit is known to be the Black-Scholes model.

### 2.2 Black-Scholes Model

The problem we now discuss is modeling possible uncertainties in the input parameters in the famous Black-Scholes formula for pricing European options. This section is based on (Buckley and Eslami 2008a). In the next section we briefly review the crisp (not fuzzy) Black-Scholes formula. Then we model uncertain future interest rates and uncertain volatility using fuzzy numbers. Reference (Buckley and Eslami 2008a) discusses how we might estimate these fuzzy numbers. In (Buckley and Eslami 2008a) we also looked at stocks that give dividends during the time horizon of interest but their value is uncertain and modeled by triangular fuzzy numbers. In all cases the call value of the stock option becomes a triangular shaped fuzzy number.

The basic assumptions are that we have an European option and we assume that the stock pays no dividends during contract period. We purchase the stock option at time $t = 0$ for $\$C$ (the value of the option) per share and we have the option to buy the

348     J.J. Buckley and E. Eslami

stock at $\$K$ (strike price) per share, or cancel the contract, at time $t = T$. Time is measured in years.

### 2.2.1 Crisp Black-Scholes

Let $S$ be the price of one share of the stock at time zero and let $K$ be the strike price, the amount per share we may pay at the termination of the contract at time $t = T$. Also $r$ is the ``risk-free" interest rate for this period and $\sigma$ is the historical volatility of this stock. Then (Durbin 2006) the call value ($C$) of this contract is

$$C = F(S,K,r,\sigma,t) = S \int_{-\infty}^{d_1} N(0,1)dx - K\exp(-rt)\int_{-\infty}^{d_2} N(0,1)dx, \tag{25}$$

where

$$d_1 = [\ln(S/K) + rt + \sigma^2(t/2)]/(\sigma\sqrt{t}), \tag{26}$$

and

$$d_2 = d_1 - \sigma\sqrt{t}, \tag{27}$$

and $N(0,1)$ denotes the standard normal probability density. All inputs $S$, $K$, $r$, $\sigma$ and $t$ are crisp (not fuzzy). We continue to use continuous interest.

### 2.2.2 Fuzzy Black-Scholes

Now we allow $\overline{S}$, $\overline{r}$ and $\overline{\sigma}$ to be fuzzy numbers. So

$$\overline{C} = F(\overline{S},K,\overline{r},\overline{\sigma},t), \tag{28}$$

using the extension principle (Buckley and Eslami 2002). Let

$$\Pi(x,y,z) = min\{\overline{S}(x),\overline{r}(y),\overline{\sigma}(z)\}. \tag{29}$$

Then

$$\overline{C}(w) = sup\{\Pi(x,y,z)\,|\,F(x,K,y,z,t) = w\}, \tag{30}$$

which gives the membership function for $\overline{C}$ by the extension principle. Let $\overline{C}[\alpha] = [c_1(\alpha), c_2(\alpha)]$, $\overline{S}[\alpha] = [s_1(\alpha), s_2(\alpha)]$, $\overline{r}[\alpha] = [r_1(\alpha), r_2(\alpha)]$, $\overline{\sigma}[\alpha] = [\sigma_1(\alpha), \sigma_2(\alpha)]$, for $0 \leq \alpha \leq 1$. Then, since $F$ is a continuous function, we may find the alpha-cuts of $\overline{C}$ as follows (Buckley and Qu 1990)

$$c_1(\alpha) = min\{F(x,K,y,z,t)\,|\,\mathbf{S}\}, \tag{31}$$

and

$$c_2(\alpha) = max\{F(x,K,y,z,t)\,|\,\mathbf{S}\}, \tag{32}$$

where $\mathbf{S}$ is the statement ``$x \in \overline{S}[\alpha], y \in \overline{r}[\alpha], z \in \overline{\sigma}[\alpha]$".

It is well known (Durbin 2006) that $C$ is an increasing function of $s$, $r$ and $\sigma$. So we can evaluate equations (31) and (32) as

$$c_1(\alpha) = F(s_1(\alpha), K, r_1(\alpha), \sigma_1(\alpha), t), \qquad (33)$$

and

$$c_2(\alpha) = F(s_2(\alpha), K, r_2(\alpha), \sigma_2(\alpha), t). \qquad (34)$$

**Example 4**

Let $\overline{S} = (32/33/34)$, $\overline{r} = (0.048/0.05/0.052)$, $\overline{\sigma} = (0.08/0.10/0.12)$, $T = 0.25$ and $K = 30$. $\overline{C}$ is shown in Figure 6 with selected alpha-cuts in Table 5. A Maple (2008) program was used for these calculations and it is given in (Buckley and Eslami 2008a). We used the following alpha-cuts $\overline{S}[\alpha] = [32 + \alpha, 34 - \alpha]$, $\overline{r}[\alpha] = [0.048 + 0.002\alpha, 0.052 - 0.002\alpha]$ and $\overline{\sigma}[\alpha] = [0.08 + 0.02\alpha, 0.12 - 0.02\alpha]$, $0 \le \alpha \le 1$, in the evaluation of $\overline{C}[\alpha]$.

The fuzzy numbers give an automatic sensitivity analysis for each alpha-cut. In fact the fuzzy analysis contains interval analysis. Choose an $\alpha \in [0,1]$. Given any $S \in \overline{S}[\alpha]$, $r \in \overline{r}[\alpha]$ and $\sigma \in \overline{\sigma}[\alpha]$, then the call value of the option from equation

**Table 5.** Alpha-Cuts of the Fuzzy Call Value $\overline{C}$ in Example 4

| Alpha | $\alpha$ – Cut |
|---|---|
| 0 | [2.37, 4.39] |
| 0.50 | [2.88, 3.89] |
| 1 | [3.38, 3.38] |

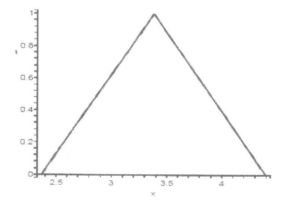

**Fig. 6.** Fuzzy Call Value in Example 4

350    J.J. Buckley and E. Eslami

(25) is in $\overline{C}[\alpha]$. Conversely, given any $c \in \overline{C}[\alpha]$, there are values of $S \in \overline{S}[\alpha]$, $r \in \overline{r}[\alpha]$ and $\sigma \in \overline{\sigma}[\alpha]$ that then give this call price $c$.

Also the alpha-cuts display the uncertainty. The maximum uncertainty in $C$, produced by the uncertainty in $S$, $r$ and $\sigma$, is the interval [2.37,4.39] its $\alpha = 0$ cut.

## 3  Pricing Forwards and Futures

This section is concerned with using trapezoidal fuzzy numbers to model the uncertainty in future interest rates in the pricing of forwards and futures. In the next section we discuss forwards and futures and their fuzzy prices/values based on certain fuzzy interest rates. Now we will use both simple interest and continuous interest. Let $r$ denote the current interest rate quoted as a percent per annum. So $r = 5\%$ pa means $r = 0.05$ per annum. Time $t$ is measured in years. If $FV$ means future value and $PV$ stands for present value, the simple interest formula is

$$FV = PV(1 + rt), \tag{35}$$

and

$$PV = FV(1 + rt)^{-1}. \tag{36}$$

For continuous interest the equations are

$$FV = PV\exp(rt), \quad PV = FV\exp(-rt). \tag{37}$$

For future interest rates we will use the LIBOR Spot rate and the LIBOR Forward rate (Durbin 2006). Consider the three month rates in Table 6 obtained from the LIBOR three month yield curves. The $3m$, $6m$,... means three months in the future, six months in the future, We will only go one year into the future so we can use the LIBOR rates. Going further out into the future would require Eurodollar Spot rates and other calculations which we will not need (Durbin 2006).

**Table 6.** LIBOR Spot and 3m Forward Rates

| Term | Spot | Forward |
|------|------|---------|
| 3m   | 3.55 | 3.62    |
| 6m   | 3.57 | 3.79    |
| 9m   | 3.62 | 3.92    |
| 12m  | 3.67 | 4.01    |
| 15m  | 3.71 |         |

The first $t = 0.25$ is for three months and we use the $3m$ Forward rate, and the second $t = 0.50$ is six months in the future and we use the $6m$ Spot rate. For continuous interest the calculation is

$$M(\exp(0.0362 * 0.25) - 1)(\exp(-0.0357 * 0.50)). \tag{38}$$

Pricing Options, Forwards and Futures Using Fuzzy Set Theory    351

The Libor Spot rates are published each day as the London Interbank Offer Rate. From the Spot rates we can determine the Forward rates. Consider the following example (Durbin 2006).You borrow \$10,000 6 months in the future for a period of 3 months. The 6 month Spot rate in Table 6 is $0.0357$ and the 9 month Spot rate is $0.0362$. What is the 3 month Forward rate for \$10,000 for the loan commencing in 6 months? We will use simple interest and assume that interest is paid in advance. Consider two equivalent senarios. Senario A is you borrow for 9 months at the $9m$ Spot rate $0.0362$. Senario B is you borrow for 6 months at the $6m$ Spot rate $0.0357$ and then take out a new loan for $3$ months at the then current Spot rate $X$. Solve for $X$ the three month Forward rate starting in six months. The equation is ($M = \$10,000$)

$$M(0.0362(9/12)) = M(0.0357(1/2)) + M(X(1/4))(1+0.0357(1/2))^{-1}, \quad (39)$$

or $X = 0.0379$ the $6m$ Forward rate in Table 6. The other Forward rates are computed in a similar way.

There are a number of ways fuzziness can enter these future interest rates. We will consider two ways and use the first method. The following example is continued in the next section. ABC corporation wants to buy 100 shares of GIMTEX in one year. The contract is to be between ABC and a futures dealer who now holds the shares in GIMTEX. The contract calls for using the LIBOR three month yield curve. Now ABC corporation will sign the contract in the next two weeks and they think the LIBOR Spot rate is more likely to decrease, than increase, over the next two weeks. Consider the $3m$ LIBOR Spot rate in Table 6. ABC thinks it will be in the interval $[3.52, 3.56]$. That is, they think it can drop three basis points or increase one basis point. One basis point is $0.01$. To complete their thinking about this rate they construct the trapezoidal fuzzy number $(3.50/3.52, 3.56/3.57)$ for the $3m$ LIBOR Spot rate over the next two weeks. They do the same for the $6m$, $9m$ and $12m$ LIBOR Spot rates. The results are shown in Table 7. Let these fuzzy numbers be called $\overline{r}_{s3}$, $\overline{r}_{s6}$, $\overline{r}_{s9}$ and $\overline{r}_{s,12}$. For notation define $\overline{r}_{sj} = (r_{sj1}/r_{sj2}, r_{sj3}/r_{sj4})$, $j = 3, 6, 9, 12$. Now we need to compute the fuzzy numbers for the $3m$ Forward rates. Let the fuzzy Forward rates be called $\overline{r}_{f3}$, $\overline{r}_{f6}$, $\overline{r}_{f9}$ and $\overline{r}_{f,12}$. So $\overline{r}_{fj} = (r_{fj1}/r_{fj2}, r_{fj3}/r_{fj4})$, $j = 3, 6, 9, 12$. We compute the $r_{fjl}$ from the $r_{sjk}$ as discussed above, $l = 3, 6, 9, 12$, $k = 3, 6, 9, 12$. All results are in Table 7. The rest of this example is in the next section. The fuzziness is in the planning stage, or the "what if" stage, for the ABC corporation. Once they sign the contract almost all the fuzziness disappears since all the interest rates become crisp (not fuzzy).

The LIBOR Spot rate is the rate at which top banks in the London market lend funds to each other. Each day these values are obtained and aggregated to a crisp number like those shown as Spot rates in Table 6. Let us take for the $3m$ fuzzy Spot rate the minimum rate given as $r_{s31}$ and the maximum rate as $r_{s34}$. Then the support (base) of the trapezoidal fuzzy number for $\overline{r}_{s3} = [r_{s31}, r_{s34}]$. Also, from the data, we determine

## Table 7. Fuzzy LIBOR Spot and $3m$ Forward Rates

| Term | Spot | Forward |
|------|------|---------|
| $3m$ | (3.50/3.52,3.56/3.57) | (3.57/3.59,3.63/3.64) |
| $6m$ | (3.52/3.54,3.58/3.59) | (3.73/3.76,3.80/3.81) |
| $9m$ | (3.57/3.59,3.63/3.64) | (3.87/3.89,3.93/3.94) |
| $12m$ | (3.62/3.64,3.68/3.69) | (3.96/3.98,4.02/4.03) |

### 3.1 Crisp Pricing Forwards/Futures

Although forwards and futures are different they can be priced the same way. These sections, and the following section, are based on (Buckley and Eslami 2008b). A forward is a contract between two parties (individuals, companies, corporations,...) to buy something at a given price at some future date. These contracts are called "over the counter" agreements because the two parties get together and design the contract to meet their needs. One problem with a forward is that one of the parties could default on the contract at the due date. To guard against default the parties may put forth collateral (stocks, bonds,..)in the contract to reduce the risk of default. A futures contract is a standardized forward contract between one party and a futures dealer at an "exchange". Two parties do not need to come together because the futures dealer is the intermediary. On a futures contract there is no risk of default because the corporation behind the exchange guaranties the contract. However, in a futures contract there is a "daily settlement" where money may be exchanged each day depending on the current price of the "underlier" in the contract. We will ignore these differences (default, daily exchanges,...) and price both futures and forwards the same way. We also ignore contract fees, taxes, etc. in the pricing. Now we continue the example started in the previous section. ABC corporation agrees to buy 100 shares of GIMTEX in one year. The contract is between ABC and a futures dealer who now holds the shares in GIMTEX. Today GIMTEX is selling for $26.32 per share. The contract calls for using the LIBOR three month yield curve. GIMTEX is expected to pay around $0.08 per share at the end of each three month period. We may handle other "cost of carry" items like storage costs, etc. in a similar manner to dividends. The pricing of this financial instrument is to first find the present value of the future cash flow (dividends) and then determine the future value at the end of the contract. We first do this using crisp (non-fuzzy) values. We will use continuous interest. Find the present value of the future dividends. Use $r_{s3},...,r_{s,12}$ from Table 6. Then;

$$PV = 8\exp(-0.0355*0.25) + 8\exp(-0.0357*0.50) +$$
$$8\exp(-0.0362*0.75) + 8\exp(-0.0367*1.00), \tag{40}$$

which equals $\$31.28$. Then the price (also called the forward price) of this contract is

$$P = (S - PV)\exp(r*1), \tag{41}$$

## Pricing Options, Forwards and Futures Using Fuzzy Set Theory    353

where $S = 100 * 26.32$ is the cost of the contract today and $r = 0.0350$ which is todays (risk-free) interest rate. The risk-free (Durbin 2006) rate is essentially a guaranteed rate of return on an investment with no risk of loss (treasury/government securities). The value is $P = \$2693.36$. The idea is that ABC does not have to spend $100 * 26.32$ today and could invest it accumulating to $A_1 = 100 * 26.32\exp(0.035 * 1)$ in one year and the futures dealer can not do that. However the futures dealer gains $\$PV$ today in dividends which accumulates to $A_2 = PV\,exp(0.035 * 1)$ in one year, so the fair forward price would be $A_1 - A_2$.

### 3.2 Fuzzy Pricing Forwards/Futures

Now we fuzzify the example using fuzzy interest rates from ABC corporation $\overline{r}_{s3}, ..., \overline{r}_{s,12}$ in Table 7 and a fuzzy number for the dividends since GIMTEX gives approximately \$0.08 per share quarterly. The fuzzy number ABC corporation takes for the dividend is $\overline{D} = 100 * (0.07/0.075, 0.085/0.09)$. The rest of the parameters are all crisp and known exactly. Let

$$F(\overline{D}, \overline{r}_{s3}, \overline{r}_{s6}, \overline{r}_{s9}, \overline{r}_{s,12}) = \overline{D}\exp(-\overline{r}_{s3}/4) + \overline{D}\exp(-\overline{r}_{s6}1/2) +$$

$$\overline{D}\exp(-\overline{r}_{s9}3/4) + \overline{D}\exp(-\overline{r}_{s,12}1.00), \tag{42}$$

and

$$\overline{P} = G(\overline{D}, \overline{r}_{s3}, \overline{r}_{s6}, \overline{r}_{s9}, \overline{r}_{s,12}) = (2632 - F(\overline{D}, \overline{r}_{s3}, \overline{r}_{s6}, \overline{r}_{s9}, \overline{r}_{s,12}))\exp(0.035 * 1). \tag{43}$$

The fuzzy forward price $\overline{P}$ is given by equation (44). We will evaluate the fuzzy function $G$ using the extension principle (Buckley and Eslami 2002). Let

$$\Pi(x_1, x_2, x_3, x_4, x_5) = min(\overline{D}(x_1), \overline{r}_{s3}(x_2), \overline{r}_{s6}(x_3), \overline{r}_{s9}(x_4), \overline{r}_{s,12}(x_5)).$$

Let $\overline{P}[\alpha] = [p_1(\alpha), p_2(\alpha)]$, $\overline{D}[\alpha] = [d_1(\alpha), d_2(\alpha)]$, $\overline{r}_{s3}[\alpha] = [r_{s31}(\alpha), r_{s32}(\alpha)]$, $\overline{r}_{s6}[\alpha] = [r_{s61}(\alpha), r_{s62}(\alpha)]$, $\overline{r}_{s9}[\alpha] = [r_{s91}(\alpha), r_{s92}(\alpha)]$, $\overline{r}_{s,12}[\alpha] = [r_{s,12,1}(\alpha), r_{s,12,2}(\alpha)]$, Then

$$\overline{P}(z) = sup\{\Pi(x_1, x_2, x_3, x_4, x_5) \mid G(x_1, x_2, x_3, x_4, x_5) = z\}. \tag{44}$$

We know how to find alpha-cuts of $\overline{P}$ (Buckley and Qu 1990) as

$$p_1(\alpha) = min\{G(x_1, x_2, x_3, x_4, x_5) \mid S\}, \tag{45}$$

and

$$p_2(\alpha) = max\{G(x_1, x_2, x_3, x_4, x_5) \mid S\}, \tag{46}$$

354    J.J. Buckley and E. Eslami

where S is the statement ``$x_1 \in \overline{D}[\alpha], x_2 \in \overline{r}_{s3}[\alpha], x_3 \in \overline{r}_{s6}[\alpha], x_4 \in \overline{r}_{s9}[\alpha], x_5 \in \overline{r}_{s,12}[\alpha]$''. We may easily evaluate equations (47) and (48) because: (1) the partial derivative of $G$ with respect to $D(x_1)$ is negative; and (2) the partial derivative of $G$ with respect to $r_{s3}(x_2)$, $r_{s6}(x_3)$, $r_{s9}(x_4)$ and $r_{s,12}(x_5)$is positive. So for $p_1(\alpha)$ we use $x_1 = 9 - 0.5 * \alpha$, $x_2 = 0.0350 + 0.0002 * \alpha$, $x_3 = 0.0352 + 0.0002 * \alpha$, $x_4 = 0.0357 + 0.0002 * \alpha$ and $x_5 = 0.0362 + 0.0002 * \alpha$, $0 \le \alpha \le 1$, in function $G$. For $p_2(\alpha)$ use $x_1 = 7 + 0.5 * \alpha$, $x_2 = 0.0357 - 0.0001 * \alpha$, $x_3 = 0.0359 - 0.0001 * \alpha$, $x_4 = 0.0364 - 0.0001 * \alpha$ and $x_5 = 0.0369 - 0.0001 * \alpha$, $0 \le \alpha \le 1$, in function $G$. This gives the left and right sides of the fuzzy number $\overline{P}$. If we use $y$ for $\alpha$ we get the left (right) side of $\overline{P}$ described as $x$ a function of $y$. This is backwards since we usually have the $x-$ axis horizontal and the $y-$ axis vertical with $y$ a function of $x$. Using the implicit function notation in Maple we can graph $x$ a function of $y$ and the graph of $\overline{P}$ is shown in Figure 8. $\overline{P}$ is a trapezoidal shaped fuzzy number. The Maple program for Figure 8 is in (Buckley and Eslami 2008b). Some $\alpha-$cuts of $\overline{P}$ are shown in Table 8. Notice that the crisp price $P = \$2693.36$ is in $\overline{P}[1]$.

**Table 8.** Alpha-Cuts of the Fuzzy Price $\overline{P}$ for the Forward/Future

| $\alpha$ | $\overline{P}[\alpha]$ |
| --- | --- |
| 0 | [2689.29,2697.40] |
| 0.50 | [2690.30,2696.39] |
| 1 | [2691.32,2695.38] |

The fuzzy numbers give an automatic sensitivity analysis for each alpha-cut. In fact the fuzzy analysis contains interval analysis. Choose an $\alpha \in [0,1]$. Given any $D \in \overline{D}[\alpha]$, $r_{s3} \in \overline{r}_{s3}[\alpha]$, $r_{s6} \in \overline{r}_{s6}[\alpha]$, $r_{s9} \in \overline{r}_{s9}[\alpha]$ and $r_{s,12} \in \overline{r}_{s,12}[\alpha]$, then the forward price from equation (42) is in $\overline{P}[\alpha]$. Conversely, given any $p \in \overline{P}[\alpha]$, there are values of $D \in \overline{D}[\alpha]$, $r_{s3} \in \overline{r}_{s3}[\alpha]$, $r_{s6} \in \overline{r}_{s6}[\alpha]$, $r_{s9} \in \overline{r}_{s9}[\alpha]$ and $r_{s,12} \in \overline{r}_{s,12}[\alpha]$ that then give this forward price.

Now ABC corporation can see how the price varies because of the fuzzy dividend and the fuzzy interest rates. This is an easy result to analyze. Consider any $\alpha-$cut: (1) the minimum price occurs when we use the maximum value of $D$ and the minimum values for all the interest rates; and (2) the maximum price is when we use the minimum value of $D$ and the maximum values for all the interest rates.

Pricing Options, Forwards and Futures Using Fuzzy Set Theory    355

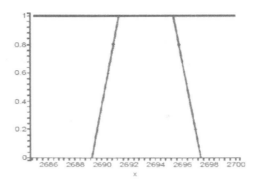

**Fig. 7.** Fuzzy Future Value of Forward/Future

Now suppose we look six months into the future and assume that at that point in time GIMTEX is trading at $27.05 per share. Let us compute the value of this contract at that point in time (six months in the future). We first do this for all crisp values and then with fuzzy interest rates. $P$ in equation (42) was the original forward price after one year. Let $K$ be the forward price of a new contract six months in the future for 100 shares at $27.05 per share to buy in six months. There will be two more dividends coming in three and six months. We find $K$ as before

$$K = (100 * 27.05 - PV)\exp(r_{f6} * 0.5), \tag{47}$$

where we use the Forward rate $r_{f6}$ from Table 6. Also

$$PV = 8 * \exp(-r_{s9} * 0.25) + 8 * \exp(-r_{s,12} * 0.50). \tag{48}$$

We find $K = 2740.67$. The forward value ``today'', six months in the future, is

$$V = (P - K) * \exp(-r_{s,12} * 0.5), \tag{49}$$

which equals $\$-46.45$. We discount the future values $P$ and $K$ one year in the future back to six months in the future.

ABC corporation now applies its fuzzy interest rates in Table 7 which it uses over the next two weeks. Let

$$\overline{PV} = \overline{D}\exp(-\overline{r}_{s9} * 0.25) + \overline{D}\exp(-\overline{r}_{s,12} * 0.50), \tag{50}$$

and

$$\overline{K} = (2705 - \overline{PV})\exp(\overline{r}_{f6} * 0.5). \tag{51}$$

Then the fuzzy value is the present value of $\overline{P} - \overline{K}$, or

$$\overline{V} = (\overline{P} - \overline{K})\exp(-\overline{r}_{s,12} * 0.5). \tag{52}$$

The fuzzy forward value $\overline{V}$ will be a function of $\overline{P}$, $\overline{D}$, $\overline{r}_{s9}$, $\overline{r}_{s,12}$ and $\overline{r}_{f,6}$. We can argue that the partial derivative of V with respect to $P$ and $D$ are positive, the partial derivative on $r_{s9}$ is negative, the partial derivative on $r_{f6}$ is also negative but it is not clear the result for $r_{s,12}$. As before we may find the alpha-cuts of $\overline{V}[\alpha] = [v_1(\alpha), v_2(\alpha)]$ as

$$v_1(\alpha) = min\{(P-K)\exp(-r_{s,12}*0.5) | S\}, \qquad (53)$$

and

$$v_2(\alpha) = max\{(P-K)\exp(-r_{s,12}*0.5) | S\}, \qquad (54)$$

where the statement **S** is ``$p \in \overline{P}[\alpha]$, $D \in \overline{D}[\alpha]$, $r_{s9} \in \overline{r}_{s9}[\alpha]$, $r_{f6} \in \overline{r}_{f6}[\alpha]$, and $r_{s,12} \in \overline{r}_{s,12}[\alpha]$''. Now we will use "Solver" in Excel (Frontline Systems 2008) to solve equations (55) and (56) for selected values of alpha. The results are in Table 9. The program we used for Table 9 is in (Buckley and Eslami 2008b). An approximate graph of $\overline{V}$, only through the three alpha-cuts in Table 9, is shown in Figure 8. $\overline{V}$ is a trapezoidal shaped fuzzy number.

The fuzzy sets produce a sensitivity analysis and an interval analysis. As the price $P$ varies in its alpha-cut, and the dividends $D$ are in their alpha-cut and the interest rates $r$ change throughout their alpha-cuts, the value will vary in its alpha-cut for all $\alpha \in [0,1)$. ABC corporation now wants to analyze these results. Consider any $\alpha$ −cut: (1) the minimum value occurs when we use the minimum values for $P$ and

**Table 9.** Alpha-Cuts of the Fuzzy Value $\overline{V}$ After Six Months

| $\alpha$ | $\overline{V}[\alpha]$ |
|---|---|
| 0 | [−52.70, −39.69] |
| 0.50 | [−51.14, −41.38] |
| 1 | [−49.58, −43.07] |

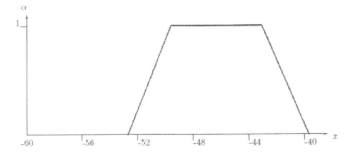

**Fig. 8.** Fuzzy Value after Six Months

$D$ and the maximum values for all the interest rates except $r_{s,12}$; and (2) the maximum price is when we use the maximum values of $P$ and $D$ and the minimum values for all the interest rates except $r_{s,12}$. This can be verified by taking the partial derivatives of $V$ with respect to these variables. Looking at the Solver solution of this example we found that the minimum (maximum) value is for the minimum (maximum) value of $r_{s,12}$. Now ABC corporation knows how uncertainty in future interest rates, and dividends, effect the future value $V$. We see that the crisp solution $V = -46.45 \in \overline{V}[1]$.

## 4 Summary and Conclusions

There are usually a number of uncertain parameter values that go into the pricing models for options and derivatives (Futures/Forwards). These uncertainties may be handled using probability theory or fuzzy set theory. We illustrated using fuzzy numbers in this chapter. We claim it is easier to model this uncertainty with fuzzy sets than with probabilities. Suppose a future interest rate is the value a random variable. We need to know this random variable's probability density function. Suppose it is the normal distribution. Now we need to estimate the mean $\mu$ and standard deviation $\sigma$ of the normal. But point estimators for $\mu$ and $\sigma$ are themselves uncertain. How are we to handle the uncertainty in $\mu$ and $\sigma$? The use of fuzzy sets is more straight forward and does not have this second order uncertainty.

## References

Buckley, J.J., Eslami, E.: Pricing stock options using black-scholes and fuzzy sets. Iranian Journal of Fuzzy Systems (to appear) (2008a)

Buckley, J.J., Eslami, E.: Pricing forwards/futures and swaps using fuzzy numbers. J. Advances in Fuzzy Sets and Systems (to appear) (2008b)

Buckley, J.J., Eslami, E.: Pricing stock options using fuzzy sets. Iranian J. Fuzzy Systems (to appear) (2008c)

Buckley, J.J., Eslami, E.: Introduction to fuzzy logic and fuzzy sets. Springer, Heidelberg (2002)

Buckley, J.J., Qu, Y.: On using -cuts to evaluate fuzzy equations. Fuzzy Sets and Systems 38, 309–312 (1990)

Durbin, M.: All about derivatives. McGraw-Hill, New York (2006)

Frontline Systems (2008), http://www.frontsys.com

Maple 9, Waterloo Maple Inc., Waterloo, Canada (2008)

# Fuzzy Capital Rationing Models

Esra Bas and Cengiz Kahraman

Department of Industrial Engineering, Istanbul Technical University

**Abstract.** Addressing uncertainty in Lorie-Savage and Weingartner capital rationing models has been considered in the literature with different approaches. Stochastic and robust approach to Weingartner capital rationing problem are examples of non-fuzzy approaches. In this chapter, we provide examples of fuzzy approach to Lorie-Savage problem, and illustrate the models with numerical examples. The solution of the generic models requires evolutionary algorithms; however for the models with triangular or trapezoidal fuzzy numbers, branch-and-bound method has been suggested to be sufficient.

## 1 Introduction

Lorie-Savage defined a capital rationing problem as a 0-1 linear programming (LP) model with fixed capital budget, and no lending and borrowing for each period, while Weingartner applied LP to Lorie-Savage problem in Weingartner (1967). Although Weingartner model has been analyzed and criticized extensively, it remained a basic capital rationing model, and its drawbacks have been reduced by several extensions to the model (for example, see Park and Sharp-Bette 1990). One of the drawbacks that Weingartner (1967) excluded uncertainty in future cash flows, and interest rates, has been analyzed only in recent years. Non-fuzzy and fuzzy approaches to capital rationing problem in the literature are remarkable.

Stochastic approach to capital rationing problem has been proposed by Kira and Kusy (1990), and Kira et al (2000). Robust optimization approach was applied to capital rationing and capital budgeting problems by Kachani and Langella (2005). Soyster (1973) originally proposed robust optimization, and assumed the worst-case scenario for the realization of the coefficients of an LP model. Ben-Tal and Nemirovski (2000) alleviated the over-conservatism of Sosyter (1973) by proposing the partitioning of the coefficients in each constraint into two sets as certainty set and uncertainty set. Bertsimas and Sim (2004) proposed the adjustment of the robustness of an LP model by predefined parameters.

Several fuzzy capital rationing models appeared in the literature. Chiu and Park (1998) modelled Lorie-Savage capital rationing model as a fuzzy 0-1 LP model by considering budget limits, periodic net cash flows, periodic cash outflows, and discount rates as triangular fuzzy numbers, and proposed branch-and-bound method for the solution. Chiu and Park (1998) also proposed so-called "weighted method" to compare the fuzzy numbers, and discussed other fuzzy ranking methods in the literature. For the details of fuzzy ranking methods, the interested reader is referred to Chen and Hwang (1992). Huang (2007a) applied chance-constrained programming to Lorie-Savage problem with fuzzy cash outflows and net cash flows, and Huang (2007b) applied chance-constrained programming to Lorie-Savage problem with random fuzzy cash outflows and net cash flows, and proposed so-called "fuzzy simulation based genetic algorithm"

C. Kahraman (Ed.): Fuzzy Engineering Economics with Appl., STUDFUZZ 233, pp. 359–380, 2008.
springerlink.com © Springer-Verlag Berlin Heidelberg 2008

and "hybrid intelligent algorithm" for the solution, respectively. Huang (2008) proposed fuzzy mean-variance model for Lorie-Savage problem, and fuzzy simulation based genetic algorithm for the solution of the model.

The literature for fuzzy LP models is also extensive. Several articles have appeared with either fully fuzzified or partially fuzzified models. Buckley and Feuring (2000) studied a fully fuzzified LP model, discussed fuzzy inequality and maximization of an objective function, and proposed an evolutionary algorithm for the solution. Partially fuzzified models include models with fuzzy objective function coefficients or fuzzy constraints or constraint matrix with fuzzy coefficients. Zimmermann (1978) proposed a fuzzy LP model with several objective functions. Delgado et al (1989) proposed a general fuzzy LP model with fuzzy constraints, constraint matrix with fuzzy coefficients, and fuzzy right-hand sides.

The structure of this chapter is as follows. In section 2, example models of non-fuzzy and fuzzy approaches to capital rationing will be provided, and fuzzy capital rationing models will be illustrated with the numerical examples. Finally, in section 3, we will give the conclusions.

# 2 Uncertainty-Based Approaches to Capital Rationing Models

In section 2.1, we provide Kira and Kusy's (1990) stochastic capital rationing model, and Kachani and Langella's (2005) robust capital rationing model. In section 2.2, we provide Chiu and Park's (1998) fuzzy capital rationing model, Huang's (2007a) and Huang's (2007b) chance-constrained programming models, and Huang's (2008) fuzzy mean-variance model, and illustrate the models with the numerical examples.

## 2.1 Stochastic and Robust Approaches to Capital Rationing Model

Kira and Kusy's (1990) stochastic capital rationing model and Kachani and Langella's (2005) robust capital rationing model are two examples of non-fuzzy approaches to model uncertainty.

### 2.1.1 Kira and Kusy's (1990) Stochastic Capital Rationing Model

Kira and Kusy (1990) proposed the following model as a stochastic capital rationing model:

$$
\begin{aligned}
&\max \\
&\sum_n \left\{ \sum_j \left[ c_{nj}(1+r)^{-n} \right] x_j - \min \sum_l (p_n^+ y_{nl}^+ + p_n^- y_{nl}^-) P(\beta_{nl}) \right\}
\end{aligned}
$$

$$
\begin{aligned}
&s.t. \\
&\sum a_{hj} x_j = b_h && \forall h \in \{1, 2, \dots\dots\dots, m_1\} \\
&\sum t_{nj} x_j + y_{nl}^+ - y_{nl}^- = \beta_{nl} && \forall n \in \{1, 2 \dots\dots\dots, m_2\} \\
&&& \forall l \in \{1, 2, \dots\dots\dots, L_i\} \quad (1) \\
&x_j \geq 0 && \forall j \in \{1, 2, \dots\dots\dots, J\} \\
&y_{nl}^+ \geq 0, y_{nl}^- \geq 0 && \forall n \in \{1, 2 \dots\dots\dots, m_2\} \\
&&& \forall l \in \{1, 2, \dots\dots\dots, L_i\}
\end{aligned}
$$

where

$c_{nj}$ : net cash flow for project $j$ at period $n$

$r$ : discount rate

$x_j$ : project selection variable for project $j$

$p_n^+$ : penalty associated with not utilizing the available funds at period $n$

$p_n^-$ : penalty per dollar borrowed at period $n$

$y_{nl}^+$ : the shortage variable for the $l$th realization of the resource $n$

$y_{nl}^-$ : the surplus variable for the $l$th realization of the resource $n$

$\beta_{nl}$ : the $l$th realization of the resource $n$

$a_{hj}$ : the coefficient for project $j$ on the $h$th deterministic constraint

$b_h$ : the resource constraint for the $h$th deterministic constraint

$t_{nj}$ : the coefficient for project $j$ on the $n$th stochastic constraint

and the first $m_1$ constraints are deterministic, whereas the constraints $(m_1 + 1)$ to $(m_1 + m_2)$ are stochastic.

### 2.1.2 Kachani and Langella's (2005) Robust Capital Rationing Model

Kachani and Langella (2005) considered the following Weingartner capital rationing model:

$$\max \quad \sum_{j=1}^{J} NPV_j x_j$$

$$s.t. \quad -\sum_{j=1}^{J} a_{nj} x_j \leq M_n \quad \forall n \in \{0,1,\ldots\ldots\ldots N\} \qquad (2)$$

$$0 \leq x_j \leq 1 \qquad \forall j \in \{1,2,\ldots\ldots\ldots, J\}$$

where $NPV_j = \sum_{n=0}^{N} \dfrac{a_{nj}}{(1+r)^n}$ is Net Present Value of project $j$, $a_{nj}$ is net cash flow for project $j$ at period $n$ for $\forall n \in \{0,1,\ldots\ldots\ldots, N\}$, $N$ is project life and budget life for all projects, $r$ is discount rate, $M_n$ is budget limit at period $n$ for $\forall n \in \{0,1,\ldots\ldots\ldots, N\}$, and $x_j$ is project selection variable for project $j$, where $x_j = 1$ if project $j$ is fully funded, $0 < x_j < 1$ if project $j$ is partially funded, and $x_j = 0$ if project $j$ is not funded for $\forall j \in \{1, 2,\ldots\ldots\ldots, J\}$. Since $\tilde{a}_{nj} \in [a_{nj} - \hat{a}_{nj}, a_{nj} + \hat{a}_{nj}]$ for $j \in J_n$ is assumed to be a symmetric and bounded random variable, and $\Gamma_n \in [0, |J_n|]$ is defined as a predetermined protection level to adjust the robustness for $\forall n \in \{0,1,\ldots\ldots\ldots, N\}$, model (2) is converted into its robust counterpart as follows (Kachani and Langella 2005):

$$\max \quad \sum_{n=0}^{N} \frac{x_{J+n+1}}{(1+r)^n}$$

*s.t.*

$$-\sum_{j=1}^{J} a_{nj}x_j + z_n\Gamma_n + \sum_{j\in J_n} p_{nj} + x_{J+n+1} \leq M_n \qquad \forall n \in \{0,1,\ldots\ldots,N\}$$

$$z_n + p_{nj} \geq \hat{a}_{nj}x_j \qquad\qquad \forall n \in \{0,1,\ldots\ldots,N\} \qquad \forall j \in J_n \quad (3)$$

$$z_n \geq 0 \qquad\qquad \forall n \in \{0,1,\ldots\ldots,N\}$$

$$p_{nj} \geq 0 \qquad\qquad \forall n \in \{0,1,\ldots\ldots,N\} \qquad \forall j \in J_n$$

$$x_{J+n+1} \geq 0 \qquad\qquad \forall n \in \{0,1,\ldots\ldots,N\}$$

$$0 \leq x_j \leq 1 \qquad\qquad \forall j \in \{1,2,\ldots\ldots,J\}$$

where Kachani and Langella (2005) transformed model (2) into an equivalent model with objective function coefficients without $a_{nj}$ for $\forall n, \forall j$ before formulating its robust counterpart. They also formulated the dual model of the primal model (3), and analyzed the complementary slackness results. They applied the proposed model to a three-project example and a two hundred-project example. The interested reader is referred to the article for the details of the robust optimization approach to capital rationing and capital budgeting.

## 2.2 Fuzzy Capital Rationing Models

Chiu and Park (1998), Huang (2007a), Huang (2007b), and Huang (2008) proposed fuzzy capital rationing models, and their solution.

### 2.2.1 Chiu and Park's (1998) Fuzzy Capital Rationing Model

$$\max \quad PW_{(x_1,x_2,\ldots\ldots,x_J)}$$

$$s.t. \quad \sum_{j=1}^{J} -P_{nj}x_j \preceq B_n \qquad \forall n \in \{0,1,\ldots\ldots,N\} \qquad (4)$$

$$x_j \in \{0,1\} \qquad \forall j \in \{1,2,\ldots\ldots,J\}$$

where $P_{nj} = (p_{njo}, p_{nj1}, p_{nj2})$ is a triangular fuzzy number (TFN) representing periodic cash flow for project $j$ at period $n$ for $\forall n \in \{0,1,\ldots\ldots,N\}$ and $\forall j \in \{1,2,\ldots\ldots,J\}$ where $N$ is budget life, and $J$ is the number of the projects; $B_n$ is the fuzzy budget limit at period $n$ such that $B_{n1}$ is the most possible budget, and $B_{n0}$ is the largest possible budget for $\forall n \in \{0,1,\ldots\ldots,N\}$; $x_j$ is project selection variable for project $j$, where $x_j = 1$ if project $j$ is funded, and $x_j = 0$ if project $j$ is not funded for $\forall j \in \{1,2,\ldots\ldots,J\}$. $\preceq$ is a fuzzy extension of the binary relation $\leq$. $PW_{(x_1,x_2,\ldots\ldots,x_J)}$ is a TFN representing present worth of the combined cash flow of funded projects, and $r_n = (r_{n0}, r_{n1}, r_{n2})$ is a TFN representing discount rate at period $n$ for $\forall n \in \{1,2,\ldots\ldots,L\}$ where $r_0 = 0$, and $L$ is project life for all projects. The calculation of present worth of project $j$ $PW_j$ is based on Chiu and Park (1994) as follows:

$$PW_j = \left( \begin{array}{c} \sum_{n=0}^{L} \left[ \dfrac{\max\{p_{nj0}, 0\}}{\prod_{n'=0}^{n}(1+r_{n'2})} + \dfrac{\min\{p_{nj0}, 0\}}{\prod_{n'=0}^{n}(1+r_{n'0})} \right], \sum_{n=0}^{L} \dfrac{p_{nj1}}{\prod_{n'=0}^{n}(1+r_{n'1})}, \\ \sum_{n=0}^{L} \left[ \dfrac{\max\{p_{nj2}, 0\}}{\prod_{n'=0}^{n}(1+r_{n'0})} + \dfrac{\min\{p_{nj2}, 0\}}{\prod_{n'=0}^{n}(1+r_{n'2})} \right] \end{array} \right) \quad (5)$$

Note that Chiu and Park (1998) considered different budget life and project life for all projects. Chiu and Park (1998) also emphasized that the present worth calculation proposed in Chiu and Park (1994) yields a non-linear fuzzy number, however in "reasonable discount rate ranges such as ±4% of these rates (Chiu and Park 1998)", the present worth calculation in equation (5) yields a TFN.

Chiu and Park (1998) proposed the following branch-and-bound flow chart in Figure 1 for the solution of the model (4).

The distinctive features of Chiu and Park's (1998) branch-and-bound method can be summarized as the so-called "weighted method" to rank the projects according to

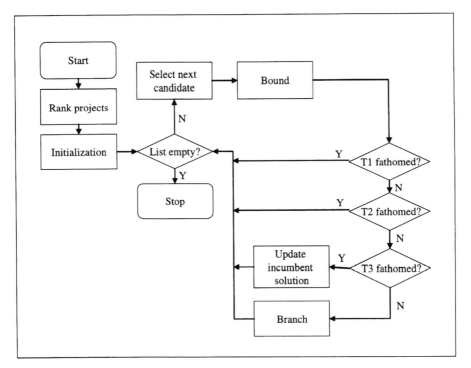

**Fig. 1.** Branch-and-bound flow chart for Chiu and Park's (1998) model

364     E. Bas and C. Kahraman

their present worth, and fathoming tests to fathom a current candidate solution. The interested reader is referred to the article for the details of the branch-and-bound method proposed in Chiu and Park (1998).

*2.2.1.1  Weighted method to Rank the Projects.* If we define present worth of project $j$ as $PW_j = (pw_{j0}, pw_{j1}, pw_{j2})$, then

$$w_1 \left( \frac{pw_{jo} + pw_{j1} + pw_{j2}}{3} \right) + w_2 pw_{j1} \tag{6}$$

is the so-called "dominance indicator" for the comparison of present worth of project $j$, where $w_1$ and $w_2$ are relative weights of each criterion (Chiu and Park 1998). If we set $w_1 = 1$ as in Chiu and Park (1998), then equation (6) is equivalent to the following:

$$\left( \frac{pw_{jo} + pw_{j1} + pw_{j2}}{3} \right) + w pw_{j1} \tag{7}$$

where $w = 0.3$ may be recommended if $pw_{j1}$ is fairly important, and $w = 0.1$ may be recommended if $pw_{j1}$ is slightly important (Chiu and Park 1998).

*2.2.1.2  Fathoming Tests to Fathom a Current Candidate Solution*

**Fathoming test 1 (T1)**
If the current objective function is inferior to the updated lower bound of the so-called "incumbent solution (Chiu and Park 1998)", then the current candidate solution is fathomed. Chiu and Park (1998) proposed a TFN with extreme negative numbers such as $(-9999999, -9999998, -9999997)$ as a lower bound for the beginning of the problem.

**Fathoming test 2 (T2)**

$$\sum_{j=1}^{J'} - P_{nj} x_j \succ B_n + \sum_{j=J'+1}^{J} \max\{P_{nj}, 0\} \tag{8}$$

where the left hand side represents the net expenditures of the funded projects with $J'$ as the lastly considered project in the subset, and the right hand side represents the budget limit and revenues possibly generated by the subsequent projects for the $n$th period.

Equation (8) can be transformed into the following two equations with crisp parameters:

$$\sum_{j=1}^{J'} - p_{nj1} x_j > B_{n1} + \sum_{j=J'+1}^{J} \max\{p_{nj1}, 0\} \tag{9}$$

$$\sum_{j=1}^{J'} -\min\{p_{nj0},0\}x_j > B_{n0} + \sum_{j=1}^{J'} \max\{p_{nj2},0\}x_j + \sum_{j=J'+1}^{J} \max\{p_{nj2},0\} \qquad (10)$$

where equation (9) checks the feasibility of "the most possible budget" case, and equation (10) checks the feasibility of "the largest possible budget" case. If either equation (9) or equation (10) is satisfied for any $n \in \{0,1,\ldots,N\}$, then the current candidate solution is fathomed, since the current candidate solution proves to be infeasible.

**Fathoming test 3 (T3)**

$$\sum_{j=1}^{J'} -P_{nj} x_j - P_{n,J'+1}(1-x_{J'}) + \sum_{j=J'+2}^{J} -P_{nj} \preceq B_n \qquad (11)$$

where $x_{J'} = 1$ or $x_{J'} = 0$.

Equation (11) can be transformed into the following two equations with crisp parameters:

$$\sum_{j=1}^{J'} -p_{nj1} x_j - p_{n,J'+1,1}(1-x_{J'}) + \sum_{j=J'+2}^{J} -p_{nj1} \le B_{n1} \qquad (12)$$

$$\sum_{j=1}^{J'} \max\{-p_{nj0},0\}x_j + \max\{-p_{n,J'+1,0},0\}(1-x_{J'}) +$$

$$\sum_{j=J'+2}^{J} \max\{-p_{nj0},0\} \le B_{n0} + \sum_{j=1}^{J'} \max\{p_{nj2},0\}x_j + \qquad (13)$$

$$\max\{p_{n,J'+1,2},0\}(1-x_{J'}) + \sum_{j=J'+2}^{J} \max\{p_{nj2},0\}$$

where equation (12) checks "the most possible budget" case, and equation (13) checks "the largest possible budget" case. If equations (12) and (13) are satisfied for $\forall n \in \{0,1,\ldots,N\}$, then the current candidate solution is recorded as the current incumbent solution.

*2.2.1.3 A numerical Example of Chiu and Park's (1998) Fuzzy Capital Rationing Model.* The nominal values for a three-project and three-period example from Kachani and Langella (2005) are given in Table 1.

**Table 1.** A three-project and three-period example with nominal values

|        | Project 1 | Project 2 | Project 3 | Budget |
|--------|-----------|-----------|-----------|--------|
| Year 0 | -15       | -5        | -12       | 25     |
| Year 1 | 5         | 5         | 4         | 25     |
| Year 2 | 25        | 5         | 20        | 25     |
| NPV    | 6.53      | 2.64      | 5.22      |        |
| r=20%  |           |           |           |        |

366     E. Bas and C. Kahraman

Table 1 with nominal periodic cash flows, budget limits, and discount rate is transformed into Table 2 with fuzzy periodic cash flows, budget limits, and discount rates. Note that we consider ±10% range for symmetric triangular fuzzy cash flows, and +10% range for fuzzy budget limits, while we consider ±4% range for symmetric triangular fuzzy discount rates as suggested in Chiu and Park (1998).

**Table 2.** A three-project and three-period example with fuzzy cash flows, budget limits, and discount rates

|  | Project 1 | Project 2 | Project 3 | Budget |
|---|---|---|---|---|
| Year 0 | (-16.5, -15, -13.5) | (-5.5,-5,-4.5) | (-13.2, -12, -10.8) | (25, 27.5) |
| Year 1 | (4.5, 5, 5.5) | (4.5, 5, 5.5) | (3.6, 4, 4.4) | (25, 27.5) |
| Year 2 | (22.5, 25, 27.5) | (4.5, 5, 5.5) | (18, 20, 22) | (25, 27.5) |
| r= (19.2%, 20%, 20.8%) | | | | |

### "Weighted method" to rank the projects

According to the present worth calculation in equation (5), the present worth of each project $\forall j \in \{1, 2, 3\}$ is calculated as follows:

$$PW_1 = \left( \begin{bmatrix} -16.5 + \dfrac{4.5}{(1+0.208)^1} + \dfrac{22.5}{(1+0.208)^2} \end{bmatrix}, -15 + \dfrac{5}{(1+0.2)^1} + \dfrac{25}{(1+0.2)^2}, \\ \begin{bmatrix} -13.5 + \dfrac{5.5}{(1+0.192)^1} + \dfrac{27.5}{(1+0.192)^2} \end{bmatrix} \right)$$

$$= (2.643897197, 6.5277778, 10.46851493)$$

$$PW_2 = \left( \begin{bmatrix} -5.5 + \dfrac{4.5}{(1+0.208)^1} + \dfrac{4.5}{(1+0.208)^2} \end{bmatrix}, -5 + \dfrac{5}{(1+0.2)^1} + \dfrac{5}{(1+0.2)^2}, \\ \begin{bmatrix} -4.5 + \dfrac{5.5}{(1+0.192)^1} + \dfrac{5.5}{(1+0.192)^2} \end{bmatrix} \right)$$

$$= (1.308912, 2.638889, 3.984978)$$

$$PW_3 = \left( \begin{bmatrix} -13.2 + \dfrac{3.6}{(1+0.208)^1} + \dfrac{18}{(1+0.208)^2} \end{bmatrix}, -12 + \dfrac{4}{(1+0.2)^1} + \dfrac{20}{(1+0.2)^2}, \\ -10.8 + \dfrac{4.4}{(1+0.192)^1} + \dfrac{22}{(1+0.192)^2} \end{bmatrix} \right)$$

$$= (2.115118, 5.222222, 8.374812)$$

Fuzzy Capital Rationing Models    367

If we set $w = 0.3$, then according to equation (7):

$$\left( \frac{pw_{1o} + pw_{11} + pw_{12}}{3} \right) + 0.3\, pw_{11} = 8.505063302$$

$$\left( \frac{pw_{2o} + pw_{21} + pw_{22}}{3} \right) + 0.3\, pw_{21} = 3.435926$$

$$\left( \frac{pw_{3o} + pw_{31} + pw_{32}}{3} \right) + 0.3\, pw_{31} = 6.804051$$

are the dominance indicators of the projects. Finally, the ranking of the projects will be $1 \succ 3 \succ 2$. According to Chiu and Park (1998), Last In First Out (LIFO) method is used to decide the order for fathoming the current candidate solutions.

**Fathoming tests to fathom a current candidate solution**

**Current candidate solution-** $x_2 = 1, x_3 = 1$
**(T1)** is not satisfied, since the lower bound (LB) is determined as a TFN with extreme negative numbers as in Chiu and Park (1998).
**(T2)** is not satisfied for "the most possible budget" case and "the largest possible budget" case at any period.
**(T3)** is satisfied, and $(3.42403, 7.861111, 12.35979)$ is set as the lower bound of the current incumbent solution. According to equation (7) where $w = 0.3$ is assumed,

$$\left( \frac{3.42403 + 7.861111 + 12.35979}{3} \right) + 0.3 * 7.861111 = 10.23998 \text{ is the dominance}$$

indicator for the lower bound of the current incumbent solution.

**Current candidate solution -** $x_2 = 1, \ x_3 = 0, \ x_1 = 1$
**(T1)** is not satisfied, since the new objective function $(3.952809, 9.166667, 14.45349)$ dominates the current incumbent solution with $11.94099 > 10.23998$.
**(T2)** is not satisfied for "the most possible budget" case and "the largest possible budget" case at any period.
**(T3)** is satisfied, and $(3.952809, 9.166667, 14.45349)$ is set as the lower bound of the current incumbent solution.

**Current candidate solution -** $x_2 = 0, \ x_3 = 1, \ x_1 = 1$
**(T1)** is not satisfied, since the new objective function $(4.759015, 11.75, 18.84333)$ dominates the current incumbent solution with $15.30911 > 11.94099$
**(T2)** is satisfied for "the most possible budget" case at $n = 0$.

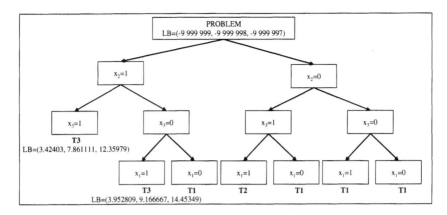

**Fig. 2.** Branch-and-bound method for the numerical example of Chiu and Park's (1998) model

Finally, $x_2 = 1$, $x_3 = 0$, $x_1 = 1$ is the optimal solution with a fuzzy objective function $(3.952809, 9.166667, 14.45349)$. The branch-and-bound procedure for the numerical example of Chiu and Park's (1998) model is summarized in Figure 2.

### 2.2.2 Huang's (2007a) and Huang's (2007b) Chance-Constrained Programming Models For Capital Rationing

Huang (2007a) and Huang (2007b) considered the following Lorie-Savage problem:

$$\max \quad \sum_{j=1}^{J} \sum_{n=1}^{L} \frac{d_{nj}}{(1+r)^n} x_j$$

$$s.t. \quad \sum_{j=1}^{J} a_{nj} x_j \leq M_n \quad \forall n \in \{1, 2, \ldots, N\} \quad (14)$$

$$x_j \in \{0, 1\} \quad \forall j \in \{1, 2, \ldots, J\}$$

where $d_{nj}$ is the net cash flow of project $j$ at the end of the period $n$ for $\forall n \in \{1, 2, \ldots, L\}$, and $\forall j \in \{1, 2, \ldots, J\}$ where $L$ is project life for all projects, and $J$ is the number of the projects; $r$ is the marginal cost of capital; $a_{nj}$ is the cash outflow of project $j$ at the beginning of the period $n$ for $\forall n \in \{1, 2, \ldots, N\}$, $\forall j \in \{1, 2, \ldots, J\}$, where $N$ is budget life; $M_n$ is the budget limit at the beginning of the period $n$ for $\forall n \in \{1, 2, \ldots, N\}$; and $x_j$ is project selection variable for $\forall j \in \{1, 2, \ldots, J\}$ where $x_j = 1$ if project $j$ is funded, and $x_j = 0$ if project $j$ is not funded. Note that budget is assumed to be depleted at the beginning of a period, while net cash flow is assumed to occur at the end of a period.

Huang (2007a) and Huang (2007b) applied chance-constrained programming to model (14).

### Definition 1

Let $X$ and $Y$ be two non-empty sets, and $F(X)$ and $F(Y)$ be all fuzzy subsets $\tilde{A}$ and $\tilde{B}$ of $X$ and $Y$, where the fuzzy subsets $\tilde{A}$ and $\tilde{B}$ are determined by their membership functions $\mu_{\tilde{A}}: X \to [0,1]$, and $\mu_{\tilde{B}}: Y \to [0,1]$, respectively. Then, *possibility, necessity, and credibility* indices are defined as follows (Ramik 2006, Huang 2007a):

$$Pos(\tilde{A} \preceq \tilde{B}) = \sup\{\min(\mu_{\tilde{A}}(x), \mu_{\tilde{B}}(y)) | x \leq y, x, y \in \Re\} \quad (15)$$

$$Nec(\tilde{A} \prec \tilde{B}) = \inf\{\max(1 - \mu_{\tilde{A}}(x), 1 - \mu_{\tilde{B}}(y)) | x > y, x, y \in \Re\} \quad (16)$$

$$Cr\{\tilde{A} \preceq \tilde{B}\} = \frac{1}{2}(Pos(\tilde{A} \preceq \tilde{B}) + Nec(\tilde{A} \prec \tilde{B})) \quad (17)$$

In other words, *possibility, necessity* of a fuzzy event $\tilde{A} \leq k$ and *credibility* of a fuzzy event $\tilde{A} \leq k$ and $\tilde{A} \geq k$ can be defined as follows, where $\tilde{A} = (x_1, x_2, x_3)$ is a TFN as in Figure 3, and $k$ is a real number (Huang 2007a, Zheng and Liu 2006):

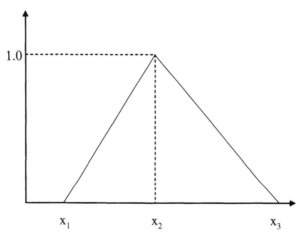

**Fig. 3.** A triangular fuzzy number $\tilde{A}$

$$Pos\{\tilde{A} \leq k\} = \begin{cases} 0, & k \leq x_1 \\ \dfrac{k - x_1}{x_2 - x_1}, & x_1 \leq k \leq x_2 \\ 1, & k \geq x_2 \end{cases} \quad (18)$$

$$Nec\{\tilde{A} \leq k\} = \begin{cases} 0, & k \leq x_2 \\ \dfrac{k - x_2}{x_3 - x_2}, & x_2 \leq k \leq x_3 \\ 1, & k \geq x_3 \end{cases} \quad (19)$$

$$Cr\left\{\tilde{A} \le k\right\} = \begin{cases} 0, & k \le x_1 \\ \dfrac{k - x_1}{2(x_2 - x_1)}, & x_1 \le k \le x_2 \\ \dfrac{x_3 - 2x_2 + k}{2(x_3 - x_2)}, & x_2 \le k \le x_3 \\ 1, & k \ge x_3 \end{cases} \tag{20}$$

$$Cr\left\{\tilde{A} \ge k\right\} = \begin{cases} 1, & k \le x_1 \\ \dfrac{2x_2 - x_1 - k}{2(x_2 - x_1)}, & x_1 \le k \le x_2 \\ \dfrac{x_3 - k}{2(x_3 - x_2)}, & x_2 \le k \le x_3 \\ 0, & k \ge x_3 \end{cases} \tag{21}$$

**Definition 2**
$(X, F(X), Pos)$ is a *possibility space* if $X$ is a non-empty set, $F(X)$ is the power set of $X$, $Pos\{\emptyset\} = 0$, $Pos\{X\} = 1$ and $Pos\{\cup_k A_k\} = \sup_k Pos\{A_k\}$ for any arbitrary collection $\{A_k\} \in F(X)$ (Huang 2007b).

**Definition 3**
Huang (2007b) defined a *random fuzzy variable* as a function from the possibility space $(X, F(X), Pos)$ to the set of random variables, while Liu (2002) defined a *random fuzzy variable* as a function from a set of random variables to $[0, 1]$.

**Definition 4**
Let $\xi = (\xi_1, \xi_2, \ldots\ldots\ldots, \xi_n)$ be an $n$ – dimensional random fuzzy vector, and $f : \Re^n \to \Re$ be a real-valued function over the $n$ – dimensional Euclidean space. Then, *chance of a random fuzzy event* characterized by $f(\xi) \le 0$ is defined as follows (Huang 2007b):

$$Ch\left\{f(\xi) \le 0\right\}(\alpha) = \sup_{Cr\{A\} \ge \alpha} \inf_{\theta \in A} Pr\left\{f(\xi(\theta)) \le 0\right\} \tag{22}$$

where $A \in F(X)$, and equation (22) represents "the random fuzzy event holds with probability Ch$\{f(\xi) \le 0\}(\alpha)$ at credibility $\alpha$" (Huang 2007b).

Huang (2007b) introduced chance-constrained programming model with random fuzzy cash outflows and net cash flows, chance-constrained programming model with random cash outflows and net cash flows, and chance-constrained programming model with fuzzy cash outflows and net cash flows, while Huang (2007a) introduced chance-constrained programming model with fuzzy cash outflows and net cash flows.

### 2.2.2.1 Chance-Constrained Programming Model With Random Fuzzy Cash Outflows and Net Cash Flows

$$\max \quad \bar{f}$$

$$\text{s.t.} \quad \text{Ch}\left\{\sum_{j=1}^{J}\sum_{ln=1}^{L}\frac{\eta_{nj}}{(1+r)^n}x_j \geq \bar{f}\right\}(\alpha) \geq \delta$$

$$\text{Ch}\left\{\sum_{j=1}^{J}\xi_{nj}x_j \leq M_n\right\}(\beta_n) \geq \gamma_n \qquad \forall n \in \{1,2,\ldots\ldots,N\} \tag{23}$$

$$x_j \in \{0,1\} \qquad \forall j \in \{1,2,\ldots\ldots,J\}$$

where $\eta_{nj}$ and $\xi_{nj}$ are random fuzzy variables representing net cash flow of project $j$ at the end of the period $n$ for $\forall n \in \{1,2,\ldots\ldots,L\}$, and $\forall j \in \{1,2,\ldots\ldots,J\}$; and cash outflow of project $j$ at the beginning of the period $n$ for $\forall n \in \{1,2,\ldots\ldots,N\}$, and $\forall j \in \{1,2,\ldots\ldots,J\}$, respectively. $\alpha$, $\beta_n$ for $\forall n \in \{1,2,\ldots\ldots,N\}$, $\delta$, and $\gamma_n$ for $\forall n \in \{1,2,\ldots\ldots,N\}$ are predetermined confidence levels, and $\max \bar{f}$ is the $(\alpha, \delta)$ − return to the objective function $\sum_{j=1}^{J}\sum_{ln=1}^{L}\frac{\eta_{nj}}{(1+r)^n}x_j$ (Huang 2007b).

### 2.2.2.2 Chance-Constrained Programming Model with Random Cash Outflows and Net Cash Flows

$$\max \quad \bar{f}$$

$$\text{s.t.} \quad \text{Pr}\left\{\sum_{j=1}^{J}\sum_{ln=1}^{L}\frac{\eta_{nj}}{(1+r)^n}x_j \geq \bar{f}\right\} \geq \delta$$

$$\text{Pr}\left\{\sum_{j=1}^{J}\xi_{nj}x_j \leq M_n\right\} \geq \gamma_n \qquad \forall n \in \{1,2,\ldots\ldots,N\} \tag{24}$$

$$x_j \in \{0,1\} \qquad \forall j \in \{1,2,\ldots\ldots,J\}$$

where $\eta_{nj}$ for $\forall n \in \{1,2,\ldots\ldots,L\}$, $\forall j \in \{1,2,\ldots\ldots,J\}$ and $\xi_{nj}$ for $\forall n \in \{1,2,\ldots\ldots,N\}$, $\forall j \in \{1,2,\ldots\ldots,J\}$ are random variables (Huang 2007b).

### 2.2.2.3 Chance-Constrained Programming Model with Fuzzy Cash Outflows and Net Cash Flows

$$\max \quad \bar{f}$$

$$\text{s.t.} \quad \text{Cr}\left\{\sum_{j=1}^{J}\sum_{ln=1}^{L}\frac{\eta_{nj}}{(1+r)^n}x_j \geq \bar{f}\right\} \geq \alpha$$

$$\text{Cr}\left\{\sum_{j=1}^{J}\xi_{nj}x_j \leq M_n\right\} \geq \beta_n \qquad \forall n \in \{1,2,\ldots\ldots,N\} \tag{25}$$

$$x_j \in \{0,1\} \qquad \forall j \in \{1,2,\ldots\ldots,J\}$$

where $\eta_{nj}$ for $\forall n \in \{1, 2, ............, L\}$, $\forall j \in \{1, 2, ............, J\}$ and $\xi_{nj}$ for $\forall n \in \{1, 2, ............, N\}$, $\forall j \in \{1, 2, ............, J\}$ are fuzzy variables; $\alpha$, and $\beta_n$ for $\forall n \in \{1, 2, ............, N\}$ are predetermined credibility levels (Huang 2007a, Huang 2007b).

Huang (2007a) and Huang (2007b) proposed fuzzy simulation based genetic algorithm and hybrid intelligent algorithm for the solution of the generic models (25) and (23), respectively, while branch-and-bound method is recommended for the solution of the model (25) with triangular or trapezoidal fuzzy numbers. In Huang's (2007b) model, fuzzy simulation and stochastic simulation are used to calculate the $(\alpha, \delta)$ − return to the objective function $\sum_{j=1}^{J}\sum_{n=1}^{L}\frac{\eta_{nj}}{(1+r)^n}x_j$, while genetic algorithm is used to find the optimal solution. The interested reader is referred to Huang (2007a), Huang (2007b), and Liu (2002) for the details of fuzzy simulation based genetic algorithm, and hybrid intelligent algorithm.

*2.2.2.4 A Numerical Example of Huang's (2007a) Model.* The following model will be considered as an example:

$$\max \quad \bar{f}$$

$$s.t. \quad \text{Cr}\left\{\sum_{j=1}^{3}\sum_{n=1}^{3}\frac{\tilde{a}_{nj}}{(1+r)^n}x_j \geq \bar{f}\right\} \geq 0.9$$

$$\text{Cr}\left\{-\sum_{j=1}^{3}\tilde{a}_{nj}x_j \leq M_n\right\} \geq 0.95 \qquad \forall n \in \{1, 2, 3\}$$

$$x_j \in \{0, 1\} \qquad \forall j \in \{1, 2, 3\}$$

where $\tilde{a}_{nj} = (a_{nj0}, a_{nj1}, a_{nj2})$ is a triangular, normal, and symmetric fuzzy number for $\forall n \in \{1, 2, 3\}$, $\forall j \in \{1, 2, 3\}$. We consider $(0.9 - 0.95)$ predetermined credibility levels as in Huang (2007a), and Table 3 which is a slight modification of the data in the numerical example of Chiu and Park's (1998) model in Table 2.

Table 3. A three-project and three-period example with fuzzy cash flows

| | Project 1 | Project 2 | Project 3 | Budget |
|---|---|---|---|---|
| Year 1 | (-16.5, -15, -13.5) | (-5.5,-5,-4.5) | (-13.2, -12, -10.8) | 25 |
| Year 2 | (4.5, 5, 5.5) | (4.5, 5, 5.5) | (3.6, 4, 4.4) | 25 |
| Year 3 | (22.5, 25, 27.5) | (4.5, 5, 5.5) | (18, 20, 22) | 25 |
| r=20% | | | | |

We calculate $Cr\left\{-\sum_{j=1}^{3}\tilde{a}_{nj}x_j \leq 25\right\}$ from (20) as follows:

$$Cr\left\{-\sum_{j=1}^{3}\tilde{a}_{nj}x_j \leq 25\right\} =$$

$$\begin{cases} 0, & 25 \leq -\sum_{j=1}^{3}a_{nj2}\,x_j \\[3ex] \dfrac{25 + \sum_{j=1}^{3}a_{nj2}\,x_j}{2\left(\sum_{j=1}^{3}a_{nj2}\,x_j - \sum_{j=1}^{3}a_{nj1}\,x_j\right)}, & -\sum_{j=1}^{3}a_{nj2}\,x_j \leq 25 \leq -\sum_{j=1}^{3}a_{nj1}\,x_j \\[3ex] \dfrac{-\sum_{j=1}^{3}a_{nj0}\,x_j + 2\sum_{j=1}^{3}a_{nj1}\,x_j + 25}{2\left(\sum_{j=1}^{3}a_{nj1}\,x_j - \sum_{j=1}^{3}a_{nj0}\,x_j\right)}, & -\sum_{j=1}^{3}a_{nj1}\,x_j \leq 25 \leq -\sum_{j=1}^{3}a_{nj0}\,x_j \\[3ex] 1, & 25 \geq -\sum_{j=1}^{3}a_{nj0}\,x_j \end{cases}$$

where $-\tilde{a}_{nj} = (-a_{nj2}, -a_{nj1}, -a_{nj0})$ for $\forall n \in \{1,2,3\}$, $\forall j \in \{1,2,3\}$.

Since we determine a $0.95-$credibility level for the constraint $-\sum_{j=1}^{3}\tilde{a}_{nj}x_j \leq 25$ $\forall n \in \{1,2,3\}$, we consider the following:

$$Cr\left\{-\sum_{j=1}^{3}\tilde{a}_{nj}x_j \leq 25\right\} =$$

$$\begin{cases} \dfrac{25 + \sum_{j=1}^{3}a_{nj2}\,x_j}{2\left(\sum_{j=1}^{3}a_{nj2}\,x_j - \sum_{j=1}^{3}a_{nj1}\,x_j\right)}, & -\sum_{j=1}^{3}a_{nj2}\,x_j \leq 25 \leq -\sum_{j=1}^{3}a_{nj1}\,x_j \\[3ex] \dfrac{-\sum_{j=1}^{3}a_{nj0}\,x_j + 2\sum_{j=1}^{3}a_{nj1}\,x_j + 25}{2\left(\sum_{j=1}^{3}a_{nj1}\,x_j - \sum_{j=1}^{3}a_{nj0}\,x_j\right)}, & -\sum_{j=1}^{3}a_{nj1}\,x_j \leq 25 \leq -\sum_{j=1}^{3}a_{nj0}\,x_j \\[3ex] 1, & 25 \geq -\sum_{j=1}^{3}a_{nj0}\,x_j \end{cases}$$

where right-hand sides imply mutually exclusive cases for $Cr\left\{-\sum_{j=1}^{3}\tilde{a}_{nj}x_j \leq 25\right\}$ $\forall n \in \{1,2,3\}$.

We next calculate $\mathrm{Cr}\left\{\sum_{j=1}^{3}\sum_{n=1}^{3}\dfrac{\tilde{a}_{nj}}{(1+0.2)^n}\,x_j \geq \bar{f}\right\}$ from (21) as follows:

$$\mathrm{Cr}\left\{\sum_{j=1}^{3}\sum_{n=1}^{3}\frac{\tilde{a}_{nj}}{(1+0.2)^n}\,x_j \geq \bar{f}\right\} =$$

$$\begin{cases}
1, & \bar{f} \leq \sum_{j=1}^{3}\sum_{n=1}^{3}\dfrac{a_{nj0}x_j}{(1+0.2)^n} \\[4ex]
\dfrac{2\sum_{j=1}^{3}\sum_{n=1}^{3}\dfrac{a_{nj1}x_j}{(1+0.2)^n} - \sum_{j=1}^{3}\sum_{n=1}^{3}\dfrac{a_{nj0}x_j}{(1+0.2)^n} - \bar{f}}{2\left(\sum_{j=1}^{3}\sum_{n=1}^{3}\dfrac{a_{nj1}x_j}{(1+0.2)^n} - \sum_{j=1}^{3}\sum_{n=1}^{3}\dfrac{a_{nj0}x_j}{(1+0.2)^n}\right)}, & \sum_{j=1}^{3}\sum_{n=1}^{3}\dfrac{a_{nj0}x_j}{(1+0.2)^n} \leq \bar{f} \leq \sum_{j=1}^{3}\sum_{n=1}^{3}\dfrac{a_{nj1}x_j}{(1+0.2)^n} \\[4ex]
\dfrac{\sum_{j=1}^{3}\sum_{n=1}^{3}\dfrac{a_{nj2}x_j}{(1+0.2)^n} - \bar{f}}{2\left(\sum_{j=1}^{3}\sum_{n=1}^{3}\dfrac{a_{nj2}x_j}{(1+0.2)^n} - \sum_{j=1}^{3}\sum_{n=1}^{3}\dfrac{a_{nj1}x_j}{(1+0.2)^n}\right)}, & \sum_{j=1}^{3}\sum_{n=1}^{3}\dfrac{a_{nj1}x_j}{(1+0.2)^n} \leq \bar{f} \leq \sum_{j=1}^{3}\sum_{n=1}^{3}\dfrac{a_{nj2}x_j}{(1+0.2)^n} \\[4ex]
0, & \bar{f} \geq \sum_{j=1}^{3}\sum_{n=1}^{3}\dfrac{a_{nj2}x_j}{(1+0.2)^n}
\end{cases}$$

Since we determine a $0.90-\text{credibility}$ level for the constraint $\sum_{j=1}^{3}\sum_{n=1}^{3}\dfrac{\tilde{a}_{nj}}{(1+0.2)^n}\,x_j \geq \bar{f}$, we consider the following:

$$\mathrm{Cr}\left\{\sum_{j=1}^{3}\sum_{n=1}^{3}\frac{\tilde{a}_{nj}}{(1+0.2)^n}\,x_j \geq \bar{f}\right\} =$$

$$\begin{cases}
1, & \bar{f} \leq \sum_{j=1}^{3}\sum_{n=1}^{3}\dfrac{a_{nj0}x_j}{(1+0.2)^n} \\[4ex]
\dfrac{2\sum_{j=1}^{3}\sum_{n=1}^{3}\dfrac{a_{nj1}x_j}{(1+0.2)^n} - \sum_{j=1}^{3}\sum_{n=1}^{3}\dfrac{a_{nj0}x_j}{(1+0.2)^n} - \bar{f}}{2\left(\sum_{j=1}^{3}\sum_{n=1}^{3}\dfrac{a_{nj1}x_j}{(1+0.2)^n} - \sum_{j=1}^{3}\sum_{n=1}^{3}\dfrac{a_{nj0}x_j}{(1+0.2)^n}\right)}, & \sum_{j=1}^{3}\sum_{n=1}^{3}\dfrac{a_{nj0}x_j}{(1+0.2)^n} \leq \bar{f} \leq \sum_{j=1}^{3}\sum_{n=1}^{3}\dfrac{a_{nj1}x_j}{(1+0.2)^n} \\[4ex]
\dfrac{\sum_{j=1}^{3}\sum_{n=1}^{3}\dfrac{a_{nj2}x_j}{(1+0.2)^n} - \bar{f}}{2\left(\sum_{j=1}^{3}\sum_{n=1}^{3}\dfrac{a_{nj2}x_j}{(1+0.2)^n} - \sum_{j=1}^{3}\sum_{n=1}^{3}\dfrac{a_{nj1}x_j}{(1+0.2)^n}\right)}, & \sum_{j=1}^{3}\sum_{n=1}^{3}\dfrac{a_{nj1}x_j}{(1+0.2)^n} \leq \bar{f} \leq \sum_{j=1}^{3}\sum_{n=1}^{3}\dfrac{a_{nj2}x_j}{(1+0.2)^n}
\end{cases}$$

Fuzzy Capital Rationing Models    375

where    right-hand    sides    imply    mutually    exclusive    cases    for

$\mathrm{Cr}\left\{\sum_{j=1}^{3}\sum_{n=1}^{3}\frac{\tilde{a}_{nj}}{(1+0.2)^{n}}x_{j}\geq \bar{f}\right\}$. As an example, we give the following problem

instance, where the credibility level is 1, and solve the model with the branch-and-bound method.

$$\max \quad \bar{f}$$

$$s.t. \quad \sum_{j=1}^{3}\sum_{n=1}^{3}\frac{a_{nj0}}{(1+0.2)^{n}}x_{j}\geq \bar{f}$$

$$-\sum_{j=1}^{3}a_{nj0}x_{j}\leq 25 \qquad \forall n\in\{1,2,3\}$$

$$x_{j}\in\{0,1\} \qquad \forall j\in\{1,2,3\}$$

If we set the upper bound of the first constraint as an objective function, then the following model is used for the branch-and-bound method:

$$\max \quad \sum_{j=1}^{3}\sum_{n=1}^{3}\frac{a_{nj0}}{(1+0.2)^{n}}x_{j}$$

$$s.t. \quad -\sum_{j=1}^{3}a_{nj0}x_{j}\leq 25 \qquad \forall n\in\{1,2,3\}$$

$$x_{j}\in\{0,1\} \qquad \forall j\in\{1,2,3\}$$

Since $\sum_{n=1}^{3}\frac{a_{n10}}{(1+0.2)^{n}}=2.39583333$, $\qquad \sum_{n=1}^{3}\frac{a_{n20}}{(1+0.2)^{n}}=1.145833$,

$\sum_{n=1}^{3}\frac{a_{n30}}{(1+0.2)^{n}}=1.916667$; then the ranking of the projects will be $1\succ 3\succ 2$. According to Chiu and Park (1998), LIFO method is used to decide the order for fathoming the current candidate solution, and the lower bound for the beginning of the problem is set as an extremely negative number such as $-9\,999\,999$.

**Current candidate solution -** $x_2 = 1, x_3 = 1$
The objective function is $3.0625$. Since the feasibility condition is satisfied, $3.0625$ is set as the lower bound of the current incumbent solution.

**Current candidate solution -** $x_2 = 1, x_3 = 1, x_1 = 1$
The feasibility condition is not satisfied.

**Current candidate solution -** $x_2 = 1, x_3 = 0, x_1 = 1$
The objective function is $3.5416667$. Since the feasibility condition is satisfied, $3.5416667$ is set as the lower bound of the current incumbent solution.

**Current candidate solution -** $x_2 = 0, x_3 = 1, x_1 = 1$
The feasibility condition is not satisfied.

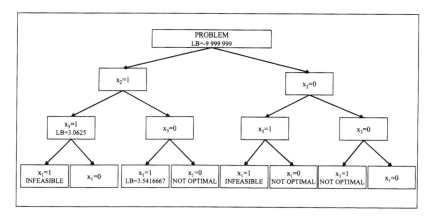

**Fig. 4.** Branch-and-bound method for the numerical example of Huang's (2007a) model

Finally, $x_2 = 1$, $x_3 = 0$, $x_1 = 1$ is the optimal solution with an objective function 3.5416667. The branch-and-bound method for the numerical example of Huang's (2007a) model is summarized in Figure 4.

### 2.2.3 Huang's (2008) Fuzzy Mean-Variance Model

Huang (2008) developed the following fuzzy mean-variance model for the Lorie-Savage problem in (14).

$$\max \quad E\left[\sum_{j=1}^{J}\sum_{n=1}^{L_j}\frac{\eta_{nj}}{(1+r)^n}x_j\right]$$

$$s.t. \quad V\left[\sum_{j=1}^{J}\sum_{n=1}^{L_j}\frac{\eta_{nj}}{(1+r)^n}x_j\right] \leq p_1$$

$$E\left[\sum_{j=1}^{J}\xi_{nj}x_j\right] \leq E[\varsigma_n] \quad \forall n \in \{1, 2, \ldots, N\} \quad (26)$$

$$V\left[\sum_{j=1}^{J}\xi_{nj}x_j\right] \leq p_2 \quad \forall n \in \{1, 2, \ldots, N\}$$

$$x_j \in \{0, 1\} \quad \forall j \in \{1, 2, \ldots, J\}$$

where $\eta_{nj}, \xi_{nj}, \varsigma_n$ are fuzzy variables representing net cash flow of project $j$ at the end of the period $n$ for $\forall n \in \{1, 2, \ldots, L_j\}$, cash outflow of project $j$ at the beginning of the period $n$ for $\forall n \in \{1, 2, \ldots, N\}$, and budget limit at the beginning of the period $n$ for $\forall n \in \{1, 2, \ldots, N\}$, respectively. $N$ is budget life, and $L_j$ is project life of project $j$. $E$ and $V$ represent the expected value and variance, respectively. $p_1$ and $p_2$ are predetermined tolerable variance level of the total fuzzy net present value, and the total fuzzy cash outflow, respectively.

## A problem instance for the generic model (26)

Let $\eta_{nj} = (\eta_{nj0}, \eta_{nj1}, \eta_{nj2})$, $\xi_{nj} = (\xi_{nj0}, \xi_{nj1}, \xi_{nj2})$ and $\varsigma_n = (\varsigma_{n0}, \varsigma_{n1}, \varsigma_{n2})$ be symmetric TFNs. Then, model (26) is equivalent to the following model as in Huang (2008):

max

$$\frac{1}{4}\left(\sum_{j=1}^{J}\sum_{n=1}^{L_j}\frac{\eta_{nj0}}{(1+r)^n}x_j + \sum_{j=1}^{J}\sum_{n=1}^{L_j}\frac{2\eta_{nj1}}{(1+r)^n}x_j + \sum_{j=1}^{J}\sum_{n=1}^{L_j}\frac{\eta_{nj2}}{(1+r)^n}x_j\right)$$

s.t.

$$\left(\sum_{j=1}^{J}\sum_{n=1}^{L_j}\frac{\eta_{nj2}}{(1+r)^n}x_j - \sum_{j=1}^{J}\sum_{n=1}^{L_j}\frac{\eta_{nj0}}{(1+r)^n}x_j\right)^2 \le 24p_1$$

$$\sum_{j=1}^{J}\xi_{nj0}x_j + \sum_{j=1}^{J}2\xi_{nj1}x_j + \sum_{j=1}^{J}\xi_{nj2}x_j \le \varsigma_{n0} + 2\varsigma_{n1} + \varsigma_{n2} \qquad \forall n \in \{1, 2, \dots\dots\dots, N\}$$

$$\left(\sum_{j=1}^{J}\xi_{nj2}x_j - \sum_{j=1}^{J}\xi_{nj0}x_j\right)^2 \le 24p_2 \qquad \forall n \in \{1, 2, \dots\dots\dots, N\}$$

$$x_j \in \{0,1\} \qquad \forall j \in \{1, 2, \dots\dots\dots, J\}$$

$$(27)$$

Huang (2008) proposed fuzzy simulation based genetic algorithm for the solution of the generic model (26), while branch-and-bound algorithm has been found to be sufficient for a model with triangular or trapezoidal fuzzy numbers.

*2.2.3.1 A numerical Example of Huang's (2008) Model.* We assume $p_1 = p_2 = 1.5$ as tolerable variance level of the total fuzzy net present value, and the total fuzzy cash outflow for the model (27). We consider Table 4 for the numerical example which is a slight modification of the data in the numerical example of Chiu and Park's (1998) model in Table 2.

**Table 4.** A three-project and three-period example with fuzzy cash flows and budget limits

|  | Project 1 | Project 2 | Project 3 | Budget |
|---|---|---|---|---|
| Year 1 | (-16.5, -15, -13.5) | (-5.5, -5, -4.5) | (-13.2, -12, -10.8) | (22.5, 25, 27.5) |
| Year 2 | (4.5, 5, 5.5) | (4.5, 5, 5.5) | (3.6, 4, 4.4) | (22.5, 25, 27.5) |
| Year 3 | (22.5, 25, 27.5) | (4.5, 5, 5.5) | (18, 20, 22) | (22.5, 25, 27.5) |
| r=20% | | | | |

Let $\tilde{a}_{nj} = (a_{nj0}, a_{nj1}, a_{nj2})$ be a triangular, normal, and symmetric fuzzy number for $\forall n \in \{1, 2, 3\}$, $\forall j \in \{1, 2, 3\}$ and let $\tilde{M}_n = (M_{n0}, M_{n1}, M_{n2})$ be a triangular, normal, and symmetric fuzzy number for $\forall n \in \{1, 2, 3\}$. Then the following model is equivalent from the model (27):

max

$$\frac{1}{4}\left(\sum_{j=1}^{3}\sum_{n=1}^{3}\frac{a_{nj0}}{(1+r)^n}x_j + \sum_{j=1}^{3}\sum_{n=1}^{3}\frac{2a_{nj1}}{(1+r)^n}x_j + \sum_{j=1}^{3}\sum_{n=1}^{3}\frac{a_{nj2}}{(1+r)^n}x_j\right)$$

s.t.

$$\left(\sum_{j=1}^{3}\sum_{n=1}^{3}\frac{a_{nj2}}{(1+r)^n}x_j - \sum_{j=1}^{3}\sum_{n=1}^{3}\frac{a_{nj0}}{(1+r)^n}x_j\right)^2 \le 36$$

$$-\left(\sum_{j=1}^{3}a_{nj0}x_j + \sum_{j=1}^{3}2a_{nj1}x_j + \sum_{j=1}^{3}a_{nj2}x_j\right) \le M_{n0} + 2M_{n1} + M_{n2} \qquad \forall n \in \{1,2,3\}$$

$$\left(\sum_{j=1}^{3}a_{nj0}x_j - \sum_{j=1}^{3}a_{nj2}x_j\right)^2 \le 36 \qquad\qquad\qquad \forall n \in \{1,2,3\}$$

$$x_j \in \{0,1\} \qquad\qquad\qquad\qquad\qquad\qquad\qquad \forall j \in \{1,2,3\}$$

Since all fuzzy numbers are triangular, branch-and-bound method can be used for the solution. Since

$$\frac{1}{4}\left(\sum_{n=1}^{3}\frac{a_{n10}}{(1+r)^n} + \sum_{n=1}^{3}\frac{2a_{n11}}{(1+r)^n} + \sum_{n=1}^{3}\frac{a_{n12}}{(1+r)^n}\right) = 5.43981481,$$

$$\frac{1}{4}\left(\sum_{n=1}^{3}\frac{a_{n20}}{(1+r)^n} + \sum_{n=1}^{3}\frac{2a_{n21}}{(1+r)^n} + \sum_{n=1}^{3}\frac{a_{n22}}{(1+r)^n}\right) = 2.1990741$$

$$\frac{1}{4}\left(\sum_{n=1}^{3}\frac{a_{n30}}{(1+r)^n} + \sum_{n=1}^{3}\frac{2a_{n31}}{(1+r)^n} + \sum_{n=1}^{3}\frac{a_{n32}}{(1+r)^n}\right) = 4.351852$$

the ranking of the projects will be $1 \succ 3 \succ 2$.

According to Chiu and Park (1998), LIFO method is used to decide the order for fathoming the current candidate solutions.

**Current candidate solution -** $x_2 = 1, x_3 = 1$
The current objective function is $6.5509259$. But, the feasibility condition is not satisfied for the first constraint, since $48.67646176 > 36$.

**Current candidate solution -** $x_2 = 1,\ x_3 = 0,\ x_1 = 1$
The current objective function is $7.6388889$. But, the feasibility condition is not satisfied for the first constraint, since $67.14891975 > 36$.

**Current candidate solution -** $x_2 = 1,\ x_3 = 0,\ x_1 = 0$
The current objective function is $2.1990741$, and all the feasibility conditions are satisfied. Then, $2.1990741$ is set as the lower bound of the current incumbent solution.

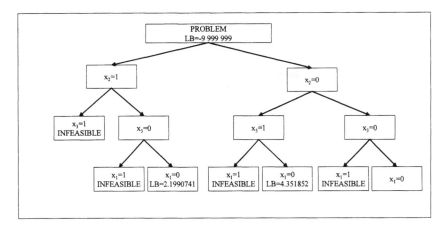

**Fig. 5.** Branch-and-bound method for the numerical example of Huang's (2008) model

**Current candidate solution -** $x_2 = 0$, $x_3 = 1$, $x_1 = 1$
The current objective function is $9.7916667$. But, the feasibility condition is not satisfied for the first constraint, since $120.0850694 > 36$.

**Current candidate solution -** $x_2 = 0$, $x_3 = 1$, $x_1 = 0$
The current objective function is $4.351852$, and all the feasibility conditions are satisfied. Then, $4.351852$ is set as the lower bound of the current incumbent solution.

**Current candidate solution -** $x_2 = 0$, $x_3 = 0$, $x_1 = 1$
The current objective function is $5.43981481$. But, the feasibility condition is not satisfied for the first constraint, since $37.06329304 > 36$.

Finally, the optimal solution is $x_2 = 0$, $x_3 = 1$, $x_1 = 0$ with an objective function $4.351852$. The branch-and-bound method for the numerical example of Huang's (2008) model is summarized in Figure 5.

## 3 Conclusions

Fuzzy modelling is one of the methods to encompass uncertainty in Lorie-Savage and Weingartner capital rationing models, whereas stochastic and robust approaches are also available in the literature. In this chapter, several fuzzy models proposed in the literature have been provided, and illustrated with the numerical examples. While dominance method to compare the present worth of each project determines the results in Chiu and Park's (1998) model, predetermined credibility levels, confidence levels, and tolerable variance levels in Huang's (2007a), Huang's (2007b), and Huang's (2008) models determine the results, respectively. Branch-and-bound method has been suggested to be sufficient for the models with triangular or trapezoidal fuzzy numbers, while evolutionary methods such as genetic algorithm has been suggested to be inevitable in case of non-triangular or non-trapezoidal fuzzy numbers.

# References

Ben-Tal, A., Nemirovski, A.: Robust solutions of linear programming problems contaminated with uncertain data. Mathematical Programming 88(3), 411–424 (2000)

Bertsimas, D., Sim, M.: The price of robustness. Operations Research 52(1), 35–53 (2004)

Buckley, J.J., Feuring, T.: Evolutionary algorithm solution to fuzzy problems: Fuzzy linear programming. Fuzzy Sets and Systems 109(1), 35–53 (2000)

Chen, S.-J., Hwang, C.-L.: Fuzzy multiple attribute decision making: Methods and applications. Springer, New York (1992)

Chiu, C.-Y., Park, C.S.: Fuzzy cash flow analysis using present worth criterion. The Engineering Economist 39(2), 113–138 (1994)

Chiu, C.-Y., Park, C.S.: Capital budgeting decisions with fuzzy projects. The Engineering Economist 43(2), 125–150 (1998)

Delgado, M., Verdegay, J.L., Vila, M.A.: A general model for fuzzy linear programming. Fuzzy Sets and Systems 29(1), 21–29 (1989)

Huang, X.: Chance-constrained programming models for capital budgeting with NPV as fuzzy parameters. Journal of Computational and Applied Mathematics 198(1), 149–159 (2007a)

Huang, X.: Optimal project selection with random fuzzy parameters. International Journal of Production Economics 106(2), 513–522 (2007b)

Huang, X.: Mean-variance model for fuzzy capital budgeting. Computers & Industrial Engineering (2008) doi: 10.1016/j.cie.2007.11.015

Kachani, S., Langella, J.: A robust optimization approach to capital rationing and capital budgeting. The Engineering Economis 50(3), 195–229 (2005)

Kira, D.S., Kusy, M.I.: A stochastic capital rationing model. The Journal of the Operational Research Society 41(9), 853–863 (1990)

Kira, D., Kusy, M., Rakita, I.: The effect of project risk on capital rationing under uncertainty. The Engineering Economis 45(1), 37–55 (2000)

Liu, B.: Random fuzzy dependent-chance programming and its hybrid intelligent algorithm. Information Sciences 141(3-4), 259–271 (2002)

Park, C.S., Sharp-Bette, G.P.: Advanced engineering economics. John Wiley & Sons, Inc., New York (1990)

Ramik, J.: Duality in fuzzy linear programming with possibility and necessity relations. Fuzzy Sets and System 157(10), 1283–1302 (2006)

Soyster, A.L.: Convex programming with set-inclusive constraints and applications to inexact linear programming. Operations Research 21(5), 1154–1157 (1973)

Weingartner, H.M.: Mathematical programming and the analysis of capital budgeting problems. Markham Publishing Company, Chicago (1967)

Zheng, Y., Liu, B.: Fuzzy vehicle routing model with credibility measure and its hybrid intelligent algorithm. Applied Mathematics and Computation 176(2), 673–683 (2006)

Zimmermann, H.-J.: Fuzzy programming and linear programming with several objective functions. Fuzzy Sets and System 1(1), 45–55 (1978)

# Future Directions in Fuzzy Engineering Economics

Cengiz Kahraman

Department of Industrial Engineering, İstanbul Technical University, 34367, Maçka, İstanbul

The discounted cash flow techniques are the economic techniques for analyzing alternatives. These economic analysis techniques taking the time value of money into account are the present worth analysis, the future worth analysis, the rate of return analysis, the benefit/cost ratio analysis, the payback period analysis, and the annual worth method.

Under risky conditions, each action leads to one of a set of possible specific outcomes, each outcome occurring with a known probability. And under uncertainty, where actions may lead to a set of consequences, but where the probabilities of these outcomes are completely unknown. A risky situation is thus a situation where the outcome is unknown to the decision-maker, i.e. he/she is not sure which outcome will occur and the uncertainty may lead to erroneous choices.

The conventional economic analysis methods under certainty are based on the exact numbers of parameter estimations like project life, interest rate, and cash flows. Under certainty, the project life is exactly 10 years and it can not be defined as "around 10 years". In reality, it is not practical to expect that certain data, such as the future cash flows and discount rates, are known exactly in advance. Usually, based on past experience or educated guesses, decision makers modify vague data to fit certain conventional decision-making models. The modified data consist of vagueness such as *approximately between 9% to 12%* or *around $25,000*.

Fuzzy set theory can be used in an uncertain economic decision environment to deal with the vagueness of human thought. Fuzzy numbers can capture the difficulties in estimating these inputs, such as cash amounts and interest rates in the future, for conventional decision-making models. When we have vague data such as interest rates and cash flows to apply discounted cash flow techniques, the fuzzy set theory can be used to handle this vagueness. The fuzzy set theory has the capability of representing vague data and allows mathematical operators and programming to apply to the fuzzy domain. The theory is primarily concerned with quantifying the vagueness in human thoughts and perceptions.

A rational approach toward decision-making should take into account human subjectivity, rather than employing only objective probability measures. This attitude towards the uncertainty of human behavior led to the study of a relatively new decision analysis field: Fuzzy decision-making. Fuzzy systems are suitable for uncertain or approximate reasoning, especially for the system with a mathematical model that is difficult to derive. Fuzzy logic allows decision-making with estimated values under incomplete or uncertain information.

A major contribution of fuzzy set theory is its capability of representing vague data. Fuzzy set theory has been used to model systems that are hard to define precisely. As a methodology, fuzzy set theory incorporates imprecision and subjectivity

---

C. Kahraman (Ed.): Fuzzy Engineering Economics with Appl., STUDFUZZ 233, pp. 381–382, 2008.
springerlink.com © Springer-Verlag Berlin Heidelberg 2008

into the model formulation and solution process. Fuzzy set theory represents an attractive tool to aid research in industrial engineering (IE) when the dynamics of the decision environment limit the specification of model objectives, constraints and the precise measurement of model parameters. This book provided a survey of the applications of fuzzy set theory in engineering economics.

Almost every research area, after the fuzzy set theory is introduced by L.A. Zadeh (1965), is handled by two ways of logic: Classical one and fuzzy one: classical (crisp) linear programming (LP) and fuzzy LP; classical probability and fuzzy probability; classical analytic hierarchy process (AHP) and fuzzy AHP; classical confidence intervals and fuzzy confidence intervals; classical engineering economics and fuzzy engineering economics, etc. The main reason of this is that the science of mathematics is divided into classical mathematics and fuzzy mathematics by many researchers. The scientists should be in agreement that mathematics is a unique science which can not be viewed by two different ways. Any book entitled *mathematics* should anymore involve the topics taking both binary and multiple-valued logic into account. Hence, in the future, it is expected that the word *fuzzy* will not be put on the cover page of the books entitled like "fuzzy engineering economics" and "fuzzy optimization techniques".

The engineering economics books of the future should naturally be integrated by fuzzy cases of the topics. For instance, the present worth analysis should be handled by both probabilistic and possibilistic points of view. Because of the way of human's thinking, fuzziness is a component of real life and science even the rules of physics and nature are not fuzzy.

In classical engineering economics books, the decision environments or conditions are divided into three parts: deterministic conditions, risky conditions, and uncertainty conditions. An engineering economics book of the future should include a new section: *"Investment Decisions under Fuzziness."*

The recent editions of engineering economics books give more importance to evaluate investment projects using multiple criteria. Analytic hierarchy process, TOPSIS, VIKOR, and outranking methods are some of the multiple-criteria evaluation methods. The crisp and fuzzy models of these methods should be included to evaluate investment projects under fuzziness in the engineering economics books of the future. Some topics of financial engineering like real options under fuzziness should also be included.

# Index

Accelerated cost recovery system  171
After-tax cash flow  166
Aggregated importance weights  255
Aggregating modes  265
Aggregation function  183, 199
Aggregation methods  266
Analysis of stock markets  VII, 289
Annual depreciation  161
Annual revenues  38
Annuity  132

Balance depreciation  159
Before-tax cash flow  166
Benefit-cost analysis  VI, 129
Benefit/cost ratio  15
Benefit cost ratio  VI, 129
Binary relation  76
Binomial model  340
Black-scholes model  VII, 339, 348
Black-scholes pricing formula  333
Branch-and-bound method  359, 363
Break-even  84

Capital budget  359
Capital budgeting decision  1
Capital budgeting decisions  1
Capital recovery calculations  180
Capital recovery factor  3, 14
Cash flow  14, 75
Cash flow analysis  VI, 145
Center of gravity  23
Centralized interval arithmetic  110
Challenger  148
Chance-constrained programming model  369, 371

Choquet integral  VI, 183, 186
Classical replacement analysis  148
Combining fuzzy  60
Common multiple of the alternative lives  133
Comparing complex fuzzy numbers  VII, 217
Comparison criteria  40
Compensation index  174
Compound interest factor  13
Compound interest rates  3
Constant depreciation  161
Constraint satisfaction problem  114
Consumer price index  173, 175
Continuous compounding  2, 4, 133
Continuous interest  341, 350
Continuous uniform payments  135
Conversion to Straight Line  163
Converting a fuzzy number  75
Convex subset  323
Correlated  33
Cost-of-living index  174
Cost minimization  36
Cost push inflation  174
Costs escalation  37
Covariance  315
Credit risk analysis  VII, 183, 209
Crisp black-scholes  348
Crisp decision trees  234
Crisp pricing forwards/futures  352

Data mining  199, 289
Data mining techniques  291
Data-mining  183

384    Index

DB factor    163
Debt rating model    209
Decision tree    231
Declining balance    162
Declining balance method    164
Demand-pull inflation    174
Depreciation    159
depreciation and tax    160
depreciation methods    161
depreciation table    164
Differentiation of the choquet integral    185
Discounted cash flow    330
Discount factor    13
Discount rates    359
Discrete compounding    131
Discrete payments    136
Double declining balance    163, 164

Effective interest rate    4
Engineering economic decision-making    1
Equivalent annual cash flow    81
Equivalent annual worth    15, 80, 84
Equivalent annual worth analysis    V, 71
Equivalent uniform annual benefit    81, 142
Equivalent uniform annual cash flow analysis    148
Equivalent uniform annual cost    81, 148
Expected return of a portfolio    317
Expected value    233
Expert's opinions aggregation    248
Extension principle    19, 21, 313, 342

Fathom a current candidate solution    365
Fathoming tests    365
Financial investments    317
Flexible query language    293
Fulfillment thresholds    294
Functional dependencies    290
Future options    V, 43, 64
Future worth    15, 177
Future worth calculations    177, 179
Fuzzy after-tax cash flow analyses    VI, 159
Fuzzy B/C ratio    136
Fuzzy B/C ratio analyses    142

Fuzzy benefit/cost analysis    129
Fuzzy black-scholes    348
Fuzzy capital rationing models    359, 362
Fuzzy cash flow    VII, 217
Fuzzy cash flow project    225
Fuzzy cash flows    VI
Fuzzy cash outflows    58, 372
Fuzzy classification    V, 43, 66
Fuzzy comparators    294
Fuzzy compound    135
Fuzzy continuous compounding    135
Fuzzy continuous payments    VI, 129, 133
Fuzzy decision analysis    291
Fuzzy decision trees    231, 233
Fuzzy depreciation    VI, 159
Fuzzy depreciation payments    171
Fuzzy deterioration rate    174
Fuzzy discount rate    156
Fuzzy discrete payments    VI, 129, 135
Fuzzy discrete uniform series continuous compounding    136
Fuzzy discrete uniform-series    136
Fuzzy equivalent annual-worth analysis    71
Fuzzy equivalent uniform annual worth    74
Fuzzy equivalent uniform annual worth analysis    147
Fuzzy EUAW    149
Fuzzy extended dependencies    291, 292
Fuzzy extension    109
Fuzzy functional dependencies    297
Fuzzy inflation    178
Fuzzy inflation rate    174, 178
Fuzzy integral    183
Fuzzy internal rate of return    97, 99, 105
Fuzzy interval ordering    264
Fuzzy mean-variance model    377
Fuzzy meta-knowledge base    293, 297
Fuzzy minimum attractive rate of return    VI, 159, 166
Fuzzy multi-attribute data mining    289
Fuzzy multi-attribute decision problem    225
Fuzzy multiobjective evaluation of investments    243

Index     385

Fuzzy multiple attribute decision analysis VII, 217
Fuzzy net cash flow   178
Fuzzy net present value   V, 43, 59, 62, 65, 67, 331
Fuzzy net present value analysis   307, 330
Fuzzy noise   32
Fuzzy nominal interest rate   134
Fuzzy numbers   308, 310
Fuzzy present value   43, 59, 62, 67
Fuzzy present value - concept   V, 43, 58
Fuzzy present worth   24
Fuzzy quantifiers   296
Fuzzy real option valuation   307, 332
Fuzzy replacement analysis   145, 152
Fuzzy sensitivity analysis   183
Fuzzy sets   308
Fuzzy SQL   289
Fuzzy tax rate   VI, 159
Fuzzy tree crossover   291
Fuzzy values comparison   247

Geometric payments   6
Geometric payments future worth factor   2
Geometric payments present worth factor   2
Geometric series   14
Geometric series annuity factor   15
Geometric series present worth factor   14
Giga-investments   327
Gradient payments   6
Gradient payments future worth factor   2
Gradient payments present worth factor   2
Gradual dependencies   290
Gradual functional dependencies   297
Gradual fuzzy dependencies   299

Imprecision of information   321
Income tax   159
incompleteness of information   321
Incremental benefit   130
Incremental cost   130
Incremental present value   130
Inflated interest rate   2, 5
Inflation   37, 173

Inflation-adjusted interest rate   176
inflation- free interest rate   176
Inflation and real interest   177
Inflation measurement   175
Initial investment   75
Interest factors   6
Internal rate of return   105, 243
Interval-valued possibilistic mean   315
Interval arithmetic   110
Interval arithmetic rules   113
Interval extended zero   127
Interval extensions of simplest linear equation   112
Interval Limited Choquet integral   184
Inverse factor   13
Inverse intervals   112
Investment Analysis   236
Investment Decisions   289
Investment Evaluation Problem   VII, 243
investment risk   317

Kuhn-Tucker   324

Left-hand fuzzy number   100
LIFO method   376
Linguistic weights   248
Lorie-Savage   VII, 359

Maintenance costs   156
Mean-variance methodology   307, 317
Mellin convolution   223
Mellin transform   VII, 217
Mellin transforms   223
Minimum acceptable rate of return   41
Modal interval arithmetic   111
Multi-attribute utility analysis   246
Multiple criteria decision making   244
Multiple criteria investment   274

n Step Model   346
Negative uniform cash flows   135
Net annual benefit   132
Net present value   43, 58, 60, 105, 243, 248
Neural net solutions   109
Newton's method   109
Nominal interest rate   4
Non-linear fuzzy equation   VI, 105
Non- standard interval arithmetic   110

386    Index

Normative Measures on Fuzzy Numbers 315

One step model    341
Operating system selection    148, 152
Optimistic value    109
Optimization Problems    V, 43, 65
Oscillator    292
Outranking method    75

Pareto regions    267
Partial correlation    31
Partially correlated    35
Payback period    15, 105
Periodic cash outflows    359
Period is Infinite    80
Pessimistic value    109
Portfolio Selection Problem    307, 322
Possibilistic Approach    322
Possibilistic mean value    315
Possibilistic Portfolio Selection Problem 324
Possibilistic variance    315
Possibility distributions    308, 319
Possibility theory    246
Present value of benefits    133
Present value of costs    133
Present worth    15
Present worth annuity factor    14
Present worth calculations    178
Present worth of a future amount    13
Price increase    173
Pricing options    339
Probabilistic approach    V, 43, 60, 217
Probabilistic concepts    60
Probabilistic method for fuzzy values 248, 261
Probabilistic method for fuzzy values comparison    261
Probability distributions    319
Producer price index    175
Profitability Index    105
Project selection problems    62
Proportional probability distribution 219

Random cash outflows    372
Random fuzzy cash outflows    371
ranking fuzzy numbers    75
Ranking methods    137

Ranking portfolios    317
Rate of return    15, 166
Regular annuity    132
Regular interval arithmetic    108
Relational data base management systems    299
Replacement of equipment    147
Right-hand fuzzy number    100
Robust approach    VII, 359
Robust capital rationing model    361
Robustness index    26
Robust optimization approach    359

Salvage value    75, 160
Segment interval analysis    110
Sensitivity Analysis    199
Sensitivity analysis    199
Simple moving average    292
Single declining balance    162
Single payment future worth factor    2
Single payment present worth factor    2
Sinking fund factor    3, 15
Soft decision support    317, 327
Soft decision support systems    307
Spatial games    65
Standard interval extension    113
Stochastic capital rationing model    360
Stock market analysis    289
Straight line    164
Straight line depreciation    159, 161
Straight line method    163
Strategic investment planning    327, 328
Subsethood measure    252
Sugeno integral    183, 199, 201
Sum-of-years digits depreciation    161

Tax implications    37
Tax rate    177
T-conorm integral    196, 197
Technology selection    147
The discounted rate    75
The index of optimism    76
The largest possible budget    365
The most possible budget    365
Time value of money    3
Tool steel material selection problem 248
Transparent method    248
Trapezoidal fuzzy number    218, 313
Trapezoidal fuzzy numbers    86

Triangular density function 223
Triangular fuzzy number 218, 312
Two Step Model 344

Uncertain rate 21
Uncertainty-based approaches to capital
    rationing models 360
Uncorrelated 34
Uncorrelated cash flows 11
Uncorrelated initial cost 36
Uniform payments future worth factor
    2

Uniform probability distribution 219

Variance 315
visual k-chart technical analysis 291

Weighted method 364
Weighted possibilistic mean 315
Weighting function 315
Weingartner capital rationing models
    VII, 359
Whole sale price index 175